BIG BANG PSEUDO-THEOLOGY

YUXIANG WU   Lǔ WU

BIG BANG PSEUDO-THEOLOGY

Title: Big Bang Pseudo-Theology, Second Edition
Author: Sean Yuxiang Wu, Lü Wu
Published 2023 by Two-W Object
Printed in the United States of America
First Edition: August, 2022
Second Edition: January 2023

BISAC: Science / Cosmology - Big Bang, Theology, Science, Einstein, Hawking, Relativity, Microwave Background, Red Shift, Cosmic, Dark Energy, Dark Matter, Law of Existence, Space, Time

ISBN English Book No.: 978-1-7370977-5-4
Library of Congress Control Number: 2022920107

ISBN English Hard Cover No.: 978-1-7370977-4-7
Library of Congress Control Number: 2022920113

ISBN 繁体中文 Book No.: 978-1-7370977-6-1

For information please address:

Two-W Object

2W Object
yuxiangwu@outlook.com

Following English and Chinese versions of the books are published:
English version: BIG BANG PSEUDO-THEOLOGY, SECOND EDITION
Chinese version 简体中文: 大爆炸伪神学
Chinese version 繁體中文: 大爆炸偽神學

A Few Important Notes

The Big Bang theory, which originated from a model proposed by the Belgian Catholic priest and physicist Georges Lemaître about 100 years ago, presented the formation of the universe as caused by the explosion of a primeval atom.

Such a theory cannot be termed "scientific" for the simple reason that science cannot explain how only one atom would produce the whole universe through one blasting.

In relation to religion, the aforesaid theory also does not glorify God either, and for these two reasons:

- **How the universe was created cannot be known to man, only God knows;**
- **God does not need science to explain or support his actions.**

Science is but a small part of the knowledge that God allows man to master. The vast majority of phenomena whose workings we cannot master is known only by God.

People can try to discover more of God's secrets, but people shouldn't call their own thoughts as God's thoughts, their own stories as God's stories, and label them as "Scientific". Doing so looks like praising God, but actually hurting God. Human attempts to use science to defend God often turn into harm to God. This has big lessons in history.

In mainstream astronomical cosmology, the Big Bang theory depicts the creation of the universe as the result of an atomic explosion, and arrogantly mobilizes all forces to depict it as "scientific" and beyond dispute.

Analysis of the main evidence that supports the Big Bang theory – which we consider pseudo-scientific theology (pseudo-theology) – finds that such evidence is basically composed of a series of artificial myths, which we briefly summarize below:

Myth 1: The all-sky map of cosmic microwave background was compiled by NASA at a cost of billions of taxpayer dollars. The map was based on data received by so-called detectors (are receivers) receiving data from an area of **less than one light-minute** of space near the Earth. The scientists extend such received data from such limited scope to approximately **13.7 billion light years**, and use it to support Big Bang theory.

Myth 2: Dark energy was invented 70 years later after formation of the Big Bang theory. It is used to provide the power for the continuing expansion of the universe that has already cooled and should contract from its origin explosion. The scientific community mobilized all its forces to search for it without success.

Myth 3: Turn hidden celestial bodies that belong to normal celestial bodies at the time of their discovery into mysterious invisible dark matter.

Myth 4: There are numerous authoritative NASA observational wrong reports like this one: found a **30-year-old baby** black hole at **50 million light-years** away.

Myth 5: A ray moving around a rigid rod can shorten the rod and slow the flow of time on the rod.

Myth 6: Nothing is in the empty-space. How can it be bent or twisted? Bending or twisting is caused by any matter or its properties presented in space, not by space itself.

……

This book debunks all these "scientific" myths from a real **scientific** point of view.

What should the universe be like? We'll take you out of the smoke of the pseudo-god Big Bang and experience a magnificent NEW UNIVERSE based on science.

Dedicate this book to:

Li Wang and Eugene Wu, for their excellent help on the research, writing, thesis completing and translation for this book.

Our family members.

Forever teachers: Michael Hood, Neville G. W. Cook, S. Adiga, S. E. Dreyfus, Xianceng Zhang, Jingmin Xin, Jian Wu, Naiguang He, Zigang Jin, Yuexi Chang, Yuanben Liu, Jinfan Li, Fanzhong Li, Yao He, Guangjiao Zhou, Ziqiang Zhu

Friends who have given sincere help during the publication of this book: Su Qing, Nie Leng, Li Shuoru, Shuiqing Wu, Weihong Huang, Liu Fangjian

Yichun, our root and home town.

Berkeley, our second home town.

Because of you, the starry sky journey is not lonely, and the lonely universe is full of warmth.

Special thanks to:

Thanks to the editors who selected the "The Abstruse of the Distance" (距离的奥妙) section of my book "*Who Should Talk about Cosmos* (*谁有权谈论宇宙*)" into the 7.15 Excellent Chinese Courses for Chinese Universities. Without the encouragement of these editors, whom I have not met, I would not have had enough confidence and courage to finish this book.

Thanks to the editors of Matter Regularity magazine from the Beijing Relativity Association. Your long-term, unremitting efforts have made available and maintain a safe place for valuable research not accepted by the mainstream.

Content

About Dr. Yuxiang Wu

2021 published "10 Questions To NASA"

2015 – 2019, Published 10 related papers on "Matter Regularity." Published books "Debate of Light and Dark," and "Fading Modern Cosmology."

2005, published "Chinese book: Who Should Talk about Cosmos," one section of it was selected into Chinese College Essence textbook.

2001, again won the World Journal Chinese Novella Essay Contest

1999 to 2007, he served as the dean of the School of Software, Beijing Normal University Zhuhai Campus, and the director of the Software Research Institute of the Academician Park of Dalian University; Distinguished Professor of Guizhou University; Distinguished Professor and Postgraduate Supervisor of Beijing Wuzi University. Due to acclimatization, he returned to work in the United States in 2007.

1999, won the World Journal Chinese Novella Essay Contest .

1990, received Ph.D. from U.C. Berkeley School of Engineering.

1987, won UC Berkeley Regents' Fellowship full scholarship

1981, enrolled into master's degree, got 5 major course credits before classes began.

1977, admitted to university, got English course credit before class began.

1968, after graduated from junior middle school, worked as a farmer in mountain villages for 10 years.

About Lü Wu

2022, Initiated pseudo-theology concept.

2018, Bachelor of UC Berkeley Engineering

2015, Published 6 related papers on "Matter Regularity."

Forward

Is the Big Bang Theory a "Science" or A Pseudo-Theology

This book is the sequel to two previous titles: *Debate of Light and Dark* (2016) and *10 Questions To NASA* (2021). It incorporates new perspectives proposed by Lü Wu, including: the relationship between God and science and the character of the pseudo-theology known as the Big Bang theory. Thus, readers can form a new perspective on how we understand the universe, what authentic science is, and how we can best recognize and interpret problems within the context of the relationship between science and God. We warmly invite readers, especially from religious groups, to explore our ideas with us.

The authors also collectively published six related papers together in 2015 (refer to the reference table of the book).

After years of painful consideration and discussion, we finally realized that pseudo-theology is the essence of the Big Bang theory.

You may be surprised by this statement. You may also be surprised why the word science in the title to this Foreword is enclosed in quotation marks.

About a hundred years ago, a Belgian priest, Lemaître, proposed the Big Bang theory. He argued that the entirety of the physical universe was initially blasted out from a single particle. He called it a "primeval atom." **In an explosion, the primeval atom gives rise to space and time and an ever-expanding universe. The expansion continues to this.**

Big Bang Cosmology

Excerpt from NASA web site

The Big Bang Model is a broadly accepted theory for the origin and evolution of our universe. It postulates that 12 to 14 billion years ago, the protion of the universe we can see today was only a few millimeters across. It has since expanded from this hot dense state into the vast and much cooler cosmos we currently inhabit.

We can see remnants of this hot dense matter as the now very cold cosmic microwave background radiation which still pervades the universe and is visible to microwae detectors as a uniform glow across the entire sky.

FOUNDATIONS OF THE BIG BANG MODEL

The Big Bang Model rests on two theoretical pillars:

General Relativity

The first idea dates to 1916 when Einstein developed his General Theory of Relativity which he proposed as a new theory of gravity. A key concept of General Relativity is that gravity is no longer described by a gravitational "field" but rather it is supposed to be a distortion of space and time itself. Physicist John Wheeler put it well when he said "Matter tells space how to curve, and space tells matter how to move."

The Cosmological Principle

After the introduction of General Theoru of Relativity a number of scientists, including Einstein, tried to apply the new gravitational dynamics to the universe as a whole. This required an assumption about how the matter in the universe was distributed. The simplest assumption to make is that the matter in the universe is homogeneous and isotropic when averaged over ver large scales. This is called the Cosmological Principle.The cosmic microwave background radiation, the remnant heat from the Big Bang, has a temperature which is highly uniform over the entire sky. This fact strongly supports the notion that the gas whch emitted this radiation long ago was very uniformly distributed.

These two ideas form the entire theoretical basis for Big Bang cosmology.

Figure 0.1 Excerpt from NASA's Big Bang Theory official website

From Figure 0.1 we can’t find any clue for how the “only a few millimeters across” universe was able to expand to the current huge universe. And we can't see why the general theory of relativity supports the Big Bang theory. We leave it to be discussed in later chapters.

Please look up out of the window and see the big rivers and seas, the mountains and forests, the Sun, the Moon, and the stars. Everything, according to the Big Bang theory, was exploded from one small particle. As for what the universe looked like before the explosion, and what was outside the area of the explosion upon its completion, is basically not discussed at all. As for why such an explosion only happened once and so on, scientists have not even considered it.

Of course, there is no way to discuss this issue scientifically. How can a "primeval atom," as Lemaitre proposed, produce the universe as we know it today? Is this not a miracle from Heaven, from the Creator, from God!

Nonetheless, modern mainstream cosmology directly attributes "science" to the Big Bang theory, thereby transforming Big Bang theology into accepted science. Obviously, the Big Bang generated by an atom explosion is not science, but theology. And **it is not true theology** either. It is **pseudo theology – a concept that we discuss in detail in this book.**

Since then, the Big Bang theory, which has won out in various debates about the universe, has become the foundation of today's scientific cosmology. And during these debates, the most basic, unbelievable hypothesis of the Big Bang theory (borne from Lemaitre's exploding primeval atom), which mortals can't explain, is mostly ignored. Academia is entangled in too many other details such as dark energy, expanding universe, and so on in the name of science.

Therefore, modern cosmology is a real COSMIC THEOLOGY, and is just prefixed with the term "science". Big Bang theory is a COSMIC PSEUDO-THEOLOGY.

Religious friends must be pleasantly happily surprised: Science has really become the slave of theology. Is there anything better for religion and for God? Cosmic science not a slave of cosmic theology?

However, we regret to point out that this is not a true theology of the cosmos, but a pseudo-theology. In the long run, instead of enhancing belief in God, **it may ultimately lead to the collapse of belief in God**.

Due to the congenital defects of the Big Bang model, and because there are always people who are persistent in their pursuit of true science, as we are doing now, in time we will **use science to** finally **prove** that Lemaitre's pseudo "scientific" Big Bang" universe is indeed a notion mistaken. Then, with the failure of the Big Bang model, our belief in God can be greatly affected, especially in the long run.

In this light, is it not enough to cool down the overzealous nerves of the people of God while considering the flames burning Giordano Bruno (a scientist who insists on geocentricity)?

A short story can illustrate this point.

In the last century, a priest went to preach in the deep mountains far away from my hometown. He built a divine tree that would bear luminous fruit and performed miracles. The mountain people who followed this priest were much convinced of God's powers. However, with the passage of time and the changes that time brings, people understood that the luminous fruit was not a myth but a kind of electricity, and the presence of God, which the priest revealed, vanished.

People will not forget that the church insisted on the "geocentric theory" — that is, that the Earth was the center of the universe — the resulting persecution of Copernicus, and the burning of Bruno to death (all of which had a long-term negative impact on the presence and authority of the true God).

What then constitutes the real and permanent maintenance of God's presence and authority?

We are not speaking here of the usual reverence and devotion given to God, which is what the church, priests, and believers should abide by. Here we emphasize what has been overlooked: For things we don't understand, don't arbitrarily **use the name of God to explain them through science**!

For something that everyone wants to know but is impossible to know — such as how did the universe come into being? — don't use our common human thinking to explain it, let alone the little scientific knowledge that human beings have grasped up to this point in history. On the contrary, if it can be proved that this thing is really impossible for humans to explain through science, and also it can be scientifically proved that it cannot be explained by science, then this proof has achieved its purpose of **honoring Heaven** in a certain way.

If we can prove a thing is unknowable and inexplicable to human beings, of course, only God knows it.

This is the mantra of people: only Heaven knows!

The difference here is this: when trying to explain the unknowable in the name of God through science, if doing so fails, it impairs the credibility of God. In this light, admitting and proving that human beings are limited in their ability to understand the universe exalts the credibility of God.

In the book, we will discuss this issue in detail.

Now let's briefly review the history of the invention of the Magic Theory of Big Bang Pseudo-science – the **Big Bang Pseudo-theology.**

In 1927, Father Lemaître proposed a pseudo-theology theory that the universe was formed by the explosion, or Big Bang, of a primeval atom.

In 1929, the noted astronomer, Edwin Hubble, announced that all celestial bodies are moving away from the observer. His rationale: spectral analysis of the light from celestial bodies. When that light shifted into the red area of the spectrum, it meant that the celestial body was moving away from the observer. In effect, Hubble offered proof of Father Lemaître's Big Bang theory through science. While whole academia agrees with his proof, nobody thinks or doubts that the Interpretation Method about the observed data is correct or not.

More than half a century later, in the 1980s, experts suddenly awakened that the universe, as it cooled after the Big Bang, should be contracting instead of expanding. As their interpretation of the observational data told them the opposite, they invented, with a kind of magical thinking, something they called "dark energy." How did the universe continue to expand when it should be contracting? Dark energy was the answer.

Let's look at this from another perspective:

In the 1950s, scientists proposed that if the Big Bang theory were consistent with the simplest model of the universe at their disposal, then the existence of a cosmic microwave background (CMB) uniformly distributed throughout the universe was necessary. With its detection through a tiny near-Earth observational aperture, experts postulated it as evenly distributed throughout the universe. They then used this postulate as their main evidence to support the Big Bang theory. Because NASA had spent countless amounts of money, manpower, and material resources over decades to detect and analyze the cosmic microwave background, scientists took this thing – however much it was or not so closely related to the Big Bang — as one of its primary supporting "scientific" theories.

The result: the cosmic microwave background (CMB), said to be evenly distributed throughout the universe, became one of the main theories that NASA scientists used to support the Big Bang theory (Figure 0.1). In fact, CMB is far less important than dark energy, but too much money has been spent on the CMB, they have to emphasize its importance.

Now we see that when an atom explodes, and from that explosion the universe evolves, it also needs to create, as a long-lasting result, both a microwave background and dark energy. It's really hard work.

These two things are a bit similar, with both evenly distributed throughout the universe. But one is visible and the other completely invisible. Perhaps experts will create another fairy tale to unify them? Isn't this another good chance to win a Nobel Prize?

What this book does, then, is scientifically prove the following in detail in later chapters:

1. No matter how much authority backs it up and how much money is squandered on it, the Big Bang theory is essentially a non-"scientific" or pseudo-theological theory full of myths, with many inconceivably simple errors made in its name.
    - For some NASA research results, either we point out the mythical component involved, such as dark energy, or analyze critical theories, such as applied to the redshift, which is based on the Big Bang model, or the all-sky microwave map derived from a tiny near-Earth area for observation.
    - We should recognize that any applied mathematical model derived from an irrational physical model must also be irrational.

The theory of relativity has been heavily relied on by the astronomical and cosmology circles in recent years. Be that as it may, the physical model from which relativity theory derived was wrong.

The Big Bang was produced by the explosion of an atom. The special theory of relativity is mathematically derived from a physical model that a light ray moving around a rod would change the length of the rod and the time flow on the rod.
Empty space is neither bendable nor distortable.
All of these three are mythological.

2. The universe is unknowable to human beings. The grandeur of the universe and the limitations placed on travel, by the variably slow speeds we can attain, must make human beings stop in awe when considering the countless unknown mysteries of the universe. The true measure of the universe can only be known by the Creator!
   Humans need to understand both the general scope of the universe we can know, and the correct way to understand what we can know.
3. Call for creation
   All of the above may be shocking for readers to consider. But consider: why are there so many problems that scientists devoted to studying the cosmos cannot resolve? Some of these problems are even simple and intuitive ones. **Humans need to break free from the shackles of old thinking and expand their ability to understand the true science of the universe.** More importantly, it is necessary to discuss where, or on what, does our ultimate ability to understand the universe scientifically rest? In other words, how much cosmic information is the Creator willing to reveal to us?

Of course, this kind of critical thinking is not an easy task. According to our more than ten years of experience in exploring cosmic problems, the main difficulty we have found lies not in whether our proof is correct, but whether that proof is beyond the scope of academics. It is involved in too many things that has nothing to do with academics. For example, famous cosmic scientists will think the topic violates his authority. The academic world would think how dare you criticize Einstein …

However, human cognition is always progressing. All non-scientific pseudo-theology, all despicableness, and politics will eventually be revealed by the erosive force of time.

We need to establish a correct worldview of the cosmos in which we live.

However, our efforts have been blocked by the joint efforts of scientific and technological authority and religious circles. Because they firmly believe that they are safeguarding scientific truth and the credibility of God! This kind of power is extremely powerful. Even in the Internet age, it seems that people can express their opinions freely, but it is not the case! They are suffocating!

Let's look at a simple example of how the astronomical sciences and religion have joined together, and then discuss how we feel about this union. Simply put, both science and religion are wrong!

But this is also an excellent opportunity to turn bad things into good, especially for the religious circles, this is an excellent opportunity to truly establish the supreme glory of God, don't miss it!

Now let's look at the following science news: (2023-01-25 retrieved)
https://www.iau.org/news/pressreleases/detail/iau1812/

Figure 0.2 The bulletin of the International Astronomical Union on October 29, 2018, the first sentence in the English section below the picture is as follows: "An electronic vote has been conducted among all members of the International Astronomical Union, and the resolution to recommend renaming the Hubble law as the Hubble–Lemaître law has been accepted."

Figure 0.2 tells us that the International Astronomical Union voted firmly to support the priest's pseudo-theology that an atom exploded out of the universe.

For the astronomical community, this resolution highlights the lack of imagination in the astronomy community, cornered by the limitations of the Big Bang theory, and had to turn to God for help.

But whether God will help them is a big question.

On the other hand, people in the religious circles feel the goodwill of people in the scientific circles, and are particularly satisfied with the combination of Father Lemaître's myth of an atom exploding out of the universe with science and technology, thinking that this is the use of science to prove the glory of God!

But I want to say to religious people: But, have you thought about it carefully? If Hubble's law is proved to be wrong, if Hubble's law is proved to be against the expansion of the universe, and if Hubble's law is proved to be basically independent of the motion of celestial bodies, at that time, bond Hubble's law with God, is glorifying God or hurting God? How much does this behavior hurt God?

Dear reader, we will present all these proofs to you in the fourth chapter of this book. Turn ifs into facts! At that time, you can judge rationally: How wrong it is to bind science and God together! How hard that behavior hurts God!

Repeat it again:

Science is just a little bit of knowledge that God allows man to master!

God does not need the support of science!

One should not presumptuously support God with plausible science!

On the contrary, everything that people do not know, Heaven (God) knows. The most obvious example is which came first, the chicken or the egg? Science can't explain this simple thing at all, only Heaven knows!

At this time, according to this way of thinking, friends in religious circles, can you find the right way to truly glorify God?

Part I

The Big Bang Theory – Why is Not Science but Pseudo-Theology?

Looking back at the history of the Big Bang can evoke some interesting ideas for us. Let's take a look here with a perspective that is not exactly the same as that of the mainstream academic world:

In 1927, Catholic priest and physicist Georges Lemaître put forward the hypothesis that the universe originated from an original atomic explosion, and believed that today's universe was produced by the explosion of an atom;

In 1929, Edwin Hubble published an article in the Proceedings of the National Academy of Sciences (PNAS) on the observed relationship between the distance of galaxies and the speed of receding galaxies—Hubble's law—unveiled the expanding universe and forever changed human's understanding of the universe. In the following decades, the astronomy community relied on the power of Hubble's theory to make the Big Bang theory popular. The theory of an atom-exploded produced universe mythic has become the orthodox theory of mainstream astronomical science and academic circles!

Until the 1980s, someone (don't know who the researcher was, and we were tough to look it up, anyway we guess, this researcher is a well-deserved Nobel Prize winner, but was academia silenced by academic tycoons because of opposing the Big Bang theory) suddenly discovered: the cooled universe should not expand, but contract!

Therefore, the expansion of the universe deduced by Hubble's redshift theory is completely contrary to the Big Bang, and runs counter to the theory that the space of the Big Bang should shrink after cooling. But academia seems to be quietly forgetting this. Have you seen any critiques about it?

Another important point is that if considered from the perspective of geometric scale, Hubble's law not only does not support the theory of cosmic expansion, but on the contrary denies the theory of the Big Bang!

Unless the Earth is at the center of the universe, the light emitted by celestial bodies received by the telescope on Earth from different directions should reach the telescope at totally different speeds. Since the Earth is in the expanding Big Bang universe, even if all the light propagated at a constant speed as Einstein said, when we observe around, the light information received from different directions should have completely different speeds, instead of like the description of Hubble's law. We'll discuss this later.

At this time, the astronomical community should reflect: Is the myth Big Bang theory that an atom explodes out of the universe a science? Can it be supported by scientific evidence?

But the leaders of this generation of astronomy and cosmology do not have enough scientific innovation spirit. They obsessively follow the path of the Big Bang theology, racking their brains to fabricate a pseudo-mythical substance of dark energy to fight against the contracting universe!

It's a pity that almost half a century has passed, and the astronomy and academic leaders have not found a trace of dark energy. But they are still researching and publishing the form, nature, and method of finding dark energy... It seems that tomorrow they can find the dark energy that they say accounts for 68% of the total matter in the universe!

Let us tell you, dear reader, they will never find dark energy! The reasons are detailed in Chapter 2, which discusses dark energy. You only need to consider what the purpose of inventing dark energy is, and according to this purpose to test the so-called dark energy they found, you can easily identify whether the found thing is really dark energy that can maintain the accelerated expansion of the universe!

NASA has also spent more than half a century doing something that wasted resources and money. This is the continuous launch of 4 generations of satellite probes to depict the cosmic microwave all-sky map. This project has spent hundreds of millions of dollars. The WMAP satellite probe proposed in 1995 alone has a budget of 150 million US dollars. The key problem is that this is a project that has no scientific basis. What is used is a so-called detector that can only receive signals in situ. What it detects is a group of uniformly distributed (that is, non-flowing) waves near the earth. What it really sees is a space that does not exceed 1 light minute. Based on the detected results, an all-sky map of tens of billions of light-years across the universe was drawn, which is extremely strange.

Well, let us listen to our "rebellious" question remarks against the pseudo-theology of the Big Bang!

Correctly understanding the universe is a magnificent dream.

Can this dream come true? What if it can't be done? Is it to maintain a false "scientific" theory, or to admit that human beings themselves are not capable enough to see the overall true face of the Creator?

Greek philosophers have discussed the possibility of an eternal universe, and legends or myths, such as Pangu's creation of the world from ancient China, are also primitive speculations about the origin of the universe. For their part, Western religions have always believed that the universe was created by God at a particular moment in the past.

These are understandable and normal human activities of ordinary people and believers of God.

However, when someone lifts God to an absurd height, it is not only God who falls in the end, but also the person who sent God to that high, if absurd, place. Although this or that man who does so may have a sincere heart and well believe in God, his good intentions notwithstanding, he did bad things to God!

The Belgian Catholic priest Georges Lemaître is such a person. He did something inappropriate, even if with good intentions, for his, and our, relationship with God.

In 1927, based on mathematics and astronomical observations, Lemaître put forward the hypothesis that the universe originated from a primeval particle or **atom that exploded**.

His hypothesis tells us that the mountains, rivers and oceans around us, the scintillating Milky Way in the sky, the stars all over the sky... all came from the explosion of an atom!

But the creation of the universe, all that we can wonder at and more, came from, was born in a miraculous, not human, imagination. It was a miracle that human beings can never explain with normal thinking or science, and something that only God is able to do.

In this way, **the cosmic Big Bang Theory became, from the moment Lemaître proposed it, as if in counterpoint, a scientific God.**

However, whether this brand is good or bad for maintaining the reputation of God requires deeper insight. My opinion is that the short-term effect is okay, but the long-term damage to God's reputation is, quite simply, immeasurable.

Why would we say this?

If we simply say: God created the world. Then people who believe in God will believe this sentence; people who do not believe in God will only attribute it to a religious belief, without the expression of disgust or opposition.

Father Lemaitre's mistake is clear: for him, God created the universe scientifically. This has done great harm to God. Science should not be closely related to God's actions, much less when using science to explain His actions.

Lemaître's Big Bang (also known as Big Bang cosmology) proposed that the universe was formed by a dense and hot singularity after a Big Bang that occurred some 13.8 billion years ago. The universe has a history of evolution from hot to cold. During this period, the cosmic system is constantly expanding.

From such a theory, a complete scientific cosmology has developed. Father Lemaître imagined using science to prove his absurd Big Bang theory as authentic science. In Lemaître's view, the God of faith became the God of Science!

The expansion of the universe was also predicted by astronomers Friedmann and Hubble. Their arguments are the cornerstone of the Big Bang theory as developed by Lemaître, Friedmann, Robertson, and Walker.

Does labelling the Big Bang produced by an atom as scientific really deepen the brilliance of God for us? Will it make the world believe in God more?

No!

Consider the flames that burned Giordano Bruno. How much has that crazy fire burned the credibility we have given to God? Why not accept the lessons of history? Never use science to prove the rightness of GOD.

When you associate God's creation with science, you initiate a process that damages God's credibility! When we talk about the Big Bang with ordinary friends who don't study the universe, some often respond, tirelessly disdainfully: Oh, a world created by an atomic explosion!

Please understand this:

God's creation of the universe does not need to be proved!
God does not need any human's proof!

God does not need scientific proof!

God grasped the entire universe and its mysteries. Science is a very small part of the knowledge that God grants to people!

A miracle is a miracle, what cannot be explained by humans is a true miracle, and what can be explained later is not a miracle later!

If you tie a miracle too closely to science, and do not care how much money is used for research on its behalf, or how large the scale of research is, beware. If the latter does not accept that the phenomena it studies yet does not understand is not by its nature a miracle, then that research becomes pseudoscience or a pseudo miracle.

Father Lemaître's theory of the Big Bang created by an atomic explosion has been around for about a century, and the number of scientific studies and achievements tied to it is innumerable. Today, in contemporary research on the cosmos, we find a variety of easily discernible "sciences" can only be established in the name of God, such as dark matter, such as the found the 30 year-old baby black hole, etc. These blasphemes are against science itself, but also against the glory of God.

A large number of scientific researchers are devoted themselves to the research of astronomical and cosmic science, some throughout their entire professional lives. Have they ever thought that the science they present testifies to a kind of pseudo-God theology? How will they feel when the "science" they did was proved wrong?

Since you want to tie the Big Bang tightly to science, then there are always those who are obsessed with science who will tirelessly do their scientific research on the Big Bang. When the so-called Big Bang science piled up with untold taxpayer money has been irrefutably scientifically proven to be downright false, even with the accumulated academic authority and occupied voice, the falsehood might be maintained for a period of history, but it finally will be revealed. We believe that history will eventually give the true face of right and wrong!

The science that God allows man to master is limited. Take, for example, this old question: which came first, the chicken or the egg? This is a question that only God, not humans, can answer. In fact, this question is not scientific at all. The scientific question should state: If chickens are to be passed on from generation to generation, then, if there were chickens first, how many chickens should be there? If there were eggs first, how many eggs should there be? Still, only God, not science, can answer this question.

There are many questions that are similar to the chicken and egg question noted above. For example, according to the theory of evolution, why don't animals have an eye on the back of their head? Wouldn't that greatly increase their competitiveness?

Human beings are often overconfident, thinking that they have mastered or are close to mastering the ultimate truth of the universe. But what is the real truth?

There are so many things that science doesn't know.
There are so many things that only Heaven knows.

In Part I, we will clarify the errors that populate several of the main pillars of the Big Bang Theory, in order to understand how much pseudo-science and pseudo-theology can bring people both spiritual and material troubles. Then, we will **scientifically** prove that the complete, true face of the universe is unknowable! No matter how hard we try, it is in vain. But God in Heaven; yes, God knows!

Since the true face of the universe is scientifically unknowable, what is knowable is especially important to determine!!

Finally, we want to illustrate the errors inherent in the theory of relativity from the perspective of the relationship between physical models and applied mathematical models. Be that as it may, fundamentally speaking, the theory of relativity also has something of the breath of God.

Let's begin our journey toward the truth, which God alone knows and dispenses, and abandon false theologies wherever and whenever they arise.

Chapter I

The Magical Cosmic Microwave Background – Measurements from a Sesame Seed

In determining the "scientific" evidence for the myth that an atom as it exploded produced the universe, the cosmic microwave background holds a pivotal place. NASA's two-step process for proving the existence of a uniformly distributed cosmic microwave background is as follows:

First, use the pseudo-theology that the universe produced by an atom exploding to speculate on the existence of the cosmic microwave background: The Big Bang theory holds that after the Big Bang universe cools down, it will leave a uniform distribution of residual heat in the entire universe with a diameter of 93 billion light-years, that is, the microwave background radiation.

Second, there is the use of local observation data from the vicinity of the Earth to prove the existence of the cosmic microwave background. NASA has spent hundreds of millions of dollars to construct instruments on satellites that detect microwave emissions from near-Earth space, with an aperture for data reception of no more than one light-minute in diameter. From the data gained, scientists have constructed a whole sky map of the cosmic microwave background, which they use to support the Big Bang theory.

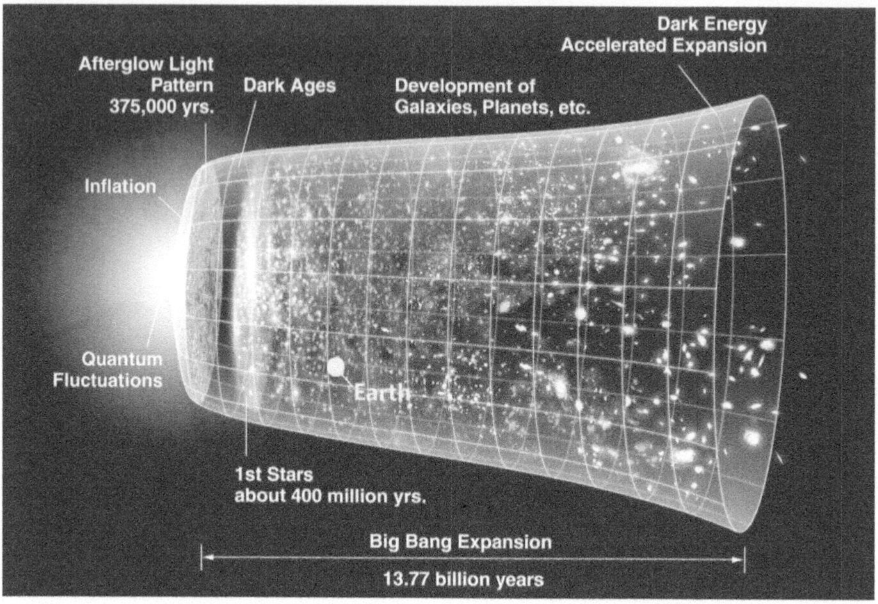

Figure 1.1 Timeline of the universe created by the NASA/WMAP science team. The author adds "Earth" in the picture. NASA used satellites in near-Earth orbit with instrumentation that can receive microwave emissions. From the data received, NASA scientists concluded that they had detected microwaves in the 13.77 billion light-years entire universe. Such microwave emissions were in an evenly distributed pattern with an intensity of less than 3k. The most important keywords here are: Near Earth, Receiving, Entire Universe, Evenly Distributed, less than 3k Intensity

Can such measurements prove the theoretical assumptions that prompted them?

Considered in another way, it's like encircling an area of one square meter of water in the Pacific Ocean, then drawing from the data obtained from that one square meter an image of the entirety of Earth's oceans

It's also similar to receiving data over decades from an area of space the size of a sesame seed, and then sorting and calculating it all and applying it to the entire Earth

Is this science, mythology, or theology?

Is this full of mythology?

Let's retrace this amazing "science" from its beginnings.

Introduction to the Cosmic Microwave Background (CMB)

Cosmologists identify the cosmic microwave background (CMB, also known as 3K background radiation) as the thermal radiation left over from the "Big Bang."

You can't see the CMB with your naked eyes, but it is everywhere in the universe. It is invisible to humans because it is so cold, just 2.725 degrees above absolute zero (minus 459.67 degrees Fahrenheit, or minus 273.15 degrees Celsius.)

The decades long period that it took NASA to draw the all-sky map of the cosmic microwave background (from 1941 to 2014), the extensive resources used (it is said that more than one trillion dollars have been used for more than 70 years), and the observation model thus constructed is not only bad science but begs the question of why this effort took place altogether. How did it happen?

For more than 70 years, NASA has been spending a lot of taxpayers' money launching space probes to detect the cosmic microwave background. Astronomers have published thousands of research articles on the cosmic microwave background and used it as a major support for the Big Bang Theory.

The compilation of perspectives on what we now know as a myth – an exploding atom forming the universe – has engaged contemporary astronomers and cosmologists completely. Be that as it may, the theories propounded are growing increasingly complex, and the narratives attached to them more and more detailed. We can sum them up, if casually, as follows:

According to the simplest Big Bang pseudo-god theory, the entire universe needs to have a uniformly distributed cosmic microwave background (CMB).

The theory holds that the cosmic microwave background is a type of electromagnetic radiation that fills the entire universe. Its characteristics are the radiation on an absolute temperature scale of 2.725K. The frequencies involved are in the microwave range; thus, the named cosmic background radiation. In effect, sensitive radiation telescopes can identify a faint background "glow" that is identical in all directions, and has nothing to do with any star, galaxy, or other distinctive object. The electromagnetic spectrum of this light is strongest in the microwave region. The cosmic microwave background also is cited as the oldest light in the universe. It shows tiny temperature fluctuations that correspond to subtle differences in local density, represents all future structures, and is the seed of today's stars and galaxies.

Because any model of the universe must account for this electromagnetic radiation, the cosmic microwave background is key to accurately measuring the universe. The temperature of the cosmic microwave background in the radiation spectrum of black bodies is 2.72548±0.00057 K. Its “glow” is uniform in all directions. Although slight residual variations exhibit anisotropy, there is a uniform distribution of hot gas that has expanded to the size of the present universe. Spectral radiation from different angles in the sky contain the same anisotropy, or irregularities, which vary with the size of the region. It **has been measured in detail.** There is a high degree of consistency **throughout the observable universe**, and the dim but measured anisotropy very broadly supports the Big Bang model.

We feel funnier to see these increasingly vivid pictures. One of the simplest questions: **What instruments** do experts use to **measure** microwaves **in the entire universe**?

Readers, please note: no matter how advanced the instruments are, no matter where they are placed – whether on mountains or satellites – they are mechanisms that detect data; that is, they are only **receivers**!

Scientists must also rely on several **receivers** in **near-Earth orbit** to complete their observations of this weak microwave signal **uniformly** distributed **throughout the WHOLE universe**, and which enables them to describe the tiny anisotropy noted. Don’t you think this is **another myth**?

Please look back at Figure 0.1. Read the last sentence in the first paragraph of the figure: “*We* ***can see*** *remnants of this hot dense matter as the now very cold cosmic microwave background radiation which still* ***pervades the universe*** *and* ***is visible to microwave detectors as a uniform glow across the entire sky.***”

Here, in addition to the cosmology community trying its best to avoid discussing the myth that the entire universe was created from an atomic explosion, another puzzling myth appears: Is it possible for a microwave detector to capture an average of 2.37K microwaves **glowing uniformly throughout the entire cosmic sky**?

Could something like this happen?

Of course, **it is impossible!**

We dare to assert that even at the edge of the solar system microwave detectors cannot detect the microwaves that behaves like NASA described; that is merely one light-year away.

Understanding this does not require advanced knowledge of the universe or the use of sophisticated detectors. It can be completely judged based on the most basic scientific principles learned in middle school physics, the two most important of which are:

All detectors, telescopes, and other instruments, including microwave

detectors, are pure receivers. They can only receive information such as light waves or microwaves that have reached the location of the instruments!

There are no stationary waves. The propagation of any wave attenuates with distance. The formula for calculating the attenuation of the wave source intensity Q with the distance R is very simple, that is: the light intensity I = $\frac{Q}{4\pi R^2}$ **at the distance R from the source.**

Now cosmic experts tell us: the intensity of microwaves at a distance of 1 km from the detector is 2.7k, and the intensity after transmission to the microwave detector is 2.7k; the intensity of microwaves at a distance of 1 light-year is 2.7k, and the intensity of the microwave at 13.7 billion light-years away is 2.7k, and NASA the detector can see such weak microwave 13.7 billion light-years away, and the intensity after being transmitted from there to the detector is still 2.7k. What formula is used to calculate this? Is this still known science? This is downright false pseudo-theology!

Whether in a laboratory or anywhere, we would like to ask experts to do a simple experiment to show how any light or wave, including microwaves or radiation, can possibly stay still (not flow at all)?

Waves must flow. If a wave does not flow then it is not a wave, but a myth.

Waves also need pressure to flow. For example, lake water does not flow on its own but only by way of the pressure exerted on it by gravity or wind. The microwaves that maintain a uniform distribution of 2.7K and are still flowing are light waves that exist in mythology, and should not bother us when considering the universe and its origins.

And a kind reminder: Einstein said that the speed of light is constant. Be careful not to contradict Einstein.

Well, we now roughly understand where the problem of the myth of the microwave lies. Let us examine this myth more closely.

If, at the edge of our solar system, there exists a cosmic microwave with an intensity of less than 3K, is it possible for it to reach detectors in near-Earth orbit? Even if it is transmitted to the Earth by an agency yet unknown to us, is its intensity still 3K after being attenuated by one light-year distance? Can dedicated space probes orbiting the Earth be able to receive it? Of course, this is only possible in mythology.

NASA has created another mind-scaring super myth with countless dollars.

Below let's first briefly introduce the advanced detection instruments and results used by NASA in various periods for decades, and then scientifically argue why the whole sky map of the microwave background, as determined by contemporary science, is a cross-century joke.

Instruments used to receive cosmic microwave emissions: What have they told us?

Scientists believe that the discovery of cosmic microwave background radiation is of great significance to modern astronomy as it provided strong evidence in support of the Big Bang theory. Together with the discovery of quasars, pulsars, and interstellar organic molecules, it completed the list of what was termed, during the 1960s, the "four great discoveries" of astronomy. A decade later, in 1978, Penzias and Wilson who both discovered the cosmic microwave background radiation, were finally awarded the Nobel Prize in Physics for it.

Figure 1.2 below depicts the equipment used in four different periods for studying the cosmic microwave background, and the achievements obtained.

Figure 1.2 Microwave background receivers and the Microwave background Whole Sky Maps in different periods (from NASA). From top to bottom are Penzias and Wilson period, COBE period, WMPA period and Plank period.

The period most important to contemporary scientists involves the data collected by the Wilkinson Microwave Anisotropy Detector (WMAP). This instrument compiles massive amounts of data on the cosmic microwave background. Such data then enables interpretations by scientists presented in numerous journal papers and related all-sky maps. In Figure 1.3 (below) a typical all-sky map image is presented.

Table 1.1 Milestones from theory to observation of the cosmic microwave background

1941 – Andrew McKellar was attempting to measure the average temperature of the interstellar medium, and used the excitation of CN doublet lines to measure that the "effective temperature of space" (the average bolometric temperature) is about 2.3 K

1946 – George Gamow calculates a temperature of 50 K (assuming a 3-billion-year-old universe),

1948 – Ralph Alphard and Robert Herman estimate "the temperature in the universe" at 5 K.

1949 – Ralph Alpher and Robert Herman re-re-estimate the temperature at 28 K

1953 – George Gamow estimates 7 K.

1956 – George Gamow estimates 6 K.

1957 – Tigran Shmaonov reports that "the absolute effective temperature of the radio emission background ... is 4±3 K".

1960s – Robert Dicke re-estimates a microwave background radiation temperature of 40 K

1964–65 – Arno Penzias and Robert Woodrow Wilson measure the temperature to be approximately 3 K. Robert Dicke, James Peebles, P. G. Roll, and D. T. Wilkinson interpret this as a signature of the Big Bang.

1966 – Rainer K. Sachs and Arthur M. Wolfe theoretically predict microwave background fluctuation amplitudes created by gravitational potential variations between observers and the last scattering surface (see Sachs-Wolfe effect)

1968 – Martin Rees and Dennis Sciama theoretically predict microwave background fluctuation amplitudes created by photons traversing time-dependent potential wells

1969 – R. A. Sunyaev and Yakov Zel'dovich study the inverse Compton scattering of microwave background photons by hot electrons (see Sunyaev-Zel'dovich effect)

1983 – Researchers from the Cambridge Radio Astronomy Group and the Owens Valley Radio Observatory first detect the Sunyaev-Zel'dovich effect from clusters of galaxies

1983 – RELIKT-1 Soviet CMB anisotropy experiment was launched.

1990 – FIRAS on the Cosmic Background Explorer (COBE) satellite measures the black body form of the CMB spectrum with exquisite precision, and shows that the microwave background has a nearly perfect black-body spectrum and thereby strongly constrains the density of the intergalactic medium.

January 1992 – Scientists that analyzed data from the RELIKT-1 report the discovery of anisotropy in the cosmic microwave background at the Moscow astrophysical seminar.

1992 – Scientists that analyzed data from COBE DMR report the discovery of anisotropy in the cosmic microwave background.

1995 – The Cosmic Anisotropy Telescope performs the first high-resolution observations of the cosmic microwave background.

1999 – First measurements of acoustic oscillations in the CMB anisotropy angular power spectrum from the TOCO, BOOMERANG, and Maxima Experiments. The BOOMERANG experiment makes higher quality maps at intermediate resolution, and confirms that the universe is "flat".

2002 – Polarization discovered by DASI.

2003 – E-mode polarization spectrum obtained by the CBI. The CBI and the Very Small Array produces yet higher quality maps at high resolution (covering small areas of the sky).

2003 – The WMAP spacecraft produces an even higher quality map at low and intermediate resolution of the whole sky (WMAP provides no high-resolution data, but improves on the intermediate resolution maps from BOOMERanG).
2004 – E-mode polarization spectrum obtained by the CBI.
2004 – The Arcminute Cosmology Bolometer Array Receiver produces a higher quality map of the high-resolution structure not mapped by WMAP.
2005 – The Arcminute Microkelvin Imager and the Sunyaev-Zel'dovich Array begin the first surveys for very high redshift clusters of galaxies using the Sunyaev-Zel'dovich effect.
2005 – Ralph A. Alpher is awarded the National Medal of Science for his groundbreaking work in nucleosynthesis and prediction that the universe expansion leaves behind background radiation, thus providing a model for the Big Bang theory.
2006 – The long-awaited three-year WMAP results are released, confirming previous analysis, correcting several points, and including polarization data.
2006 – Two of COBE's principal investigators, George Smoot and John Mather, received the Nobel Prize in Physics in 2006 for their work on precision measurement of the CMBR.
2006-2011 – Improved measurements from WMAP, new supernova surveys ESSENCE and SNLS, and baryon acoustic oscillations from SDSS and WiggleZ, continue to be consistent with the standard Lambda-CDM model.
2014 – On March 17, 2014, astrophysicists of the BICEP2 collaboration announced the detection of inflationary gravitational waves in the B-mode power spectrum, which if confirmed, would provide clear experimental evidence for the theory of inflation. However, on 19 June 2014, lowered confidence in confirming the cosmic inflation findings was reported.
2015 – On January 30, 2015, the same team of astronomers from BICEP2 withdrew the claim made on the previous year. Based on the combined data of BICEP2 and Planck, the European Space Agency announced that the signal can be entirely attributed to dust in the Milky Way.

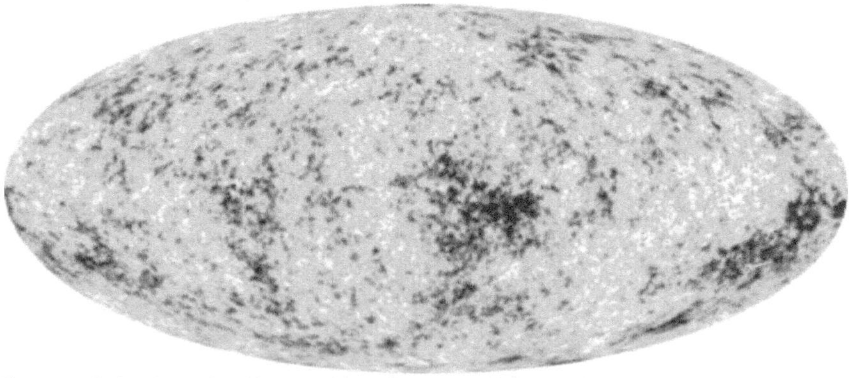

Figure 1.3 the cosmic microwave background whole sky map drawn according to WMAP detected data

With continuing data received by satellites launched by the United States, the European Union, and Russia, the whole-sky map has been expanded to include other ray wavelengths, such as the infrared, and so on.

However, after careful consideration, we have determined that if we use the methods that scientists have mastered, or other developing methods – even those methods merely imagined – it is impossible to observe the CMB (which is only 2.37k and is uniform distributed throughout the entire universe). In fact, the all-sky map of the CMB cannot be obtained directly from observations. It is the absurd result of a derivation from given theory.

In fact, it is doubtful whether we can observe the microwave value of each space point in the solar system within a distance of just about one light-year.

We describe our views from the following four aspects:

1. Use of cutting-edge but unsuitable equipment: the WMAP space instrument is a pure *RECEIVER*

The following presents a view of, and text information on, the WMAP space instrument as excerpted from NASA:

https://wmap.gsfc.nasa.gov/media/990180/index.html

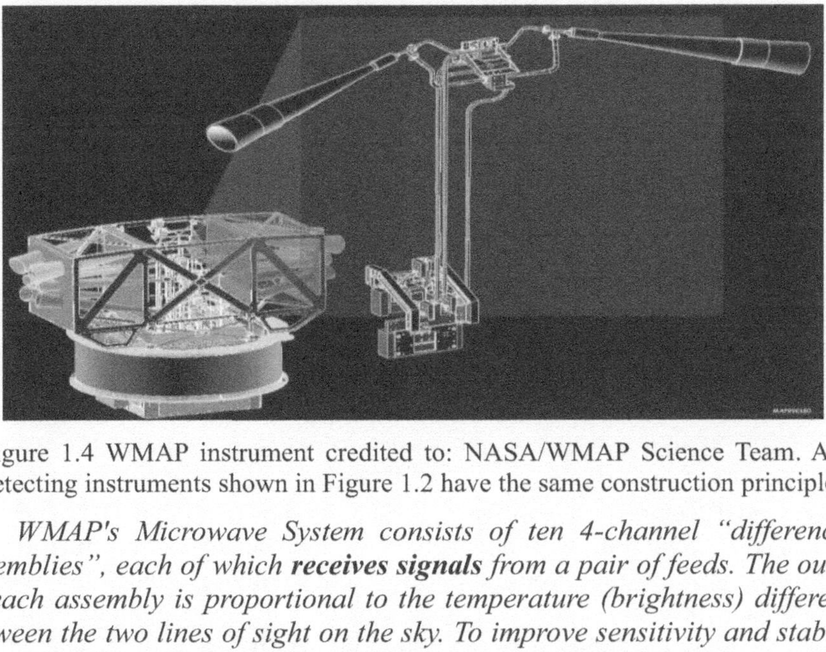

Figure 1.4 WMAP instrument credited to: NASA/WMAP Science Team. All detecting instruments shown in Figure 1.2 have the same construction principle.

WMAP's Microwave System consists of ten 4-channel "differencing assemblies", each of which ***receives signals*** *from a pair of feeds. The output of each assembly is proportional to the temperature (brightness) difference between the two lines of sight on the sky. To improve sensitivity and stability the system has a cold section where the first stages of microwave signal amplification occur, then a warm section where the remaining amplification occurs before being converted to an electrical signal.*

The above view and description clearly identify the WMPA as an **instrument designed to receive signals**. It does not have any capacity to emit and receive reflected signals like a radar from great distances. It can only receive signals that arrive on its receiver in situ!

In fact, all space telescopes, detectors, or any long-distance observing instruments can only receive data in situ. The data obtained by such instruments is limited, making it unsuitable for detection of the cosmic microwave background, which is theoretically defined as **uniformly distributed** throughout the entire universe.

2. An impossible mission: Detecting microwaves evenly distributed throughout the entire universe using data received by instruments in near-Earth orbit

So, we would like to ask NASA the following questions:

Is it possible for a microwave detector to detect microwaves with an average of 2.37K that glow uniformly in the entire cosmic sky?

It is completely impossible!

In figure 1.4, the description of the WMAP instrument clearly shows that: WMAP **only has signal-receiving components** and does not have the ability to actively observe distant celestial bodies. The WMAP detector claims to detect, but actually it can only receive information that reaches the point on the satellite orbit where it is located. In other words, the WMAP detector can only detect data from its orbit around the Earth. Detection of long-range data is beyond its capacity.

What's more mysterious is that the cosmic microwave background is defined as "**uniformly distributed throughout the universe**." Uniform distribution implies stasis, not flow compromising or negating entirely the capacity of the instrument to receive microwave data from distance. **How then can the instrument receive data on an evenly distributed microwave background one meter away**, not to mention microwaves originating from areas of space billions of light-years away? This is really a big cowhide that spans the entire universe for decades and costs billions of dollars!

This is not only the case with WMAP detectors, but also the microwave background detectors used in different periods, as shown in Figure 1.2. They will not actively detect but will passively wait for the arrival of microwaves. Microwaves have to flow from high to low, but evenly distributed microwaves have the same pressure everywhere, they can't flow from one point to another.

From their positions in near-Earth orbit, WMPA and other instruments received and continue to receive microwave data, now for more than half a century. And it can be said that they have traveled to every point in the sky near the Earth and measured the microwaves at these points around the Earth. We can conclude then that NASA has obtained a full-circle, uniformly distributed microwave map around the Earth, which extent does not even equal one light-second (or 186,000 miles) across.

What does this show? Think of the so-called 13.7 billion light-year-scale universe!

Could the data obtained from measuring the sky with a radius of less than one light second be used to identify microwaves with the same intensity throughout the universe? The quick answer is no.

Therefore, it is absurd to think that cosmic microwaves are uniformly distributed in the universe, and it is extremely unscientific and unreasonable to say that "*microwave detectors can* ***see*** *that it is* ***evenly*** *emitting light in the entire cosmic sky*".

3. A Self-circular argument: NASA's Big Bang Theory

CMB is one of the bases for proof of the Big Bang theory. Based on the above discussion, the process of how scientists use the CMB to prove the Big Bang theory is summarized in Figure 1.5.

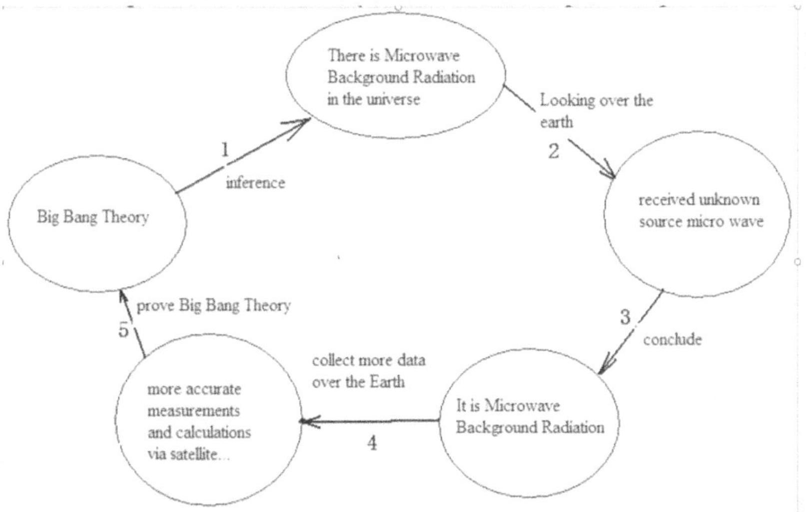

Figure 1.5 Circular proof of the Big Bang theory using microwave background

According to the Big Bang theory, an exploded atom created the universe along with the microwave background, which exhibits uniform distribution. As we have previously noted, this atom was not only tasked with creating the universe but also with generating microwaves that are evenly distributed throughout the universe.

How can this be proved?

Experts will tell us: the Big Bang theory requires an even distribution of microwaves throughout the universe. Receiving data on those microwaves in the space defined by near-Earth orbit equals to that they are uniformly distributed throughout the universe.

The problem here, as depicted in Figure 1.5, concerns two points in the figure: point 1 (inferring, from the Big Bang theory, there is microwave background radiation in the universe) and point 5 (using this inference to prove the Big Bang theory).

Let's use the following table to illustrate this confusing logical relationship:

Table1.2

We assume "the car accidents are all caused by drunkenness,"	Astronomers assume "the microwaves are all caused by the Big Bang,"
We found a lot of drunk people on the road.	Astronomers measured a lot of microwaves near the earth.
Use it as evidence that "all car accidents are caused by drunkenness"!	Use it as evidence that "all microwaves in the universe are caused by the Big Bang"!

Isn't the logic in this absurd?

Here, the most ridiculous scientific circular argument in the history of science appeared:

The Big Bang theory requires a uniformly distributed microwave background; this evenly distributed microwave background requires the Big Bang theory to prove its existence.

If it is determined that the microwave background is a product of the Big Bang, then it should determine that there is a Big Bang first, and then the microwave background is derived. Logically speaking, it is necessary to measure the microwave background in the entire universe, rather than only locally measure around the Earth and use it to say it is distributed in the whole universe. NASA experts said that they saw (Figure 0.1) the microwaves are evenly distributed in entire universe, and are the same as the vicinity of the earth, and that the microwaves of the whole universe have been measured with most advanced instruments (in fact, for decades, only microwaves near the earth have been repeatedly received,) and then those exaggerated data were used as the evidence for the Big Bang Theory.

NASA experts believe that microwaves are evenly distributed throughout the entire universe exactly the same like they are in vicinity of the Earth. They believed that if the microwaves are measured near the earth with instruments, it equals to the microwaves in the entire universe are measured. So, they actually just used the most advanced receivers near the Earth to repeatedly receive microwaves for decades, and used this locally received data as the data measured in the entire universe, telling the world that this is an image of the whole sky map, using the received local data as the data of entire universe, and taking it as "scientific" evidence for the one atom Big Bang theory.

In fact, there is no basis to assume that data measured at one small area is the same when measured at another area, and certainly no basis to assume that such data can be found as evidence of a uniform distribution of microwaves throughout the universe. This only becomes possible within the context of the Big Bang theory. In truth, using the microwave background as supporting evidence for the Big Bang theory has trapped scientists in a mythical loop of self-circular argumentation.

Let us take a closer look at how the most powerful combination of theories, experts, and satellites in contemporary times has turned decay into magic, and made the impossible possible. Let's see how our experts have falsely used the data measured from a small point (in near-Earth orbit) and mythically evolved it into an all-sky map of the universe.

It is as if experts took measurements repeatedly from instruments each the size of a sesame seed, and then broadened their field of relevance to encompass measurements for the whole Earth!

4. Dancing on a Sesame Seed

If topography were measured from results of a study done on a sesame seed, and then broadened to the entire topography of the Earth, including all of Earth's mountains, rivers, plains, valleys, and other physical features, what would happen?

Are there scientists who would ever do such thing?

Yes, there are.

For example, if we were to study the nationwide distribution of haze, of course it would require the deployment of thousands of observation points nationwide.

Imagine a well-known Chinese expert did otherwise. This expert went to YiChun, China, to collect accurate data points for measuring haze from its urban, suburban, rural, nearby mountainous and other areas. Then this expert extrapolated from the data points he measured a national haze estimate; that is, for all of China. Employed were a variety of mathematical models based on known data trends from defined directions. Should we applaud this expert for thoroughness? Can the extrapolation actually clarify how much haze exists across the entirety country of China?

Let us assume that there is another expert noted for his or her work over hundreds of kilometers of water surrounding Hawaii, including its large and small islands. With numerous observation points in this one geographical area and data analyzed from only these points, can this expert claim that such data can accurately depict a map of all the world's oceans – accurate enough for a ship captain to plot an expedition across the Pacific?

The answer is obvious.

At the same time, let us not forget the authority commonly given to experts, however farcical their results may be. Do advanced instruments and fancy mathematical models that demonstrate theoretical reasonableness make the results that experts present to us more believable?

If an expert's model derives, for example, from a certain model developed by Einstein, will you believe it or not? If most experts in the world agree with this model and have verified or are verifying this model through various means, will you believe it or not?

Think about it; do not answer too fast. After all, it is God's operation.

Over the last decade, NASA has published a variety of microwave whole-sky maps of the universe. The observational model and data sources are not essentially different from the pattern exhibited by the examples given above, despite the grander scales involved. As for the national haze map, note that it was extrapolated from an urban and surrounding areas with an expanse in relation to the entirety of the country of no more than one millionth.

But let us return to the whole-sky map of the universe. Suppose we set NASA's satellite detection range size as N, and the size of NASA's whole sky map of the universe as W, then the ratio of N and W (N/W) is

N/W > one to one hundred trillion (N/W > 1 / 100000000000000).

In scientific observation, the equipment used is as important as the analysis of the data that the equipment collects. But neither is equal to the importance of the model that shapes the analysis of the data, which must itself be scientific and reasonable. Otherwise, a state of "garbage in, garbage out" prevails with more sophisticated instruments and more perfect data perfecting greater deceptions that hide the truth. There is little doubt that NASA's whole-sky map, drawn from satellite observations, has not drawn us closer to the truth. Rather its models and styles of analysis act in violation of the laws of nature!

What's left, then, for NASA experts? Continue on as they have and expand the microwave data collected in space near the Earth to the entire universe.

Although this is a very bold idea, unfortunately, it is still wrong!

It is an act without a scientific core to arbitrarily expand the data received from the microwave background detector in one small area to the entire universe. Isn't this just like magnifying the data obtained from repeated measurements on a sesame seed to encompass the entire Earth?

How big is the universe? Let's temporarily accept NASA's estimate of 13.7 billion light years.

How big is the microwave background satellite orbit probe? The Plank satellite's orbit is about five light-seconds, equivalent to about one hundred thousandth of a light-year.

The universe radius is 7 billion light years (assuming that the Earth is at the center). The ratio of the orbit radius of a microwave background satellite detector to the radius of the universe is **1:7 million billion**.

The circumference of the Earth at the equator is 40,075 km long. The size of a sesame seed? Assume it is 1 cm in diameter. Comparing the size of a sesame seed to that of the Earth presents a ration of 1:400 million!

That relationship is far, far less than the proportion of the microwave background satellite orbit relative to the universe's radius (70 million-billionth to more than 400 million equals **17,500,000: 1**).

What if we take a sesame seed outside and imagine someone holding a precision instrument on that seed, measuring again and again from it sufficient to form a mathematical model, and then declare a result that encompasses the entire Earth? How reliable would that result be?

In fact, if we go deeper, another question arises: where did the microwaves detected near the Earth come from?

NASA experts looked for it casually, and then concluded that a source could not be found — precisely because they theorized it came from remnants of the Big Bang. The circularity.

We give here a possible source of microwaves detected near Earth.

5. Possible sources of microwaves detected near the Earth

So, where did the microwaves come from; those identified by satellite detectors like WMPA?

The question remains. For all the research done on this question,0 no precise answer has appeared. Coupled with the redshift phenomenon – which bears evidence of an expanding universe – scientists have constructed a narrative based on theory and tied it to the Big Bang as its remnant. As far as what existed before the explosion, scientists would do well to leave that to God.

The same kind of false logic surfaces with the identification of dark matter. Never having been seen, in fact, it, too, remains a theory. The logic, however, goes something like this: because HCOs cannot be seen, they must then be dark and composed of matter; that is, dark matter.

Light, of course, a particular form of radiation, which we are sensitive to – that is, light in the spectrum visible to humans – comes into play here. Is it possible to measure the size and weight of an elementary particle that also exhibits wave characteristics, such as a photon? A photon, of course, is defined as having no weight. But if the weight of a photon is only one thousandth or millionths of another elementary particle, again, do humans have the ability to measure its weight? The little weight attached to the photon causes it to travel thousands of light-years before it slows down after leaving its light source. Can humans measure this decreasing rate of speed? Perhaps it can be measured, but we have never seen an expert test a photon in this manner. We would rather believe that photons are weightless and that the speed of light is constant. For the sake of pseudo-theology, we have given up on exploring this issue further, and cling to old concepts, however "established" they are.

Light Waves Passing through a Point P in the Space

Now we discuss a general phenomenon: what is the status of light waves passing through any point in space? We call this point P point. What are the light waves passing through this point P?

In fact, we have answered this question in detail in "Who Should Talk About Cosmos" published in 2005. Readers can refer to the subsection "Superposition of Light" on page 71 of Chapter 2 "Viewing the Universe with the Naked Eyes" which is directly related to the current discussion. Here we discuss it in greater depth.

At any point P in the space where there are no celestial bodies, there are countless various light waves emitted from different celestial bodies passing through that point P from different directions! Yes, countless light waves!

For any celestial body, when its age is greater than the value of its distance to the point P in light-years, its light waves pass through that point. For example, if a star's age is 10,000 years old, and the distance between it and the Earth is 12,000 light-years, its emitted light will not reach the Earth. But if the star's age is 15,000 years old and its distance to Earth is 12,000 light years then its emitted light will reach the Earth. But there is one condition that needs to be met: the intensity of the light emitted by this or any other celestial body, after distance attenuation is factored in, does not decay to zero; or will not decay to below undetectable levels for a receiving instrument at point P.

If a receiver is put at point P, such as on the Planck satellite, all of the information from the universe it receives can only be those waves passing by point P. And all these light waves passing through space are flowing away from their light source!

There is no light in the universe that is not moving. Light flows in one direction only: away from its source. In addition, light intensity decreases with distance. As distance increases, light intensity decreases.

But it seems the light wave information received by NASA satellites is stationary because **uniformly distributed wave can't flow**. We would like to ask NASA experts to demonstrate what a stationary light wave looks like! Of course, this is a bit difficult for experts: because there are no waves in the real world that do not flow at all, whether light, sound or water!

This is something that belongs to the pseudo-theology of the Big Bang universe that exploded out from an atom.

Let's examine this more closely. From a light source, energy, or radiation, in the form of light constantly spreads into space, and as it does so its energy, or radiation level, decreases. If the target body under observation is 'A' light-years away, an observer will see the target body as it was, not as it is. That is, an observer will see light from a target body as old as when that light first flowed into space. Now numerous celestial light waves will pass through point P, superimposing waves on each other. The detecting instrument – whether on a satellite or ground-based telescope — is merely a receiver or resolver: when receiving all that light that it does, it also filters out lights from celestial bodies that are not of interest to researchers.

In Figure 1.6, Countless various types of light pass through the point P where the observer is located. We broadly classify the light rays from k1 to k7 are drawn in the figure and they contain many different types.

k1 and k2 indicate that the light of normal celestial bodies comes to the observer from different directions and continues to travel far away.

There is an 'x' at the end of k3, indicating that the light source has been destroyed at a certain time in the past, which is discussed earlier. The situation where the Sun was destroyed 8 minutes ago.

The front end of k4 has not yet reached the point P where the observer is located, indicating that the celestial body that emits the light is too young, the light it emits is still on the way to the observer, and we cannot see this light, it is only in our reasoning, its typical example is a 30-year-old black hole located 50 million light-years away.

k5 is like k3, but its light source is destroyed at some point after passing the observer.

k6 is special, it is a kind of light wave we deduced: the light source F of k6 is relatively far away from the observer. Only some special or residual wave can be detected by the observer. Microwave background wave is this kind of light wave.

Like k6, k7 is also a case where the wave source celestial body cannot be directly observed, but part of its light waves can be measured.

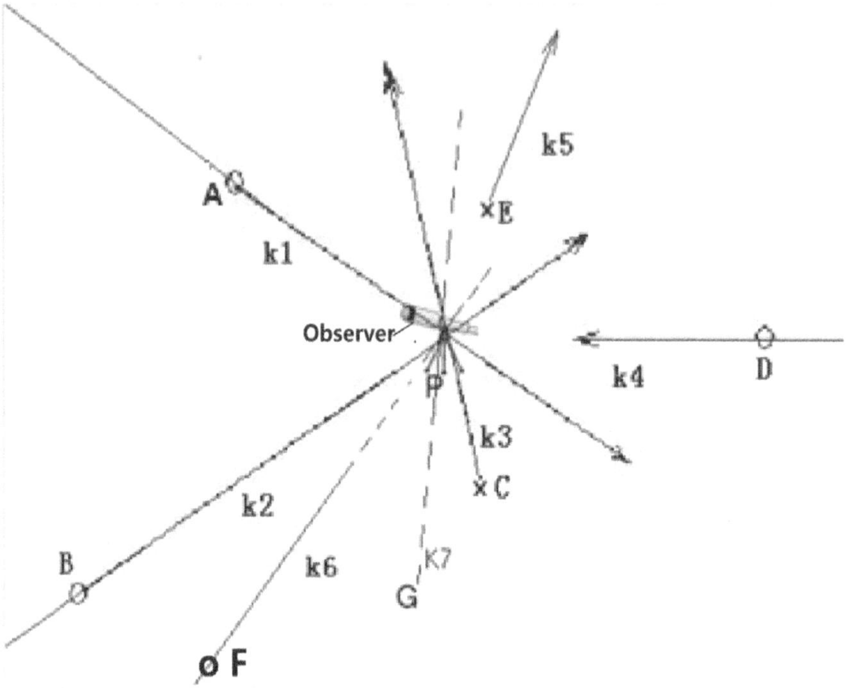

Figure 1.6 Light waves Passing through Point P. The observer is located at P. Remember that the light source celestial body is centered on the celestial body and emits light waves in all directions. We only focus on the view from the observer's point of view in the diagram, so only a straight line is used to represent the spherical light wave in the diagram. The solid line in the figure represents the normal light wave, while the dashed line represents the weaker light wave that the observer cannot clearly distinguish.

These light waves contain several lights invisible to human that would make people crazy. Let's think about how many types of light with no source can be measured.

First, we must keep in mind a basic unchangeable principle of light wave transmission: after a source emits light, it decays rapidly in inverse proportion to the 12 times of the square of the distance of transmission.

In Figure 1.6, A, B, D, F are represented by a normal celestial circle o. Light waves A and B normally pass through a detector at an arbitrary point P. D is different: it's too young, its light (represented by k4) has not yet reached P. D exists only in our reasoning. F is trickier, because its light, after travelling some distance, will exceed its saturation response point; that is, with its light attenuated badly, what is visible to the observer is not continuous but discontinuous light (indicated by a dotted line, which indicates the limited observable radius of this light source).

The light of K7 from celestial G is attenuated too much, so that the light source G cannot be identified. As a result, we can't see G. Such a situation exists in our inference. This is a class of hidden objects whose light is very weak, but not non-existent, and can be lower than the noise level from the telescope out at P. But if there is a plurality of such residual waves superimposed on hidden objects, it is possible to add these together to achieve a significant energy level.

Celestial bodies C and E, represented by X, means the light source does not exist. E only exists in our imagination, and C was discussed in the section of "Seeing is not real - reality is illusory" on page 73 of my first book "Who Should Talk About Cosmos. Suppose k3 represents the value of light-years of the distance from points P to C. Because C disappeared k3 years ago, but its light has continued to spread through the point P, it will continue to spread to point P for k3 years! If k3 = 300 million light-years, and C was born four million light years ago 5 million light-years away from Earth, and then C was destroyed 100 million years ago, its light will continue to be visible on Earth for another 300 million years on a smaller scale. The celebrated physicist, Stephen Hawking, quipped that since sunlight takes 8 minutes to reach Earth, if the Sun died now, we won't know it for 8 minutes

We can also see from Figure 1.6, there are three main types of celestial bodies in the universe represented by F, G and C, and many celestial objects that belong to these three categories: **many lights passing through point P whose source cannot be observed.**

If point P is replaced by the Earth, or the Earth is at point P, then from many unknown light sources light waves that come to the Earth very likely include the so-called cosmic microwave radiation background waves. The source of microwaves measured near the Earth possibly come from such celestial bodies like G.

We remember reading an article from NASA positing that the cosmic microwave radiation looked as if it came from two distant, non-exist celestial bodies, which is quite in line with the situation we are talking about here.

The microwaves measured near the Earth may come from one or some other invisible celestial objects. We will discuss the concept of hidden objects in detail later.

We use wisdom eyes to see through all confusion.

Chapter Summary

The microwave background theory attempts to prove that the Big Bang accurately depicts how the universe was born – with the microwave background the product of the originating explosion. As there is no provable science here, it is absurd to try, and try again, to provide the proof science requires. The behavior involved in this charade is neither scientific nor theological. It results in pure pseudo-theology.

Even if, for argument's sake, the narrative seems true, what kind of instruments can detect microwaves that are evenly distributed throughout the entire universe? No one can do this except God! And there is so much to do when this atom explodes! Don't make this atom work so hard and take on so much responsibility, please.

We would like to give NASA experts a chance to think carefully and see what kind of detectors they can devise that can measure microwaves in the entire universe near the earth? Please come up with any kind of detector you can imagine; excepting the instruments that come from God, of course. See if your dream-level detector can actually detect microwaves with an intensity of less than 3k in every place in the universe?

In this regard, NASA and their related astronomical and cosmology communities are more than a bit inauthentic. Although several generations of microwave detectors have been used, they have measured microwaves only near the earth from areas as large as a sesame dot. But the publicity generated by those experiments is also attractive, perhaps as much as the background map of the microwave distribution of the entire universe. It's so mythical, isn't it? It is not inferior to the myth of an atom blasting out the universe!

Experts used following "scientific" method: first invent the theory that the universe was generated by an atom exploding will generate the cosmic microwave background after cooling down. Then **receive** microwaves **near the Earth** to prove the theory that the **entire universe** is full of uniformly distributed microwaves. Finally use above as an argument to "scientifically" prove the Big Bang Theory. Such a wrong cycle of proof has cost taxpayers hundreds of millions of dollars. Should we ask some people whether taxpayers' money can be squandered like this?

Academically, NASA has also set a bad example to the astronomical and cosmological academic community. Why would we say so?

Please search for "UV cosmic background" on the Internet. There are not only microwave cosmic backgrounds that cost billions of dollars and decades to produce, but also various other cosmic backgrounds produced by different teams in different countries, such as X-ray cosmic backgrounds, X-ray cosmic backgrounds, ray and MeV gamma-ray cosmic background, UV/X-ray cosmic background, infrared cosmic background, etc.

Why are there so many cosmic backgrounds?
Because everyone follows the NASA model. This pattern is:

First, theoretically assumed that the ray (such as UV/X-ray) is uniformly distributed in the whole universe, and then the ray is detected (actually received) near the Earth, in space, or on the mountains, and then according to the pattern from NASA of making microwaves cosmic background, producing a map of the ray (such as UV/X-ray) distribution under the name of "cosmic", that is, the background distribution of the entire universe in which the cosmic ray was detected.

Experts are all smart people. Since you can invent a concept of cosmic microwave, and then get some microwave data near the Earth, made it "cosmic microwave background", so why can't we also use another cosmic ray, find it near the Earth, and make this ray a new "cosmic" background? Or even simpler, first find a certain kind of ray near the Earth, and then borrow the Big Bang theory to extend this ray to the entire universe!

Let's make money together! Let's make academics together. Let's "make" academic research!

Had many such kinds of cosmic background rays been made now? Sure yes!

Dear experts, scholars, academicians, and Nobel Prize winners, please come out and explain or criticize us, either admit your mistakes; or criticize what we said, and make yourself innocent, and let us taxpayers orally and heartly convinced.

Please consider this our academic challenge to NASA and other teams of scientists!

Chapter 2

The Myth of Dark Energy - It Hasn't Been Found for a Century and Never Can be Found

Introduction to Dark Energy

In general, dark energy is something that scientists have imagined in order to resolve loopholes in the Big Bang theory. More than half a century after the emergence of the pseudo-god Big Bang theory (in which an atom blasted out of itself our universe), the possibility of dark energy emerged. Recall the red-shift phenomenon, which discovered bore evidence of an ever-expanding universe, rather than an expected contracting universe. How could this happen? After the Big Bang cooled down, the universe should stop expanding and begin contracting. Thus, scientists invented dark energy as the means or motor that facilitated expansion.

Dark energy not only needs to provide power for resisting the shrinking force of the universe after cooling, but also provides continuous power for the accelerated expansion of the universe, which makes dark energy a concept that is easy for people to completely deny from a scientific point of view.

In physical cosmology, dark energy is defined as an imperceptible form of energy that fills space and thus fuels the rate of expansion of the universe. This hypothesis offers one of the most popular explanations for why the expansion of the universe is accelerating. Indeed, in the standard model of the universe that scientists have devised, dark energy occupies a majority (68.3%) of the mass-energy of the entire universe.

There are two existing models of dark energy: the cosmological constant (that is, a constant energy density that fills space uniformly) and the scalar field (that is, a dynamic field whose energy density varies with space and time). Scalar fields that have a constant effect on the universe are often included in the cosmological constant. The cosmological constant is physically equivalent to the energy in vacuum. Spatially varying scalar fields are difficult to separate from the cosmological constant because their variation is so slow.

In ΛCDM, the current best model of the Big Bang theory, dark energy is explained as a cosmological constant in terms of general relativity. However, the value of the cosmological constant, as based on general relativity and which can reasonably explain dark energy, is still surprisingly small. Comparing and choosing between the cosmological constant and other alternative theories to explain dark energy is one of the most active topics in the current field of Big Bang research.

What Is the Most Important Use of Dark Energy that Experts Have Overlooked?

Proving dark energy from a scientific point of view is very difficult despite the different narratives established in its name. In fact, dark energy does not exist other than in the imagination. It involves a stunning denial of the many achievements of scientific research, and strikes at the heart of what we know about the universe. Although scientists have yet to discover dark energy in fact, they have constructed a theoretical basis for it, ever complex and abstruse, and ever less convincing.

Here we conduct scientific criticism from several different perspectives, which begins with a question:

1. Why do cosmologists invent something like "dark energy"?
2. The most important use of dark energy that has been overlooked by experts — Accelerates the expansion of the universe that has cooled down and should be contracting according to the Big Bang theory.
3. Evidence should not be fabricated to prove a theory.

1. Why do cosmologists invent something like "dark energy"?

The Big Bang theory establishes that an explosion of energy from a primeval atom created the universe.

After the long years of the Big Bang, today the Big Bang has completely cooled down, and the universe should have stopped expanding and already started contraction a long time ago.

But at first the Big Bang Theory didn't take that into consideration.

The first factor that experts considered is the phenomenon of cosmic redshift, the theory that the universe is expanding, which provides scientific observations with what appears to be strong evidence for the Big Bang theory.

It seems that the smart experts think this way: The data I observed say that the distant celestial bodies are moving away from me, doesn't that mean that the space continues to expand? Why is the space expanded? Because of the Big Bang!

Sounds quite reasonable.

But after a few decades, experts suddenly woke up: the **universe** that has **cooled** after the Big Bang must **contracting** instead of expanding! But in this way, the scientific observation and data of the redshift of the universe obviously does not support the Big Bang theory. Not only did the redshift not support the Big Bang, but it became scientific theory that used scientific observations to oppose the Big Bang theory!

Because the scientific observations of the redshift of the universe tell us: the universe is expanding. And according to the Big Bang theory, the universe should be contracting! Therefore, the redshift scientific observation and data is against the Big Bang theory!

It was like hearing about a grenade that exploded yesterday, sending a growing cloud of explosion smoke from the center of the explosion. But today, this cloud of smoke is still expanding. Then, is it necessary to investigate: Is this the smoke really from a grenade? It is impossible for the smoke from a grenade to continue to expand for a whole day! The real reason for the expanding smoke should be reinvestigated.

The cooling universe continues to expand, is it caused by the Big Bang?

The red shift of the universe, which was originally strong “scientific” evidence to prove the Big Bang theory, has turned into evidence against the Big Bang theory.

It's so sad.

How to handle it?

Wouldn't it be okay to admit that the Big Bang Theory was wrong? Then everything would become normal.

Sadly, scientists love the Big Bang theory: Think how easy the Big Bang universe has become for scientific research, everything is within the grasp of man!

God's messengers also favor the Big Bang theory: God created the universe from a single atom! How great god is!

As a result, experts racked their brains to invent dark energy to take on the magical task of continuing to accelerate the expansion of the universe. Dark energy, they say, is what keeps the cooling, supposedly contracting universe expanding!

So, where does this omnipotent dark energy come from? Or let that Big Bang atom generate dark energy! Let that atom blast out the dark energy!

But the task of generating dark energy not only doubles the burden the atom that caused the Big Bang, it also makes it more difficult to "scientifically" prove the Big Bang.

Because it is necessary to provide the continuous power for the continuous expansion of the universe, this makes dark energy a concept that can be completely rejected by experiments from a scientific point of view. The biggest problem with not being able to make up myths to "scientifically" prove the existence of dark energy!

This invisible and intangible dark energy is amazing. It not only makes the universe that should be contracting not contract, but also makes the universe continue to expand, and the farther away it is, the faster it expands!

Magical, right?

But it's nothing.

An atom can produce a universe, so much dark energy just provides the power to expand the universe, not as magical as that atom!

Here is where dark energy enters the picture. Scientists needed a means that ensures expansion, and they found one: another theory to resolve the issue in support of the Big Bang. In fact, dark energy is a magical construct without form or detectable expression. It exists, not in science, but in mythology.

In order to illustrate the above passage, please consider this question:

Why after anything explodes, such as a grenade, does the explosion only affect an area around the center of the explosion? Why doesn't the explosion spread to places more and more distant and with greater intensity?

Classical physics dictates that the energy of the blast wave of the explosion, in fact, is similar to the expansion of light from its source: both spread in a spherical shape. With distance, **the energy produced by the explosion decreases exponentially; the farther away from the center of the explosion, the weaker the energy produced by the explosion, and the slower the diffusion of the shock wave until it collapses altogether – where effects from the explosion cease.**

In our world, the explosion, whether caused by gas cylinders or nuclear bombs, will spread into the space surrounding it. Distance attenuates the rate at which the explosion will spread. In this way, the impact of the explosion gradually diminishes and until it ceases. The explosion is confined to a certain range of space.

This basic truth should act in the same manner in the Big Bang theory with a universe whose rate of expansion should decrease after cooling.

But the Big Bang theory puts it this way: The farther celestial bodies are from the center of the explosion, the faster they will speed from the center of the explosion (their speed is proportional to the distance travelled). Only God can make such a thing happen.

The redshift tells us that space is expanding NOW! But now the whole universe has already cooled down. If there were a Big Bang, now the Big Bang created universe should be contracting! Redshift theory is exactly against the Big Bang theory!

Of course, what they say is based on the observed redshift of celestial bodies. We will discuss in detail in the fourth chapter and point out that it is not convincing to directly attribute the cause of the measured redshift to the retrograde motion of celestial bodies without proof. The dark energy we are discussing here strengthens our argument.

According to the Big Bang theory, if dark energy is involved, the explosion of a gas tank on Earth should not be contained within an area specific to the explosion and the cessation of its effects. Abetted by dark energy, such an explosion should expand infinitely.

This is obviously absurd.

However, after adding dark energy to the Big Bang theory, this conclusion is implied。

In order to complete the myth of the Big Bang, we have to answer the following question in the theory of the Big Bang pseudo-theology: how can the farther away from the center of the explosion, the faster the celestial bodies will recede. That is to answer the following question:

Where does the energy that pushes the supposedly contracted celestial bodies in the cooled universe to accelerate away from the center of the explosion come from?

Dark energy is a story derived to make up for the loopholes discovered more than 60 years after the appearance of the monatomic Big Bang pseudo-theology.

Of course, it's this mythical construct, dark energy; a narrative constructed to resolve the loopholes in the theory. In brief, it is an invention which pays a critical role in explaining how the Big Bang occurs. Dark energy is a story that was created to fill in the loopholes discovered more than 60 years after the pseudo-theology of the Big Bang.

Stuff like this is amazing. Consider, for the sake of argument, how a little bit of dark energy would affect an automobile engine. If added to the gas tank, after its first explosion in the cylinder, it would continue to expand under, effortlessly continue the engine with the force it needs to work the pistons. The engine, no longer in need of gasoline after ignition, will never stop for lack of energy.

The history of the Big Bang theory is also a bit ridiculous to review in detail. Proposed in 1927, it's been with us for more than 90 years. At first, scientists became aware that celestial bodies were accelerating and the universe expanding by way of measuring the redshift value by spectral analysis of stars. Similar to the Doppler sound wave effect, if turned on its head – where a source **grows louder the closer** it is and fainter the further away it is – scientists believed that the increase in the redshift value was caused by the movement of the celestial body away from the observer. Thus came the birth of the Big Bang theory. But after more than 60 years, in the early 1990s, scientists suddenly realized this: even if there was a Big Bang, how could this explosion make the universe that has hence cooled down continue expanding? Shouldn't it stop expanding and contract?

At this time, people should reconsider whether the Big Bang model is accurate or not. Is the observed cosmic red shift really caused by the receding movement of celestial bodies? Does the theory derived from Doppler's law, which was taken for granted, reasonably explain the redshift phenomenon of the universe? The arguments we have offered conclude in the negative.

However, the mechanized habitual thinking entraps people to move forward along the old thinking. People do not reflect on whether the previous theories derived from Doppler's law can reasonably explain the redshift phenomenon of the universe. Instead, they try to violate the basic principles of science and use "science" to forcefully explain the loopholes in the Big Bang pseudo-god theory, and turn the redshift theory against the Big Bang into a "scientific" theory for the Big Bang.

So **dark energy** was invented to fill this ridiculously large hole.

2. The important use of dark energy that experts overlooked - fighting cosmic contraction and accelerating the expansion of the universe

For more than half a century, humans have been trying to find this so-called dark energy, but never succeeded.

The latest Wikipedia entry describes dark energy as follows:

"In physical cosmology and astronomy, dark energy is an ***unknown form of***

***energy** that affects the universe on the largest scales. The first observational evidence for its existence came from supernovae measurements, which showed that the universe does not expand at a constant rate; rather, the expansion of the universe is accelerating. Understanding the evolution of the universe requires knowledge of its starting conditions and its composition. Prior to these observations, the only forms of matter-energy known to exist were ordinary matter, antimatter, dark matter, and radiation. Measurements of the cosmic microwave background suggest the universe began in a hot Big Bang, from which general relativity explains its evolution and the subsequent large-scale motion. Without introducing a new form of energy, there was no way to explain how an accelerating universe could be measured. Since the 1990s, dark energy has been the most accepted premise to account for the accelerated expansion. As of 2020, there are active areas of cosmology research aimed at understanding the fundamental nature of dark energy."*

To explain: Although Wikipedia is not an authoritative academic source, it is, after all, for public reference and revision, and reflects relevant basic information. It, therefore, is appropriate to use in this book, which speaks to general opinions and perspectives.

Cosmic experts have given us a narrative that originally included, post Big Bang, a cooling trend throughout the universe that finally works to contract matter and energy. In brief, the universe, like all other explosions, should have stopped expanding. However, the redshift theory, following the Doppler effect as measured by telescopes, identified an opposite trend: the more **intense** the redshift is under spectral analysis, the greater the speed of the celestial body moving away from the observer. And this trend has been found to be universal. Restated in a slightly different way, the farther a celestial body is from the center of an originating explosion, the faster it moves away from that explosion. Unfortunately, this narrative violates the laws of physics known to us.

As we have explained, in order to account for this phenomenon, experts imagined dark energy. Because of the dark energy produced during the Big Bang, it can make the universe, although having cooled, not contract but accelerate its expansion — and the farther away from the center of the explosion, the faster it expands.

Assuming that the lambda-CDM model of cosmology is correct, the best measurement results currently show that dark energy accounts for 69% of the total energy in the observable universe today.

From the above Wikipedia explanation provided above, we can see that the invention of the concept of dark energy involves the following logical points, however doubtful and yet unverifiable they may be：

a. Telescope observations have found that the observed celestial bodies have different redshift values.
b. Scientists have used a newly devised form of Doppler's law to explain the redshift value of the celestial body, as measured by telescope observation.

They believe that the distance of the celestial body from the observer is the only reason why the cosmic redshift occurs (despite lack of proof that this is the exclusive reason).

c. Judging from the redshift values obtained, the Big Bang Theory was born.
d. But the cooled universe should contract rather than expand;
e. And the universe whose objects are speeding away from each other also requires energy to support this acceleration.
f. So dark energy was invented. It is said that dark energy was generated during the Big Bang, which can make the cooled universe not contract but continue to accelerate its expansion. No matter how that concept is imaginary and violates the law of conservation of energy. (The explosion itself can produce energy that makes the explosion not only eternal, but increasingly accelerates the expansion of the explosion space. Where, however, is the scientific basis and logical argument for such a miraculous thing?)
g. Scientists have searched for dark energy now for decades without having found any of it; it remains a theoretical construct.

In fact, this narrative is still not accurate enough, because there is no certainty about whether something like dark energy even exists or not. Isn't it very questionable whether it really exists when a global community of scientists has tried their best to discover it but has failed for so long? Why then bypass the question – whether it exists or not – to discuss its use? In addition, doesn't endlessly discuss its form, its characteristics, various mathematical models related to it, and other efforts to study its evolution present the impression that discovering – that is, actually finding it — is not a problem at all. Seems it's not important to find it, but it's important to analyze its theoretical characteristics.

It's a pity that the global physics community doesn't have its priorities in order.

What are the important and major uses of dark energy that have been overlooked by experts? What is the role of the invention of dark energy? Hasn't it to allow an original theory about a post-Big Bang universe that is already cool down, violating the physical laws known to us? It certainly seems so. Describing the universe as originating in an explosion that expands forever is the issue. It's time that issue was put in its place.

Dark energy like this is really amazing. This kind of thing that exists everywhere is so thin that it can't be seen, and it is not even as strong as rays or light. But it not only can withstand the contraction force of the universe after cooling, but also promote the accelerated expansion of the universe!

We have to point out another headache for the experts (sorry, your problems are too many.) Assuming that dark energy was really produced by the Big Bang, is it still producing dark energy after the Big Bang cools down?

Please answer this question carefully. Dark energy supports the cooled universe from contracting after the Big Bang cools down, and a kind of dynamic equilibrium is reached at this time,

However, dark energy also promotes the accelerated expansion of cosmic space. Does the dark energy in the expanded space have greater "kinetic energy" than it in the old space without expansion? If not, how can the dark energy in the larger new space after the expansion have a greater thrust than the dark energy in the old space before the expansion? How can the expansion of space be accelerated?

It's like dividing a small part of the gas in one basketball into another deflated basketball. This deflated basketball will like what we are told, would be more elastic than the original basketball.

Magical enough, right?

So, what is the use of such magical dark energy to human beings?

The most meaningful thing about dark energy is that it can make an explosion that starts to continue indefinitely. This will solve humanity's energy problem forever!

Cosmologists have ignored this most important feature of dark energy, consciously or not!

They are silent about the characteristics of such dark energy that benefit humanity.

Think about it, even the universe can make it continue to expand, then, let the gas in a started engine continue to expand, of course, it is not a problem.

Therefore, the focus of the work of the cosmophysics community should be to find out even a little bit of dark energy as soon as possible to prove that such a magical thing really exists in the universe. Because it's so important, and it helps us use a tiny bit of starting energy to expand forever like a perpetual motion machine! Think how much that's worth! Then think about the amount of dark energy in the universe, don't make people too excited to sleep?

If anyone can find it, he will become the world's largest energy tycoon, benefiting all mankind and making himself rich! Think about the fact that after the engine is started, adding a little dark energy can make the explosive gas in the engine continue to expand, and putting it in the engine of the car can make the car run forever with almost no fuel. What a fascinating dark energy!

Of course, there are pros and cons, and the fact that dark energy can make the explosion continue to expand also makes it easier for people to tell whether experts have found real dark energy. Put a little of what you find into the blast space, and if the blast continues and spreads faster, it's really dark energy. However, such a thing that violates all the laws of physics and has obvious uses is really hard to find and hard to fool. Therefore, they felt sad and kept their mouths shut from touching the magical effect of dark energy.

If such a miraculous thing cannot be found, how can its existence be asserted? Even if there is such a thing as the Big Bang, it is hard to believe that there may be such dark energy that violates all human knowledge and all laws of physics and allows an explosion to expand forever! This kind of thing that goes against science has not been found for decades. We bet it can't be found even after hundreds of thousands of years.

Dark energy is really a good thing, it can easily solve the human energy crisis. It's a pity that this is just a dream that can't come true! Only people with super courage dare to spend a lot of money and time to search in vain for such things that are fundamentally unscientific!

Dark energy is the stuff of myth.

Human beings can never find this kind of dark energy that is produced from an explosion and can make the same explosion itself continue to expand. This kind of thing that has no scientific and technological content and is full of contradictions and cannot be justified has become the main theory of today's universe science. It is really strange! Apart from magical myths, is there any real scientific connotation in it?

3. Evidence should not be fabricated to prove a theory

As we have discussed, in order to maintain the big bang theory, both dark matter and dark energy are theoretical constructs that experts imagine exist but cannot prove.

How can one assert its existence if such a miraculous thing cannot be found? Even if there is such a thing as the Big Bang, it is hard to believe that there may be such dark energy that violates all human knowledge and all laws of physics, coming from an explosion, and allowing the explosion to expand forever! This kind of thing that goes against science has not been found for decades. We believe we can't find it for hundreds of thousands of years.

Dark energy is really a good thing, it can easily solve the energy crisis of mankind. It's a pity that this is just a good dream! This kind of thing is fundamentally contrary to science, only a person with super daring dares to spend a lot of money and time to research and search in vain!

Dark energy is the stuff of myth.

Humans will never find the kind of dark energy that generated from the Big Band and that allows the explosion itself to expand continuously. This kind of thing that has no scientific and technological content and is full of contradictions and cannot be justified has become the main theory of today's cosmic science. It is really strange! In addition to magical myths, is there any real scientific connotation in it?

Astronomic circles say that, over the next few decades, the search for dark energy will be their highest priority.

Isn't this really a waste of human intellectual and material resources?

Dark energy is something imagined to cater to the false gods of the big bang, but the task of this thing is too heavy: it must resist the shrinking force of the universe after it cools down, and it must also support the important task of expanding the space of the universe derived from the redshift theory! How can dark energy, which is so thin and weak as to be inferior to light, support such a difficult task? The mere dark energy is simply unbearable!

Most importantly, until dark energy is discovered — if it ever can be discovered — people should not use it as the foundation for a theory that describes the creation of the universe, make up various theoretical models on behalf of the theory, and discuss its various properties — **as if its discovery alone were not the issue at hand, it is more important to figure out its model, structure, use, etc.**

The absurdity of this situation was obvious to us, and we hope it is to you as well. Until it is really discovered, we call on experts to either discover dark energy, or stop hyping it, as if finding it is not a problem at all.

Now it is such a situation: experts are talking about all kinds of mysterious theories about dark energy as if dark energy has actually been found. As a result, most ordinary people believe that dark energy is already in the pockets of physicists and cosmologists.

In view of the fact that it took a lot of manpower and material resources for decades to find no dark energy, and in view of the mythical nature of dark energy, before dark energy is really found, the cosmic astronomy academic community should stop using dark energy to explain any theory! Especially the Big Bang theory of the pseudo-god of the universe.

In fact, dark energy is something that has been hitting NASA astronomy experts in the face!

And the experts have no escape this time around. No matter how good NASA telescopes and probes, no matter how authoritative, it can't make this dark energy story come true.

This is because, first of all, they made up the story too big, about 70% of the matter of the universe is dark energy; then they made up the story too close, the whole universe (including the earth of course) is distributed everywhere dark energy; and the story is so magical that dark energy can continue to expand the universe after the explosion. In this way, experts can't make use of the most advanced instruments that others don't have, find something like gravitational waves from a distance, and then make up a scripted story to continue to fool the melon eaters.

Because the magical function of dark energy to continuously expand space cannot be realized! That is, dark energy is impossible to become something in reality!

To say that dark energy accounts for more than 70% of the total matter in the universe means that the universe is full of dark energy. Such miraculous dark energy can be found near the Earth. With dark energy everywhere, we can see what their telescope can see. Although the most advanced telescope only experts own.

Of course, if there is such a magical thing, everywhere in the universe, and it can also propel the movement of the expanding universe, then the means of transportation used to propel us become an inexhaustible energy source, how cool it would be!

It's a pity that it's been missing for decades.

It is a pity that this magical dark energy can never be found!

Experts, please hurry up to find and verify this magical dark energy. Don't look for it for decades or hundreds of years. Here generation to generation you can't find it, there Nobel medals are issued piece by piece for it. Dark energy is used in this way to explain the big problem that the cooled universe continues to expand in the Big Bang pseudo-god theory.

Chapter Summary

The astronomical cosmology community has been searching for dark energy for decades, and while they've announced its discovery from time to time, their results don't amount to much, including verifiable proof. Still, however, the scientific **community has been using dark energy as a "scientific" argument** — which is absurd.

We just want to ask respected scientists to answer a question: Can you give a rough timetable for finding dark energy? You can't look for a century or a thousand years in the name of science, and at the same time use this thing that you can't find as the "scientific" basis for your various projects, theories, ... especially the Big Bang theory? Don't let people who are looking forward to seeing dark energy wait until the end of the world, generation after generation, right? Right?

Up to now, using dark energy can make a space continue to expand without any energy, and it is easier to identify the authenticity of the dark energy that is claimed to be found.

It's a pity, isn't it, that dark energy has been missing for decades, and is still missing. Can it ever be found? No。

The proofs expressed by scientists in support of the Big Bang pseudo-God theory are basically in vain!

Stop wasting taxpayer money on the Big Bang.

Young friends don't waste your intelligence or your time plunging into the smoke and banter surrounding the Big Bang. The older generations have been struggling on it for too long. With the endorsement of the Nobel Prize and the full support of the world's astronomy and cosmology circles, it looks more like a scene full of vigor and violent fire than a methodical, truthful and honesty humble scientific exploration. In fact, the bottom is empty, the heart is doubtful, the foundation is wrong, and mistakes are made at every turn. You can see that all the questions in my book are very basic, but there are a lot of errors We have pointed out that cannot be refuted. It cannot be said, for example, that the scientists and scholars who made these mistakes are not trained in the scientific and other research methods and celebrated for their training, and how they use it. But once one gets into the profound "science" of the Big Bang based on mistakes, one can't help but make naive and ridiculous mistakes, whether consciously or unconsciously.

Please excuse yourself, disavow the smoke and glitter of consensus fantasies attached to the pseudo-God Big Bang, and look outside. You will find that the universe is so vast that it can't be encompassed by one tiny Big Bang!

Chapter 3

Why Dark Matter is not Dark? What is the Real Face of the Dark Matter?

Dark matter is completely different from dark energy.

Dark energy is made up by cosmic experts in order to make up for the big loopholes in the Big Bang pseudo-god theory, and theoretically caters to the Big Bang pseudo-god theory.

Dark matter, on the other hand, started from the search for unobservable celestial bodies that can't be observed but can be discovered by gravity. It is not something that was totally invented by people like dark energy, but was deduced by people's incomplete and wrong understanding of the universe.

In astronomical and cosmological science, the concept of dark matter is another well of confusion. And **rather than examining the prospect of a Hidden Celestial Object,** our experts keep to their search for the elusive substance that dark matter is made of.

Why is dark matter not dark?

What is a hidden celestial object?

Introduction of Dark Matter

The existence of dark matter was first proposed by Dutch astronomer Jacobus Kapteyn in 1922 using stellar velocities. Since then, scientists have created dark matter based on the pseudo-theology of the single-atom Big Bang.

We have been told that dark matter is matter that does not emit any light and does not reflect any light, but has a gravitational field. Is there such a magical substance? What kind of molecular and atomic structure would such a substance have, which makes it indifferent to the universe, and without the slightest capacity to express its presence other than by gravity? So far, as scientists have yet to find dark matter of any sort, it exists only in their reasoning — Decades of profound reasoning!

According to contemporary various astrophysical observations, and cutting-edge astro-cosmological theories, both of them have shown the presence of more matter than is visible, which implies the existence of dark matter. Based on the reasoning, the experts concluded that these observations and theories all imply that there should be some unknown invisible substance, that is mystery dark matter.

In fact, this reasoning is very problematic!

How come the substances you didn't see, or can't see, then they are mystery substances? Why aren't they common celestial objects that you haven't discovered yet? Why are you making better telescopes? The purpose you are making better telescopes is to find celestial objects that are still undiscovered!

Here is the fundamental point of disagreement between us and the master experts: When discovering that there is a large amount of undiscovered matter in the universe, should we reason based on basic scientific principles to discover where these unknown substances are, or magically invent something to barely explain?

The astronomical community fabricates the magical dark energy and dark matter that they haven't found for several generations, which violates all laws of physics, to explain the large amount of unknown matter in the universe; and we believe that these unknown matters are actually ordinary unknown celestial objects. The undiscovered celestial objects are just hidden from human beings' search and discovery so far for some reasons. They may be temporarily hidden in human exploration, but more of them are forever hidden from human cognition. This is the Hidden Celestial Object we are going to discuss.

Readers are asked to keep this in mind, and it will deepen our understanding of the seemingly simple concept of Hidden Celestial Object later on.

Experts will argue: those not yet found are too small a quantity, not enough to meet the theoretical estimate of the substance amount of the universe.

This is totally wrong!

Those not yet found are normal celestial objects. The amount of them is far more than current anybody thinks! But experts' imagination is limited by the Big Bang smog, limited within the Big Bang space.

For this reason, experts believe that dark matter is abundant in the universe and has a strong influence on its structure and evolution.

Calculations have shown that numerous galaxies would behave very differently than they do if they didn't contain large amounts of unseen matter. Some galaxies wouldn't ever have formed at all, while others wouldn't move as they do. This is the main evidence for the existence of dark matter. Other evidence includes gravitational lensing observations and the cosmic microwave background, among others.

In the Lambda-CDM cosmology model, the total mass-energy content of the universe consists of 5% ordinary matter and energy, 27% dark matter, and 68% dark energy.

Therefore, dark matter is a hypothetical form of matter that experts have conjured up. Experts told us that dark matter is "dark" because it does not absorb, reflect or emit electromagnetic radiation (such as light), so it has no way to detect, and can only be discovered by gravity.

No one has directly observed dark matter except by its purported gravitational effects. To confirm the existence of dark matter, it is necessary to use the gravitational interaction between dark matter and other matter; that much is clear. But so far, no particles that make up dark matter have been found in the laboratory. Physicists have proposed a variety of particle physics theories to try to explain dark matter, and have also designed a number of detection experiments to directly observe dark matter.

From the above, we can see how naive and straightforward the introduction of the concept of dark matter is: according to observations, calculations and theories, it is speculated that there is much more matter in the universe than the observed celestial bodies. Because they can't find it, it is something mysterious, let's call it dark matter. Anyway, there are already a lot of mysteries in the universe produced by the explosion of an atom, and it's no big deal to add another one mystery, and it just happens to be paired with dark energy!

The thinking of contemporary cosmologists is so amazing! You say they have no imagination, but they can unrestrainedly imagine magical things beyond all physical science theories! If you say they have imagination, they can't deduce things that conform to basic science instead of magic according to correct scientific thinking. !

In contrast, we believe that Dark Matter is nothing less, or more, than a **Hidden Celestial Object(HCO)**. Dark matter violates basic scientific principles.

As the perspective of the new cosmic architecture discussed later, compared to dark matter, there is far more HCO than you might think. This can refer to Figure 10.5 Schematic diagram of the distribution of Hidden Celestial Objects in the universe.

What really is the dark matter that people have been searching for, for so many years?

Dark matter, unlike dark energy — the mysterious energy imagined by modern cosmologists — is yet a provable object, which cosmologists have overlooked.

We define this kind of celestial body as an HCO (Hidden Celestial Object. For different distances and different observation instruments, they are sometimes visible and sometimes invisible.

We will illustrate this from the following aspects:

1) Dark matter is something that the academic circle has been legendary for many years and has not been confirmed or discovered. It's actually something that goes against basic scientific principles.

2) An intuitive explanation of Hidden Celestial Objects (HCOs). The concept of Hidden Celestial Object inspired by a truly original history of discovering real dark matter

3) The dark matters discovered are, in fact, HCOs

Some people will say: human beings have landed on the moon; how could NASA make so many mistakes? Are you talking nonsense?

In fact, NASA's business is divided into two parts, one is aerospace, and the other is astronomy. These are two separate departments in China. The United States combined them into one organization.

Mixed departments of Astronomy and Aerospace do not seem to be working very well. They ran wildly on the path of the pseudo-God Big Bang, and everything related to the Big Bang. Dark matter became a kind of magical matter indispensable to the Big Bang theory. And the Big Bang was originally a super-fantasy created by the imagination of scientists. It is full of flaws, and there is no way to really fully prove its authenticity as it is its allies: dark matter and dark energy.

Associating any research with the Big Bang and trying to prove the theory of the Big Bang by various means, of course, reveals countless loopholes that cannot be justified, leading astray mankind's understanding of the true nature of the universe.

Maybe these people think that they are safeguarding God's credibility, but they actually do bad things with good intentions.

1. Dark matter is something that the academic circle has been legendary for many years and has not been confirmed or discovered. It's actually something that goes against basic scientific principles.

WikipediA says：

"In the standard Lambda-CDM model of cosmology, the total mass–energy of the universe contains 5% ordinary matter and energy, 27% dark matter and 68% of a form of energy known as dark energy.

Because dark matter has not yet been observed directly, if it exists, it must barely interact with ordinary baryonic matter and radiation, except through gravity. Most dark matter is thought to be non-baryonic in nature; it may be composed of some as-yet undiscovered subatomic particles. The primary candidate for dark matter is some new kind of elementary particle that has not yet been discovered, in particular, weakly interacting massive particles (WIMPs). Many experiments to directly detect and study dark matter particles are being actively undertaken, but ***none have yet succeeded.*** *Dark matter is classified as "cold", "warm", or "hot" according to its velocity (more precisely, its free streaming length). Current models favor a cold dark matter scenario, in which structures emerge by gradual accumulation of particles."*

There are several key points in the above explanation:

• There is 27% dark matter in the total matter of the universe.

• No dark matter has been observed.

• The structure of dark matter is unknown; it may be something that humans have not yet discovered.

• For many years, experts have tried and failed to find dark matter.

Although the instruments scientists have used to search for dark matter are now more sophisticated with ever greater capacities for reception of data and the models derived from that data ever more complex, what progress has been made in discovering dark matter? No progress!

No progress at all!

Dark matter is still hidden in the unknown.

Odder still is why such a substance, whose characteristics remain unknown, would prompt scientists to consider it as a key substance for the Big Bang and as a substance that helps to explain the Big Bang theory. According to the Big Bang theory, " If there were only ordinary matter in the universe, there would not have been enough time for density perturbations to grow into the galaxies and clusters currently seen. Dark matter provides a solution to this problem because it is unaffected by radiation. Therefore, its density perturbations can grow first." (Refer to astronomy text book)

We really don't understand this kind of thinking that the masters of astronomy and cosmology seem so good at: for many years, they have spent countless amounts of money and failed in their effort to find even a minute quantity of dark matter yet still use it as one of the primaries supports of the Big Bang theory. What is the logic involved? What science is this?

Conducted based on such absurd mistakes, will the research get good results?

In fact, dark matter is very simple, it is an ordinary HCO.

2. An intuitive explanation of Hidden Celestial Objects (HCOs). The concept of Hidden Celestial Object inspired by a truly original history of discovering real dark matter

The celestial objects that are invisible to the naked eye and can be seen only in a telescope are hidden HCOs to the naked eye. The celestial objects that people want to discover when constructing bigger and better telescopes are HCOs.

As the name suggests, what we call HCOs are ordinary but invisible objects that exist outside of our vision. In fact, they are everywhere, close to the Earth and far away, whether tens of millions or billions of light-years away.

Hidden celestial objects are not dark matter either. Conversely, dark matter, if it exists, belongs to a small class of HCOs.

Some HCOs can be discovered, while others, many others more, may never be discovered by humans, and exist only in correctly formed scientific reasoning.

Using gravitational fields as a primary means to discover dark matter is nothing more or less than making direct measurements of a more limited area of space: the gravitational field expressed by dark matter in that space. Unfortunately, little progress has been made thus far in this kind of research.

Another, more profitable kind of research involves the discovery of a celestial body, such as Sirius B, the companion star to Sirius; the brightest star in our night skies. Scientists discovered Sirius B by noting an erratic character to the orbit of Sirius. In order to explain it, they theorized it could only happen if a companion star, Sirius B, were present. Then visual confirmation came, and we now know that Sirius is a double star, Sirius and B orbiting around each other. In the same way, dark matter can be found by observing its impact on nearby luminous celestial objects and then extrapolating its existence. But dark matter objects found in such cases, until now, were all finally proved, without exception, to be ordinary HCOs.

No exceptions! We don't understand why cosmologists still insist on calling them mysterious dark matter, rather than normal objects that cannot be seen directly! Because these hidden objects are not dark or non-luminous, and the term, dark matter, is incorrect.

In recent years, as noted, astronomers have devoted great efforts to search for dark matter. But all the celestial bodies that have been considered as potential sources of dark matter have, so far, failed the test. With more sophisticated instrumentation, whether ground or space-based, such bodies appear as they are: not sources of dark matter but normal celestial bodies.

The most famous example of this was the discovery of Sirius' companion star Sirius B as discussed above. Figure 3.1 is a screenshot of the relevant report. This is also the first famous example of "**dark matter**" discovered through the influence of gravitational fields on the orbits of other stars.

In 1987, Virginia Trimble at the University of California, Irvine, and the University of Maryland, titled her article "the existence and nature of dark matter in the universe" (Figure 3.1). She highlighted the fact that Friedrich Wilhelm Bessel, after decades of position measurements of Sirius and Procyon, concluded by implication that their "respective trajectories were affected by invisible matter comparable to their own. "

EXISTENCE AND NATURE OF DARK MATTER IN THE UNIVERSE

Virginia Trimble

Astronomy Program, University of Maryland, College Park, Maryland 20742, and Department of Physics, University of California, Irvine, California 92717

1. HISTORICAL INTRODUCTION AND THE SCOPE OF THE PROBLEM

The first detection of nonluminous matter from its gravitational effects occurred in 1844, when Friedrich Wilhelm Bessel announced that several decades of positional measurements of Sirius and Procyon implied that each was in orbit with an invisible companion of mass comparable to its own. The companions ceased to be invisible in 1862, when Alvan G. Clark turned his newly-ground 18½" objective toward Sirius and resolved the 10^{-4} of the photons from the system emitted by the white dwarf Sirius B. Studies of astrometric and single-line spectroscopic binaries are the modern descendants of Bessel's work.

Figure 3.1 The discovery story of Sirius B. A screenshot of the first report of the discovery of "dark matter", from which the name "dark matter" comes. This type of "dark matter" is fake and is actually a hidden celestial object. Note the date the discovery happened was 1884, much earlier than the dark matter ideal

Because telescopes were relatively backward at that time, Sirius B was not directly visible by telescope observation. So it was defined as “dark matter” in the title of the report; the first time the term was used. But with the improvements of the telescope, astronomers could, by telescope observation, see Sirius B. Of course, it was not the mysterious dark matter which scientists are futilely concerned with today. However, the term "dark matter" came into being then, later to attain mythic value.

When the term "dark matter" was born, it referred to celestial bodies that were invisible but whose existence could be inferred through other factors, such as gravity. It cannot be seen because the light emitted by the object is not strong enough for a telescope that is not sensitive enough to detect it, but it can appear in the lens of a better telescope.

We don't know when and who quietly tampered with the original and correct meaning of this "dark matter" and turned it into a mythical fetish that does not emit light, absorb light, reflect light, but has gravity. Friends who are interested can search and find out which master did such an overly clever thing, and this master should also be recorded in history.

Beginning with Sirius B, all the so-called dark matter discovered through the gravitational field were hidden celestial bodies that were later reevaluated after the improvements in telescope technology. This has happened without exception.

So far, mankind has not found any magical dark matter that was later defined.

And yet, the search continues. For the community of scientists involved, cosmologists and astronomers, the search for dark matter as a primary task through the next few decades.

We think that these scientists simply cannot find their so-called "dark matter," as they define it. In my book *Debate of Light and Dark - A 100 Year Bet with NASA*, published 2016, my bet is that even after another century of dedicated searching, no dark matter will be found other than celestial objects now hidden from sight.

The concept of "dark matter" itself is a comprehensive denial of the achievements made in scientific research; it undercuts our understanding of the universe and how the universe functions – it is a non-existent mythic thing proposed by imagination to cater to the theory of the pseudo-Theology Big Bang. The concept of dark matter reflects the rigid way of thinking in the scientific world, combining invisible and normal celestial bodies with the dark matter of gods.

So far, scientific knowledge holds that any substance is composed of basic structures, such as atoms and molecules. All substances composed of atoms and molecules will absorb or reflect light, or will do so after meeting certain conditions. The current definition of dark matter – that it does not absorb or reflect light – implies that it cannot be composed of the basic material constituent elements known to us, namely molecules and atoms. In other words, all of our scientific knowledge does not help us to understand what dark matter is or to use our technology to create it.

The search for dark matter is not the bold adventure that scientists want us to believe, but an expression of an irresponsibly identified pseudo-theology that undercuts the authority of science and the methods used in science.

The number of hidden celestial bodies defined later in this chapter accounts for most of the cosmic matter of concern. When considering HCOs, we quickly find that there is no room for dark matter. Please carefully compare hidden objects and dark matter to see which concept is more reasonable and more in line with the truth!

The simplest example of an HCO follows: If a celestial body with the same brightness as the Sun is located millions light years away from the Earth, no matter how sophisticated a telescope we might use, we will not be able to observe it! It is a Hidden Celestial Object HCO. This will be discussed in detail below.

3. Dark matters discovered are Hidden Celestial Objects

The definition of dark matter as we know it is: a kind of magical matter that does not emit or reflect any light waves, but has a gravitational field. What kind of molecular and atomic structure will this substance have? We cannot say other than that its existence is implied by the gravity it exerts on other celestial bodies.

The use of a gravitational field to discover dark matter is nothing more than directly measuring that gravitational field. But little progress has been made in this respect thus far. The discovery of Sirius B, however, implied its gravitational influence on the orbit of Sirius, makes for a more reasonable research route. Recall as well that all such "dark matter objects" identified in the same manner as Sirius B were finally confirmed as ordinary celestial bodies. Before being seen, they were hidden from us. Without exception!

Without exception! Why, then, do cosmologists still insist on searching for dark matter? Must they ignore the many instances in which they've been proven wrong. Their mysterious dark matter, identified by its gravitational influence on another celestial body or bodies, finally (with improved instrumentation) identified as a normal celestial object or objects. Where does this stubbornness come from? Why not consider those HCOs that we have been looking for since the beginning of our interest in, and passion for, observing the sky? Those celestial objects that were invisible to the naked eyes but can be seen with telescopes are hidden objects relative to the naked eyes.

Dark matter is also one of the most peculiar types of mythic substances. Using what we do know, that matter is composed of atoms and molecules, what constitutes a radically different kind of matter, which atomic and molecular science cannot account for, and which does not emit light, does not reflect light, and has gravity?

Because such non-absorbent, non-reflective, but gravitational matter violates the basic structure of all matters known to us. Of equal concern is the behavior of scientists when faced with their failure to discover dark matter. Rather than admitting the error of their ways — what was once thought to be dark matter, has now been correctly identified as HCOs – they have avoided it, rarely if ever mentioning it in the published literature.

HCOs are real. On the other hand, dark matter still exists only in reasoning and imagination: this ever-strange substance that cannot be detected except by its gravitational effect. In truth, it belongs to the kind of matter that people need to "understand specifically" – especially in regard to scientific trends, foci and funding sources. And even then, it is not so convincing.

For many years, the astronomy community has spent a lot of energy searching for dark matter. But in fact, so far, all found dark matter turns out to be normal, ordinary HCOs that create or reflect light. The existence of dark matter remains "dark": an unseen substance theorized into pseudo existence, which, if it did exist somewhere in the universe – however weird that would be – would of course be very scarce.

Worrying is the behavior of scientists in the face of failure to discover dark matter. They did not admit their mistakes. They try to avoid mentioning what was once thought to be dark matter, now correctly identified as HCO. Not mentioned even in published literature.

In performing scientific research on "strange" things, an initial dose of skepticism is healthy. For something in violation of the basic principles of science—that matter consists of atoms and molecules that emit and reflect light —we should treat it with great caution, and not belief, even if supported by consensus.

A similar situation prevails with the phenomenon that we now know as the cosmic microwave background. Since experts cannot currently find its source – whether star, gas clouds or galaxy that emits such microwaves — they favor its origin in and generation by the Big Bang. An alternative possibility exists, however. As with HCOs, we have simply yet to find the star, gas cloud or galaxy that emits microwaves near to or distant from the Earth. This microwave source may be a temporarily hidden celestial body, or it may be a permanently hidden celestial body; that is to say, we may or may not ever find it.

Since experts were unable to locate the source star or galaxy that emitted the microwaves, they believed that the microwaves were produced by the Big Bang. But the concept of hidden objects offers another theoretical possibility to find stars or galaxies emitting microwaves near or far from Earth. This source of the microwave may be a temporarily hidden object, we may find it; or it may be permanently hidden; that is, we may never see it.

So, what is a **Hidden Celestial Object**? Let's discuss and define it scientifically.

Hidden Celestial Objects Overview

At first glance, the concept of a hidden celestial object is so simple that it is hardly worth mentioning. We submitted the paper about HCO in the appendix IX to the proceeding of an international conference for publication, and the reviewer said that it was "trivial" and not worthy of publication.

However, this concept, which seems to be intuitively understood by everyone, has a very important meaning.

Because the predecessors have not systematically summarized the concepts related to hidden celestial objects, the concept is also not used much in astronomical observations. Otherwise, there would not have been so many apparently erroneous observation reports. So summarizing it into a small theory is helpful for people to use the theory consciously, so as not to make various low-level mistakes due to violation of the theory in astronomical observations.

In fact, it makes sense to put this paper here alone. If dark energy does not exist and dark matter does not exist, then where does the so-called matter that accounts for 95% of the universe go? The answer lies in the hidden celestial body!

More importantly, behind the concept of hidden celestial objects is a brand-new interpretation of the distribution of celestial objects in the universe. It tells us that beyond our limited observable space, there are celestial objects that we cannot see but really exist. How the celestial objects might exist and be distributed in the universe without the Big Bang.

When we completely deny the existence of mythical dark energy and dark matter, where do these dark things which account for 95% of the matter in the universe, go? What to fill such a huge space? That is the hidden celestial objects.

With the theory of hidden celestial objects in the system, people can imagine the grand universe beyond the scope of the Big Bang, washing their thoughts from the smoke of the Big Bang, and moving to a broader new universe!

Therefore, although the concept of hidden celestial objects is very simple and intuitive, it is also worthy of application in astronomical observations full of fallacies.

What do we look for in the sky with a telescope? What are we trying to find out?

We look at known celestial bodies and for unknown celestial bodies; that is, celestial bodies that exist but have not yet been found.

What are these yet undiscovered celestial bodies?

From another perspective, what are these celestial bodies that are hidden from our senses?

In brief, cosmologists should offer us help by identifying the specific conditions needed to discover celestial bodies that have not yet been seen, and under what conditions they may not be completely visible.

Instead, they all too much request more funding for their search for dark energy and dark matter, and for the perfection of the instruments that they use in this regard.

Be that as it may, it is necessary to scientifically and accurately depict the celestial bodies being sought, so as to determine how best we can improve our capacities for observations and to discover what can be discovered; that is, **what exists but cannot be found**.

The concept of "knowing what exists but hasn't yet been found or cannot be found " is particularly important.

We would be searching for HCOs that can originate from the same microwave background and resolve the puzzle. Because there is no space for them to exist. With this concept defining the breadth of our search, we would not waste our resources searching for an imaginary duo, dark energy and dark matter, because there is no space for them to exist. Or for that matter the fictive source of the universal microwave background: the Big Bang. We would be searching for HCOs that can originate from the same microwave background and resolve the puzzle. The source of the microwave could be in one or several HCOs.

What are these invisible celestial objects3? They are the HCOs that we have been searching for and finding with our telescopes.

Of course, we will need to define them precisely with simple mathematics (algebra will do), and then analyze, from those algebraic equations, why we can't see them and how we can see them.

From this kind of simple algebraic analysis used to define HCOs, in the end we will draw three important conclusions:

• After the technology of the telescope is sufficiently sophisticated, it will not be useful to make the telescope more precise and powerful. This will be proved later.

• Although we have not yet been able to find the source of cosmic microwave radiation, other than as a remnant of a mythic Big Bang, we, for one, believe it can be found in HCOs.

• The dark matter that the entire cosmic astronomy community is frantically looking for is just an extreme case of wishful thinking. Most likely it doesn't exist. But the 12 HCOs we defined later most likely do exist.

Astronomers do not take kindly to the first conclusion above. Their reasoning here is quite clear: they always want more public funding to develop bigger and more powerful telescopes.

The latter two conclusions speak to the absence of positive proofs that the Big Bang took place.

Even more surprising is that, with such a simple and important concept as we have just expressed, the masters of the cosmic sciences even don't want to waste their intelligent to do some related research work. What's more, they think that with their telescopes they can see what they want to see, and as far as they want to see, including the weak microwave background in any corner of the one atom Big Bang universe they defined, including the cases will be discussed in chapter 6.

We can't help asking: Is this science? Is this modern science? Is this modern cosmology?

The number of HCOs is far more than the number of found objects.

Human eyes, even if aided by the power of an advanced telescope, will still be blocked at the gate of truth by a few natural barriers included in the "distance"

To figure out the sources and applications of HCOs has some significance. Here, Hubble's law pertaining to the expansion of the universe can be examined, and we will even be able to find the true source of the cosmic microwave background by it.

We have always wondered why is there no concept for HCOs in the various communities that make up the science of astronomy. Perhaps it is because of their belief that as long as the telescope is good enough, any celestial body in the universe can be seen. They need only raise the telescope to the sky, and everything in the universe will resolve on its lens?

In the following sections, we will briefly discuss the true character of HCOs.

The detailed math abstraction is written in a standard paper and attached as appendix IX in the end of this book.

Comets Whose Cycles Bring Them Near to Earth are Typical Hidden Celestial Objects

For example, we have watched Halley's Comet for many years with the naked eye. However, each time it approaches the vicinity of the Earth, before being seen by people it is an HCO; after being seen it is no longer an HCO. When it travels a certain distance from the Earth, and we can no longer see it with the naked eye, then it becomes an HCO relative to human sight once again.

This actually means that for an ordinary object, when conditions of viewing alter, it is likely to transition from being visible to invisible, or vice versa.

With the naked eye, the Andromeda galaxy appears as a weak, vague oval presence, barely visible but there nonetheless. An interesting phenomenon worth noting is that, with our naked eye, we can see the Andromeda galaxy despite its distance, 2.54 million light-years away, but cannot see the star Proxima Centauri, 4.2 light-years distant.

Why? Ordinary explanations are neither complete nor systematic. We need to carefully study the changes in related factors and find out the law of the variations involved.

Figure 3.2 involves several aspects for analysis: the light received by the eyes, the light irradiating from the comet, and the distance between the eyes and the comet. Careful study of various possible changes among the three will allow us to find some very interesting patterns.

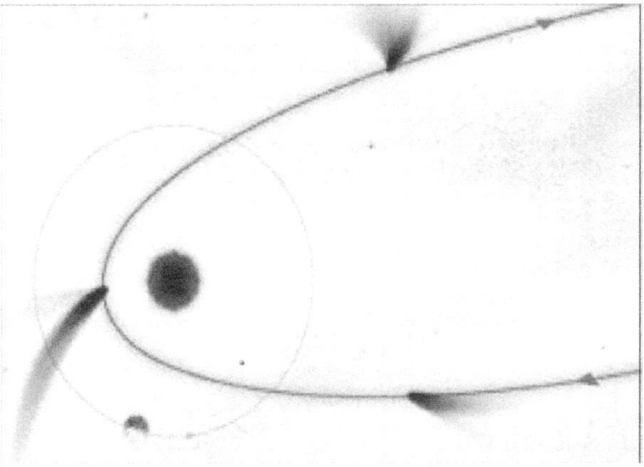

Figure 3.2 The trajectory of Halley's Comet

Several fundamental factors affecting the ability to see celestial objects

Let us first note how our eyes resolve things, including distant things, enabling us to see them. Involved is a basic law of vision expressed here with mathematical conciseness for any object or celestial body by formula (3-1):

$$Light\ eyes\ received = \frac{Light\ emitted\ from\ celestial\ object}{4\pi\ Distant^2} \quad (3\text{-}1)$$

Substituting the trajectory of Halley's Comet that is the object to be seen, we obtain a similar expression:

$$Light\ eyes\ received = \frac{Light\ emitted\ from\ the\ comet}{4\pi\ Distant^2}$$

Seeing this equation, the first thing we notice is that the light taken into eyes is inversely proportional to the square of the distance between the viewer, perhaps with a telescope, and the celestial body. In other words, it reduces very quickly with distance increases.

That is, at different distances, Halley's light irradiation on our eyes is not the same. When the distance is great, the light from Halley's Comet, which our eyes collect, is weak; in contrast, when the distance is small, the light intensifies. When the comet reaches the limits of our vision, we can't sense the light it emits and it disappears from sight.

Suppose the light intensity that Halley's Comet emits is essentially constant, without major change. So, in this case, the important factor affecting whether the Halley's Comet can be seen or not at a certain distance is the observer; and the eyes the observer has.

Good eyes, of course, can see more clearly than bad eyes; of that there is no doubt.

Then, to the same pair of eyes, under what circumstances can or cannot the light from Halley's Comet be seen?

Let's first understand how the eyes see things

The observing limit of the eyes

There was a piece on WeChat about an illiterate fellow who encountered a group of PhDs discussing why a big rain can't beat something to death. They cited formulas and reasoning until, finally, the illiterate fellow said this: Who of you have not been drenched in rain? Hearing this, the PhDs in the group suddenly scattered. The illiterate fellow was very proud that these “doctors” were clueless.

The topic being discussed was, why can't the rain beat the flying mosquitoes to death?

Some questions may seem simple but really contain a very profound truth. But if you do not stand on the height of the mountain, you cannot see the scenery on the other side of the mountain.

Now here is a silly question: can eyes see a photon?

The problem here is divided into two aspects. First, can the eyes accept a photon?

The answer is obvious: yes. If the photon goes into the eyes, the eyes cannot reject it?

Another question arises now, regarding the visual nervous system. Can it interpret the information that one photon can express? Will we eventually feel what it is like to actually "see" this photon?

After a photon enters the eye, the sensor in the eye on the retina reacts to it. However, in order to see something our eyes must receive nine or more photons within 100 milliseconds. With this quantity of photons falling on our eyes in this very brief span of time, the visual nervous system acts, producing a signal that will prompt our conscious response: we can see these photons.

In other words, if we want to see the information of light irradiated into our eyes, a certain degree of intensity of the light is required.

In fact, from the point of view of anti-interference, if the eyes transmit to the visual nervous system the information expressed by each photon, then we will be dealing with something – an image – of a mess, and the nerves in our eyes will quickly tire. Thus, minimum requirements for the light intensity we can see with the naked eye are not caused by an inherent weakness of the eye – our capacity to see – but by a very clever systems approach; in effect, a noise filter to block too weak signals.

Now it is clear, if we know the intensity of the light shining from Halley's Comet, according to formula (3-1), we can easily calculate the distance when we will be able or not be able to see Halley's Comet.

Of course, there are different eye visions. 1.5 eyes certainly see farther than 0.5. So, whether the vision of the eye itself is good or bad is an important factor affecting the observed results. This is discussed in more detail later: the important concept of observable limited radius.

Limited Observable Radius of Halley's Comet Relative to the Eyes

Limited or finite observing radius, an important concept. Using visual observation of Halley's Comet by sight will allow us to explain the concept relative to sight by the naked eye.

If, with a pair of the eyes with perfect 20/20 vision, you can see Halley's Comet from within a radius at some distance but cannot see when it is a little farther away, then this distance is called the limited observable radius of Halley's Comet relative to this pair of eyes.

Similarly, with a telescope, the greatest distance at which it can allow for observation of the comet is the limited observable radius relative to this telescope.

With improvements, telescopes can capture and resolve light from greater distances, increasing its corresponding limited observable radius. Thus, strictly speaking, the radius should be named **Relative Limited Observable Radius (RLOR)**.

Can the relatively limited observable radius be endlessly increased?

No.

Whether observed by sight with the naked eye, or by enhancements in telescope technology, including the space telescope, the RLOR relative to an object like Halley's Comet can increase, but this increase has a limitation. Eventually, no matter how much we improve our telescopes, the RLOR will not grow larger.

Why? When the intensity of the light sent by the celestial body we are observing reaches Earth, and that intensity is lower than the intensity of noise near the Earth – noise that affects seeing — no matter how advanced the telescope is, it cannot distinguish the light observed from the noise.

This effect, in relation to a celestial object, we call the **Absolute Limited Observable Radius (ALOR)**.

In the above discussion, the three mediums involved in an LOR are the observation instruments (eyes or telescopes) e, the observing target (Halley's Comet), and the distance between these two.

Changing anything in regard to these three mediums will influence the size of the observable radius, until noise is factored in. In this respect, where the ALOR comes into play, the telescope itself is not important; more important is what humans cannot control: noise. This part has not yet been discussed.

Noise Analysis

We know that during the observing process, there will be all kinds of interference noise.

The official Chinese translation of noise refers to sound as an analog, even in technical books that specifically describe astronomical telescopes.

There are various noises that influence the observation instrument. Their sources come from the natural environment and from the observation instrument itself. Whatever their source, we divide noises into two categories:

- **Controllable noises** can be controlled by humans through various means, including avoiding common atmospheric disturbances by moving the observer above them on mountains or by using a satellite in near-Earth orbit, blocking the sunlight, and so on.
- **Uncontrollable noises are composed** of two parts: those that are natural and those that are non-natural.

The uncontrollable natural noise can include diffuse light coming from the other celestial bodies when observing a planet, the background light of the sky, the zodiacal light, aurora light, and so on. Statistical properties of the photon itself also make the information received have ups and downs.

The uncontrollable non-natural noise is inherent to the observation instrument and represents insurmountable factors that optimization cannot overcome. For example, there is the size of the telescope mirror: due to materials and the processes involved in their use, it is difficult to make a large mirror with a diameter of one kilometer, and even if we could make such a large mirror, we cannot reliably collect the weak information it receives, which adds up to one photon per ten square kilometers. In addition, during the analysis of these photons, they may be absorbed and reflected, or cannot produce enough energy to record an event. Not all of the photons can be collected from the telescope to analyze either; there must be some loss.

Let us review: The noise that we have no way of controlling is

uncontrollable noise; the noise that we expect now, or in the future, to optimize out – by advances in technology — reducing its impact on observations, therefore increasing the output of observations, is **controllable noise**.

Further optimization of the telescope becomes useless after the sensitivity of the telescope reaches a certain level of the uncontrollable noise. Specifically, for example, if we increase the sensitivity of a telescope to 10 photons per square centimeter, but uncontrollable noise is 20 photons per square centimeter, the receiving level of the telescope remains at 20 photons per square centimeter. Attempting to improve its reception level to 15 or 10 photons per square centimeter is useless. Its capacity to resolve an image of received light cannot exceed the level of uncontrollable noise.

Let's use our eyes as an example. If our eyes are good enough to see in the range of four photons, but the noise level influencing our sight is in the range of eight photons, then we will not see what we want from the information provided in the range of four-photons. We can only identify the information provided by eight or more photons.

Since uncontrollable noise places restrictions on the observation of a light source, it does the same even on the most advanced of our observation instruments. The maximum observable radius is constrained by the uncontrollable noise that affects it.

Mathematical Abstract Definition of Limited Observable Radius

Limited Observable Radius (LOR) of celestial objects is such an important concept that we will define mathematically. Recall from the discussion above that it is divided into **Relative Limited Observable Radius (RLOR)** and **Absolutely Limited Observable Radius (ALOR)** of celestial objects relative to an observer. It can also simply be referred to as **Relative Observable Radius** and **Absolute Observable Radius.** (Figure 3.3)

ALOR is decided by the luminous intensity of the celestial object that is being observed and the uncontrollable noise.

RLOR is decided by the luminous intensity of the celestial object, the controllable noise plus uncontrollable noise, and the observer's capacity for optimizing the observation medium: sight, telescope, or other instrumentation.

The 'Relative' of RLOR is named because there are factors that people can contribute to.

In the equations in Figure 3.3, in the mathematical abstract qualitative formula of the relatively limited observable radius RLOR, the RLOR is determined by the luminous intensity of the celestial body, the uncontrollable noise, the controllable noise, and the sensitivity of the observation instrument.

When we consider the extreme optimization situation, that is, the controllable noise and the sensitivity of the observation instrument are reduced to the lowest level, zero, then the relatively limited observable radius RLOR becomes the absolute limited observable radius ALOR.

Abstract definition of Relative Limited Observable Radius (RLOR) is as following:

$$RLOR\ (L_C,\ NOISE_{natural},\ +\ NOISE_{control},\ SEN_{inst})$$

where: L_C: Light intensity sent by celestial c
$NOISE_{natural}$: Uncontrolable noise
$NOISE_{control}$: Controlable noise
SEN_{inst}: Instrument sensitivity

The more specific formula of RLOR is:

$$RLOR\ (L_C,\ NOISE_{natural},\ +\ NOISE_{control},\ SEN_{inst})$$

$$= \sqrt{\frac{L_C}{4\pi f(NOISE_{natural} + NOISE_{control},\ SEN_{inst})}}$$

The absolute observable limited radius ALOR:

$$ALOR(C) = RLOR\ (L_C,\ NOISE_{natural} + 0,\ 0)$$

Usually we write ALOR as LOR. That is:

$$LOR(C) = \sqrt{\frac{L_c}{4\pi f\ (NOISE)}}$$

Figure 3.3 Mathematical abstraction of the Relatively Limited Observable Radius (RLOR) and the Absolute Limited Observable Radius (ALOR)

We have determined the qualitative formula, but how to determine the specific quantitative formula is something that our readers can hopefully do.

In the end, the concept of the ALOR, or LOR, as it is also known, tells us that making bigger and better telescopes after a certain point is futile. No matter how sophisticated the telescope is, after the telescope reaches a certain quality, its scope of observation can only be within an absolute finite radius. As for the specific data that we can gain from observing the heavens, this is a subject of further study.

We have introduced several important concepts. Let's give them accurate mathematical definitions below: (you can skip the math part).

Definitions of Hidden Celestial Object (HCO)

From a mathematical perspective, we can determine the two most important simple formulas for Hidden Celestial Objects (HCOs).

Since observation instruments are located on or around the Earth, we use Earth to represent observation instruments. The distance between the Earth and the celestial object (C) is D (C, Earth). When this distance is greater than the RLOR of the celestial object C, that is:

$$\mathbf{RLOR}\,(L_C, NOISE_{natural} + NOISE_{control}, SEN_{inst}) < \mathbf{D}\,(C, Earth)$$

Then:

C, is a Hidden Celestial Object (HCO) relative to this Observation Instrument.

As can be seen, the factors that make a Hidden Celestial Object (HCO) are the same as those factors that decide the Relative Limited Observable Radius (RLOR). They are:

L_C of celestial object C, the emission intensity,
$NOISE_{natural}$, uncontrollable noise,
$NOISE_{control}$, controllable noise,
SEN_{inst}, the sensitivity of observation instruments like a telescope.

Careful analysis of the changes in the parameters of the formula, which requires a sensitivity analysis for each independent variable, gives various classifications of the hidden objects, as noted below.

No matter how advanced the telescope is, for those objects that are beyond its absolute limited observation radius, those celestial objects are **Hidden Celestial Objects (HCO).**

The celestial objects that are located outside the Relative Limited Observable Radius (RLOR) but inside the Absolute Limited Observable Radius (RLOR) are called **Temporarily Hidden Celestial Objects (THCO).** Usually, they are called **HCO**.

From an uncontrollable noise level, the Absolute Limited Observable Radius can be calculated corresponding to a celestial object. We may never see a celestial object located beyond the ALOR. The celestial objects located outside the Absolute Limited Observable Radius (RLOR) are called **Constant Hidden Celestial Objects (CHCO).**

Please note that whether a celestial object is THCO or not, it is relative to a specific observing instrument. It may be a THCO for some telescopes but can be seen by other more advanced telescopes.

The CHCO is relative to all observation instruments. If a celestial object can't be seen by the most advanced observer of the era, it is a Constant Hidden Celestial Object (CHCO).

Trying to extend the RLOR by improving the quality of the observer is the theoretical basis for turning a THCO into an observable celestial object.

Stars transform from visible to hidden objects

All the time stars burn themselves to radiate energy into space. But the star itself is a limited celestial body. When its energy is emitted to a certain moment, the energy emitted by the star will inevitably decrease. As time progresses, more and more energy is dissipated, and the star's brilliance must decrease accordingly.

When the energy emitted by the star decreases to a certain level, it becomes a hidden object to the observer. That is, the observer will never see it again.

To this observer, the star went from being visible to being hidden.

But in actual astronomical observations, humans have only used telescopes for a few hundred years. Only a few thousand years have been documented using the naked eye. And there are only a limited number of stars that can be observed and documented with the naked eye; plus the continued spread of the remnants of the star-destroying light. Therefore, it is very difficult for humans to detect a certain star disappearing from their observations.

Thus, the transition of a star from visible to hidden, which is mostly in our reasoning, is difficult to confirm by observation.

Summary of Hidden Celestial Objects (HCO)

Hidden Celestial Objects are invisible to us, but it is easy to prove their existence.

Around us, there are countless examples of real HCOs.

There is a big difference, however, in the quality of telescopes used to observe the sky. Those objects that can be seen with a high-quality telescope lens, but which are not visible with a poor-quality telescope, are known as Temporarily Hidden Celestial Objects (THCOs) to the poor-quality telescope.

All comets, for people who observe the sky with the naked eyes, or not using a high-quality telescope, are periodically Temporarily Hidden Celestial Objects. The famous Halley's Comet is a specific type of Temporarily Hidden

Celestial Object. Its periodic haunt writes a very clear footnote of Temporarily Hidden Celestial Objects.

A simpler example exists to explain this point: When we first observe the night sky with a telescope then move our eyes away from the telescope and, with our naked eyes, view the same sky, the celestial objects that were **visible with our telescope but invisible to our eyes are Temporarily Hidden Celestial Objects to our eyes. The telescope, though, proves their existence**. It's that simple.

A different example involving galaxies works in the same fashion. We can only see one galaxy in the night sky without technical aid: Andromeda (M31). This galaxy is 2.54 million light-years from Earth and is the farthest object we can see with the naked eye.

In other words, except for the Andromeda Galaxy and our own Milky Way, all other galaxies that we can see by telescope are Temporarily Hidden Celestial Objects relative to the naked eyes.

Now extend this example to those galaxies that can't be detected by the best telescopes that we currently use. They become Hidden Celestial Objects. How many are there? How many of them are Temporarily Hidden Celestial Objects and thus will possibly be found in the future because of improvements in our observing systems?

At the farthest distance we can see, we have only been able to identify a small number of particularly large celestial objects, such as supernovae. But, according to the law of large structures in the universe, the celestial distribution of such objects should be substantial. That is to say, in the space our galaxy contains there are many star clusters, individual stars and their planets, immense gas clouds, and so on. These are just normal objects that we cannot see with the naked eyes; in other words, Hidden Celestial Objects to our eyes.

Perhaps a brief summary will help, not only to clarify these concepts, but also to figure out how to define them.

1. Why can the naked eyes see the distant Andromeda Galaxy but not see the nearby star Proxima Centauri? Or why can we sometimes see Halley's Comet and sometimes not see it? ->

2. What factors influence the results of astronomical observations? What variations limit these factors? ->

3. Do our observations have limits? -> Limited Observable Radius。

4. There are two extreme limits changing in different circumstances: the analysis of absolute factors and of non-absolute factors.

5. About changes of non-absolute factors, give a definition to describe it and name it as -> Relative Limited Observable Radius。

6. About changes of absolute factors, give a definition to describe it and name it as -> Absolute Limited Observable Radius。

7. What are the corresponding changes in non- absolute observation factors?

Give a definition to describe it -> Temporarily Hidden Celestial Object

8. What are the corresponding changes in absolute observation factors? Give a definition to describe it -> Constant Hidden Celestial Object

In such a way, we defined our terms and obtained the following simple Figure 3.4, by which you can more clearly see the thought process:

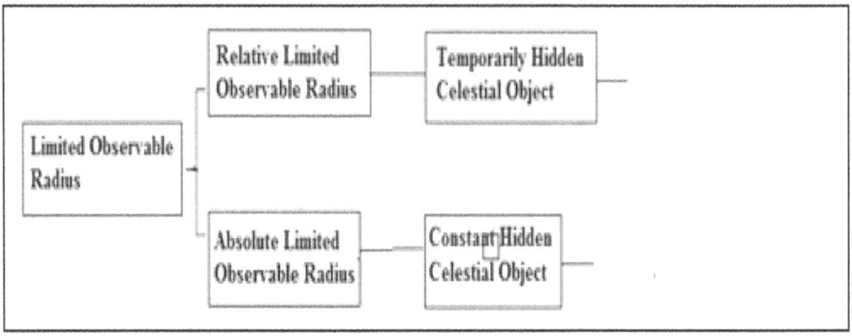

Figure 3.4 The HCO definitions extended from Limited Observable Radius

Usually in scientific research, I like to first try to abstract and generalize complex phenomena as much as possible. Analyze changes in the most common case; analyze possible changes in extreme cases. I hope this simple method of scientific research can inspire you, so I wrote the following paragraph.

Categorize the Hidden Celestial Objects

This categorization follows the classification concepts in Figure 3.4, analyzing the sensitive relationship between all relevant factors of Hidden Celestial Objects in its definition formula and considering increases or decreases in these factors caused by change, resulting in the following Figure 3.5. Please refer to Appendix 2 for the detailed mathematical classification process.

Please refer to my books for the detailed classification process. I'm here to specifically talk about the Hidden Celestial Objects of Class 10 at the lower right corner in Figure 3.5, whose light intensity is constant 0.

This situation is depicted in the class of phenomena known as "dark matter and dark energy." The official definition of dark matter and dark energy are objects that do not emit any light. Zero light intensity of dark matter and dark energy is in its definition, and their ALOR = 0; thus, dark matter and dark energy, to any observer at any distance, is a hidden object.

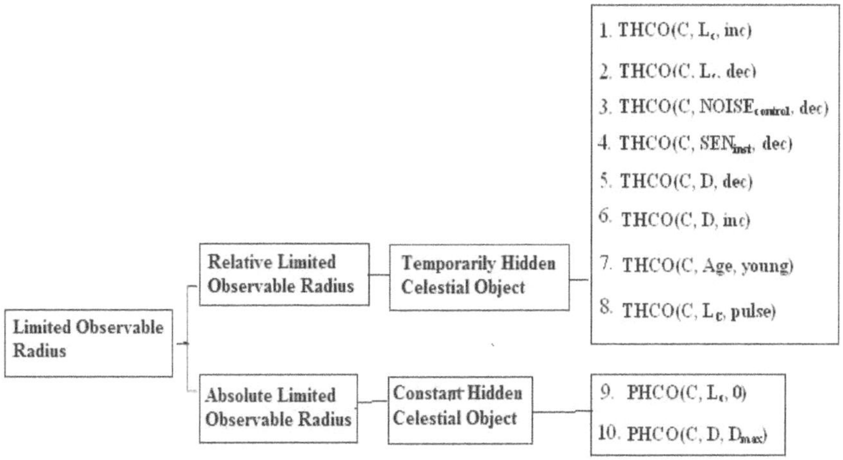

Figure 3.5 Illustrates the theoretical system of Hidden Celestial Objects

Chapter Summary

The concept of the hidden celestial object can provide us with new feasible ideas for discovering real immigrant planets that will be discussed later. It also theoretically provides for a possible hidden object emitting microwaves near Earth whose source cannot be found. This microwave source may be a temporarily hidden celestial object, or it may be a permanently hidden celestial object; that is to say, we may or may not find it.

Hidden celestial bodies are, by definition, hidden from our sight but cannot hide from our intelligence, and the wisdom we sometimes possess, when considering them with open minds.

Chapter 4

Why Cosmic Redshift Theory is Against Big Bang? What is the Real Reason for the Cosmic Redshift Produced?

A New Understanding of the Cosmic Redshift Theory – It doesn't Support Big Bang Theory

Here are 2 new insights that we just thought of when the manuscript was submitted.

1. Finding a new explanation for the redshift phenomenon of celestial bodies is not the most important issue, although before we thought it was.

Because, as we have already discussed in chapter 3 on dark energy, the redshift of celestial bodies is not evidence for the Big Bang theory, but is a strong opposition to the Big Bang theory!

As mentioned earlier, experts and scholars agree that the redshift of celestial bodies is caused by the movement of celestial bodies away from the telescope - and thus caused by the Big Bang. However, this is a naive idea.

When an explosion just explodes, it will expand rapidly, and the explosion space will continue to increase, and red shift will be a perfect theory at that period. However, when the energy of the explosive is exhausted, the explosion stops expanding, instead it begins to contract, and the explosion space begins to shrink and shrink, the redshift theory now is against the explosive theory. All forms of explosions known to man obey this law.

Then, considering that the Big Bang universe has cooled and solidified into celestial bodies, this cooled explosion space will only shrink and will not continue to expand!

The cosmic redshift theory, on the other hand, says that the explosion space of this cooling universe is expanding at an accelerated rate, and the farther away it is, the faster it expands. This is the exact opposite of what should happen after a normal explosion.

In this way, didn't the theory of cosmic redshift become thoroughly opposed to the Big Bang theory with scientific observations, data and theories?

The Big Bang theory proposed in 1927 seems to have been strongly supported by the redshift theory in 1929. So many scientists have been confused, thinking that the expansion of space is the specific manifestation of the Big Bang, which supports the Big Bang theory. It was not until decades later that the academic world woke up: it turns out to recognize that the cooled universe should not be expanding and even should be contracting.

However, the experts still continue to be sloppy, basically not mentioning the fact that the redshift theory of the universe is a strong opposition to the Big Bang theory with scientific observations and data, and they hurriedly fabricate a dark energy concept to patch up the loopholes!

Of course, the real cause of the observed redshift phenomenon, either irrelevant or not to support and even against the Big Bang theory, is worth exploring. Because in astronomical observations, the redshift theory is widely used for data processing, but in those applications, there are a large number of obvious errors (refer to Chapter 5.)

Therefore, it is of positive significance to explore the real cause of the cosmic redshift.

How terrifying is inertial thinking! With such an obvious loophole in the redshift theory (cooled down universe not expanding), due to the strong inertia of old knowledge inheritance, it took us nearly 20 years to figure out and wake up; meanwhile we have been thinking about and writing about the problems of the universe before we woke up.

In the following text, if the point of view is different from that here, the one here shall prevail.

2. View the Fallacy of Hubble's Law from the Geometric Scale

Hubble's law uses the following formula (4-1) to tell us that the velocity v of the galaxy moving away from the observer is the same in ALL directions, from which the Big Bang theory can be deduced.

However, this seemingly good reasoning is actually wrong. If the Earth was at the center of the Big Bang Universe, in that case possibly telescopes on Earth looking around would see all galaxies moving away at the same speed. But the earth is 4.2 billion years old, and it should be about 1/3 away from the center of the big bang universe, as shown in Figure 4.0.

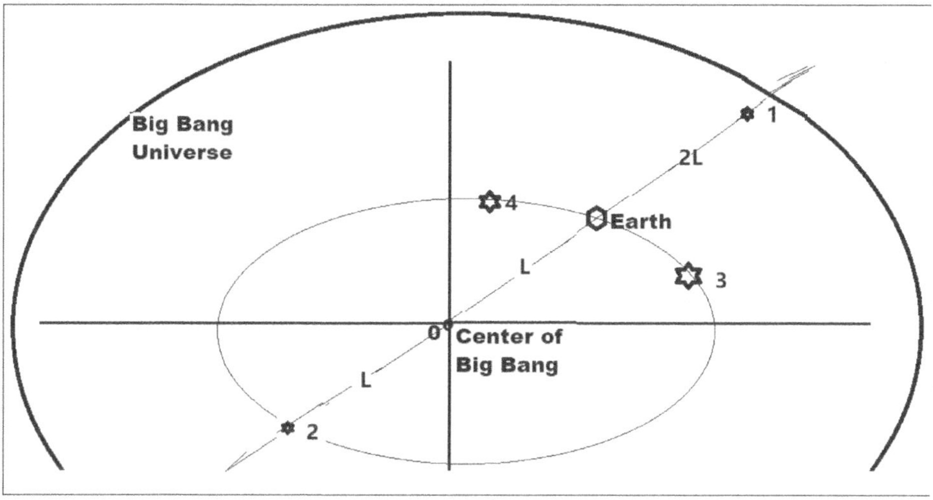

Figure 4.0 Earth and celestial bodies in different positions 1, 2, 3, 4, 5.

According to Hubble's theory, from the center of the Big Bang to the outside, the universe continues to expand in every direction, and the farther away from the center of the Big Bang, the faster the speed of celestial bodies moving away, As equation (4-1) shows.

Now, we draw a straight line, connect the Earth and the explosion center 0 to , set the distance between 0 and the earth (0-Earth) as L, and extend it to both ends, so that the distance between celestial body 1 and the earth is 2L. Velocity V1 of celestial body 1 can be calculated according to Hubble's law:

$$V1 = H_0 2L$$

Draw a celestial body 2 on the other side of the extension line center 0, so that the distance between 0-2 is L. Observing the celestial body 2 from the telescope on the earth, if the velocity V2 of the celestial body 2 away from the Earth is calculated according to Hubble's law, it should be:

$$V2 = H_0 2L$$

But, viewed from the center 0 of the Big Bang, the speed of the Earth away from the center of the Big Bang and the speed of the celestial body 2 away from the center should be basically the same, they are both H_0 L.

There is an even more serious problem, which can be demonstrated with crude calculations.

When celestial body 2 is observed from the earth, the light information of celestial body 2 is transmitted to the telescope in two different modes. From the celestial body 2 to the center of the Big Bang, the celestial body 2 moves away from the Earth, and the light information of the celestial body travels to the Earth at the speed of light, so the propagation speed of the light information of the celestial body 2 to the Earth at this distance is: (Speed of light - H_0 L;) in addition, the propagation speed of the light of celestial body 2 in the distance from the center 0 of the Big Bang to the Earth is: (Speed of light + H_0 L.)

Of course, such calculations are too rough and not perfect, but the general meaning is very clear, that is: if there is really a Big Bang universe, then when applying Hubble's law to observe and calculate in different directions of the universe, the obtained results in each direction should be very different from each other.

Introduction to the cosmic redshift

As we have discussed, in 1927 Father Lemaître proposed a mythical theory that the universe formed from the explosion of a primeval atom: the Big Bang. Two years later, in 1929, Hubble released his astronomical data. Through spectral analysis of celestial bodies in deep space, he measured a phenomenon – the redshift in light waves – that was to have a lasting impact on astronomy. Following the earlier discovery of Doppler's law of frequency changes of sound waves, Hubble interpreted the redshift in light wave changes as evidence that the celestial bodies he observed were receding from his position. In fact, most distant celestial bodies have measured redshifts; that is, they are all receding from the observer. This means the universe is expanding.

How to correctly explain the redshift phenomenon of celestial light waves has become a key factor in explicating the Big Bang theory. Should people directly apply Doppler's law (which describes how sound waves vary) to light waves, as the scientists have done? Or do we follow our systematic method and carefully consider details, such as arise in the transmission of light waves through space and telescope observations, to better explain the redshift phenomenon without considering cosmic expansion?

After analyzing in some detail, the latter holds our attention. The various issues attached to measuring the redshift as if it imitated sound waves via Doppler's law, we deduce a new method for calculating the redshift.

But in any case, let's not forget that the Big Bang was a pseudo-god universe bombed out from an atom, and the whole point of discussing in this chapter is to demonstrate from a scientific point of view that what people have always thought of as "scientific" theory is not that convincing.

Doppler Shift

As a siren approaches us, its sound waves are squeezed together and we hear them higher in pitch. After the siren has passed us, the sound waves lengthen and we hear them lower in pitch.

This apparent change in the pitch (or frequency) of a sound is called a Doppler Shift.

Cosmic Red Shift

Without any further consideration, scientists have agreed that light waves imitate sound waves, and that light from distant stars and galaxies behaves as sound wave's do. They believe that the frequency of a star's light wave would become higher as a star moves towards the Earth, and its tone, or color spectrum, becomes higher and transfers toward blue. This is called the "blue shift." If a star is moving away from the Earth, however, its tone, or color spectrum, shifts to lower frequencies and transfers toward red. This is called the "red shift" (Figure 4.1).

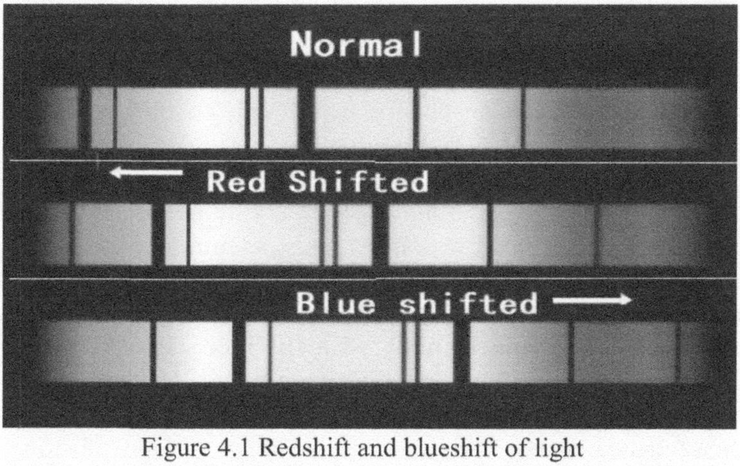

Figure 4.1 Redshift and blueshift of light

Scientists naturally take this as a fact, and describe the motion of a light source moving away from the Earth as the **main** explanation for the redshift evidenced in its light.

Hubble's Law

The continuous expansion of cosmic space is the main support for the Big Bang theory. The theory that depicts the redshift as a marker of expansion supports the assumption that the universe is expanding. For its part, dark energy, which was invented many years later, enters the discourse as yet another “scientific” support for the "continuous" expansion of space.

If a star is moving towards the Earth, its light is shifted to higher frequencies on the color spectrum, which is called a "blue shift". In contrast, if a star is moving away from the Earth, its light is shifted to lower frequencies on the color spectrum, which is called a "red shift".

Objects observed in deep space—extragalactic space, 10 megaparsecs (Mpc) or more distant—are found to have a redshift, interpreted as a relative velocity away from the Earth. More specifically, scientists have found that redshifts occur only for celestial bodies observed in deep space (that is, the distance must be greater than 32,620,000 light years for the shift to appear). The speed hypothesized by way of the redshift (for example, of various galaxies receding from the Earth) is approximately proportional to their distance from the Earth.

Since redshift occurs in most observations of the deep space universe, and experts believe that the only cause of the redshift is celestial bodies receding from the Earth, it is natural to conclude that the universe is continuously expanding.

What is particularly suspicious here is the claim that the farther the celestial body is from the Earth, the faster it will recede. Even if we imagine the existence of something like dark energy to help explain the redshift phenomenon, it's still hard to make sense of it. When is dark energy produced? Is its quantity fixed? Is it an energy from God, enabling an ever-cyclical process of replication: the more the energy is produced and consumed, the greater its quantity becomes? How can dark energy fuel the expansion of space at a faster rate in an expanding space? Try to imagine what this is like? Ask the experts to describe how it all happens, assuming dark energy exists.

With the increase of distance, the distribution of dark matter **decays with the square of distance increase**; while Hubble's law says that the **expansion of space accelerates proportionally with the increase of distance**. Please carefully compare the differences between the two.

The above is expressed in simple mathematics like this:

When the dark energy radiates from the center of the explosion, the dark energy intensity G of the unit sphere decays with the square of the distance D, which can be expressed by the following formula:

$$\frac{G}{4\pi D^2} \qquad (4\text{–}0)$$

In Hubble's law derived from the Doppler effect, Hubble ended up with the following formula:

$$v = H_0 D \qquad (4\text{-}1)$$

Where D is the distance between the observer and the galaxy, H_0 is the

Hubble constant, and galaxy speed away from the observer is v. The "appropriate distance" of D **has to be greater than 32,620,000 light years**.

Readers are asked to compare carefully, and we will not repeat them.

Because light is a kind of wave, the astronomers naturally apply the Doppler effect to explain the propagation of celestial light waves, and all too hastily marginalize other possible causes.

This very natural conversion and promotion, of course not, does not seem to require rigorous certification.

But the problem may be exactly in this "natural"!

"Natural" reasoning, which leads to such no proof, often makes people go astray. This unproven extension from sound to light often is the source of possible errors.

With comprehensive and deep analysis, however, Hubble's Law reveals some unexpected and interesting things.

Let's review the process that Astro-cosmologists have used in measuring the presumed expansion of the universe. After determining that celestial objects have different redshift values, astronomers and cosmologists directly use Doppler's law to explain those values. But is this the only way to explain the cosmic redshift? Apparently, that is what they did. -cosmologists conclude that the Big Bang occurred at some far distant point in the past, and brought into the Big Bang theory as its primary support. From the calculated redshift values, they thought that the movement of celestial objects away from the observer is the only reason for explaining the cosmic redshift, used it to establish the Big Bang theory, and concluded that a Big Bang has occurred. Have they forgotten the mythic element involved: explosion of a simple, primeval atom as the source of the Big Bang?

Doppler's law provides a method to calculate the frequency of **sound** waves: as sound waves move away or toward from their source, their frequency changes. However, does that law exclusively explain how the cosmic redshift is produced? The question stands without further verification. Nonetheless, imitating Doppler's law, scientists have attributed the cause of the cosmic redshift to the movement of celestial bodies away from the Earth. Generalizing this phenomenon as explanatory of an expanding universe then provides the "scientific" basis for the Big Bang theory.

In response, searching for alternate reasons to explain the redshift should become the most important issue of Astro-cosmological research, if for no other reason than this: most all modern astronomical observations and interpretations are related to how to interpret redshift values.

When scientists rashly use Doppler's law as they have, interpreting sound and light waves as similar phenomena, many fatal and irreconcilable theories and observational contradictions arise. This includes dark energy as we just discussed it, and what we will reveal, in the next part of this book, about the Color–Magnitude Relation (CMR) theory, and the observation results based on it that were found to be wrong. In brief, there are sufficient reasons to question whether using Doppler's law to directly explain the redshift is scientifically correct or not.

For example, when applying the K-Correction method, adding a color filter to the telescope will change the redshift value, which then needs to be corrected according to the value calculated by the K-Correction. This means that when the relative speed between the celestial body being observed and the observer does not change, the measured redshift value of the celestial body will change as the intensity of light from that celestial body changes. This also means, the observed celestial body's redshift phenomenon is not necessarily caused by the receding motion of the celestial body relative to the observer on Earth.

So, **is it possible that the observed redshift phenomena of the celestial body are mainly caused by the changing light intensities** of the celestial body? In this way, the redshift value is not solely determined by the speed of the celestial body moving away from the observer.

If changes in light intensity can cause a change in the redshift value, then different causative factors arise: changes in the distance between the celestial body and the observer; changes in the quality of the telescope, , and changes in the environment where the telescope sits. All such factors can cause the light intensity of the celestial body entering the telescope to change, thereby causing the redshift value of the celestial body to change. Of course, the relative speed between the celestial body and the observer's telescope can also cause changes in the light intensity of the celestial body.

Since the interpretation of the redshift of celestial bodies is related to the fundamental issue of whether the universe has expanded or not, **searching for the real cause and new explanation of the cosmic redshift has become a major issue related to the validity of the Big Bang theory**. If the cosmic redshift is really **ONLY** caused by the movement of celestial bodies moving away from the observer, then the Big Bang theory has a more solid foundation; if it is proved that the cosmic redshift is not mainly caused by celestial bodies moving away from the observer, then the scientific foundation of the Big Bang theory will suffer.

In order to resolve these issues, we must study all relevant factors.

1. Questions about converting from sound waves to light waves, and from Doppler effect to Hubble's law

History review: The Long-term Debate and Its Abrupt End about the Value of the Hubble Constant

It has long been recognized that the redshift is caused by the Doppler effect.

Although many scientists have sought non-Doppler explanations of the redshift mechanism, few have noticed this: the redshift value of the same celestial body varies when observed by different telescopes. Due to differences in our telescopes, the variation in redshift values climbs to ten times as much! This is completely contrary to the nature of a wave. **But it is not contrary to using different telescopes that reveal major variations in light intensities, exclusive of major variations of changing the frequency of light waves!**

According to the principle of constant light speed, the intensity of light does not affect the speed of light waves. Therefore, the change in the celestial object emitted light intensity in (4-1) (The amount of starlight collected by the telescope) should not cause a large change in the Hubble constant.

In Figure 4.2, however, we note the following: "Due to the difference in the telescopes, the redshift values were different." This kind of thing that shouldn't happen but happened, which is worth investigating carefully and scientifically.

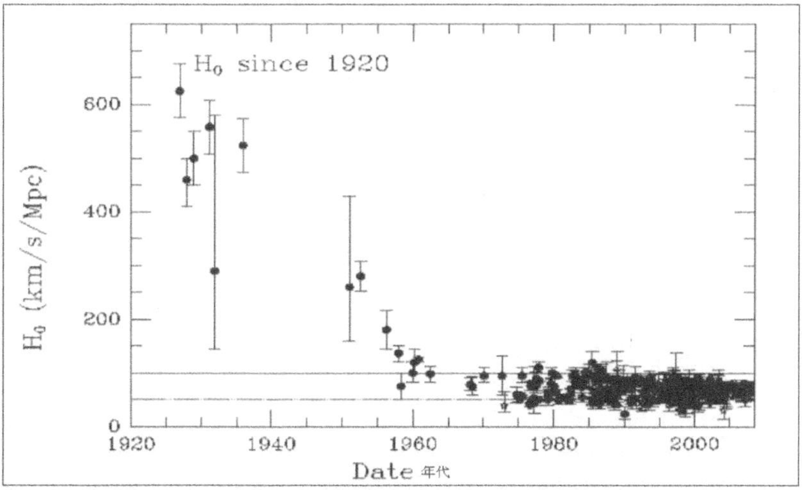

Figure 4.2 different values of Hubble Constant H0 measured in different years
https://www.cfa.harvard.edu/~dfabricant/huchra/hubble/

Historical values of the Hubble Constant show the 1920s measured values were up to ten times greater than those measured in the 1980s.

The value of the Hubble Constant was once a long and intense topic of debate. After the first half of the twentieth century, the value of the Hubble Constant H_0 was estimated between about 50 to 90 (km/s)/Mpc. The French astronomer, Gérard de VA couleurs claimed it should be 80 while the American astronomer, Allan Sandage, believed it should be 40.

In 1996, a debate between Gustav Tammann and Sidney van den Bergh was held, presided over by John Bahcall, the theme of which was the two competing values of the Hubble Constant.

We mention this to note that this fierce dispute continued for more than half a century. In such a situation, wouldn't even a few scientists eventually propose a novel solution: to interpret the phenomenon from another point of view or theory. However, with the powerful Hubble Space Telescope, NASA has thus far marginalized dissenting voices by using the enhanced capacities of the telescope to support given explanations!

In 2003, using the results of the highest precision WMAP cosmic microwave background radiation detector, the measured value was 71.4 (km/s)/Mpc, and then in 2006, 70.4 +1.3/-1.4 (km/s)/Mpc was the measured value.

Thus NASA suppressed the voices of all the different arguments, of course, it also killed the possibility that people may finally find the real cause of the redshift phenomenon at the end of the dispute. Using advanced instrumentation, **NASA obliterates the possibility of discovering the truth.**

Although without men's brains, the advanced instruments can't think, disagreements and anomalies persist, and NASA's space telescope cannot explain them.

When we notice large variations in redshift values mainly due to different observation instruments, we naturally want to explore the phenomenon by incorporating the performance of the observation instruments into the theoretical analysis of the observation results. But this does not happen generally. Instead, scientists use methods, such as the k-Correction calibration, correction theory, and others to modify the measured results of the observed data.

From this vantage point, we can discuss extant scientific theories based on analyzing and summarizing various observation results derived from different instruments with different capacities.

Is this the golden key to unlocking the true mystery of the cosmic redshift phenomenon?

As noted, until 1996, there was still debate over whether the Hubble constant was 80 or 40. This major difference cannot be simply explained by observation errors.

We believe that the scientific quality of the scientists who make observations is unquestionable. In a long-term collective debate divided into two schools, everyone, on each side of the debate, must have made adequate preparations and made the most serious and scientific observations. Just because of different observation instruments and different methods of interpreting the received information, it will cause more than double the difference, which is an incredible thing.

Let's change the perspective for a moment:

What if everyone's observations were correct? Is this possible?

What if everyone's observations were correct, but the theory used to explain their results was wrong?

If no one has considered it before, it is time for us to think about it now.

On the one hand, according to Hubble's law, light waves in motion will change their frequency, resulting in the red shift phenomenon exhibited by celestial bodies. On the other hand, the light waves reaching the Earth from the same celestial body are the same, when collected by any telescope used by anyone on Earth. All scientists who operate telescopes are extremely cautious with minimal errors involved in their calculations. Now doesn't this tell us that the different values given to the Hubble constant are likely to be related to the different telescopes used?

Here is a simple example to clarify this point. When, listening to a faint singing voice at an opera performance, you cover your ears with your hands, does the singing change in pitch? In the same way, when the opera moves into a recording studio and the singing exhibits differences in sound intensity, will that intensity change the pitch of the voice? In both instances, the answer is no. But when different machines play the same recording, the pitch is often changed.

In September 2017 I played flute solo with an electronic accompaniment at a party in San Francisco. After the accompaniment music started, I found that the music pitch out by this machine was half of one scale lower than what it normally should be, which made me very embarrassed. The music in my accompaniment DVD is of course fixed and was used many times, but there may be huge differences when played by different machines.

The Hubble constant related to a celestial body reflects the change of the frequency of its light waves. However, if the red shift is not only or mainly caused by the motion of a celestial body receding from an observer, then what is the reason for the red shift phenomenon? Is there any telescope in which systematic affecting factors have caused the red shift? Is it even the main factor that affects the observed redshift value change?

Of course, historically there have been scientists with different views on the Hubble redshift. In 1929, Fritz Zwicky proposed the "tired light" theory. Zwicky believed that photons would slowly lose their energy when moving over great distances in steady state space due to continuing interaction with each other or with other substances, or due to the action of some unknown physical mechanism. Because photon loss of energy corresponds to an increase in the wavelength of light, this effect will make the redshift increase with increasing distance.

But this theory and others like it have not gained recognition by mainstream scientists, who refer to them as non-standard cosmology.

Our opinions differ from mainstream scientists. We do not deny that the movement of celestial bodies away from an observer using a telescope causes the redshift, but we do not believe that this is the only, or most significant cause.

Ten reasons why we should be looking for a new way to correctly reinterpret and calculate the cosmic redshift ?

The main reasons are as follows:

1) There is no apparent proof that a celestial body receding from an observer is the exclusive cause of its redshift. As a result, it is necessary to expand the scope of our research and determine other salient factors that may be involved.

2) As we will discuss later in more detail, low-level, erroneous results in a large number of long-term simple astronomical observation reports have caused some scientists to doubt the basis of the Color-Magnitude Relationship (CMR) theory: Hubble's law. These errors, which were made when applying the redshift theory to easy verifiable fields of cosmic objects, leads us to reconsider the validity of Hubble's law.

3) When applying the K-Correction method, adding a color filter to the telescope will change the received redshift value, which needs to be corrected according to the value calculated by K-Correction. The noted astrophysicist, David Hogg, puts it this way: "the adjustment of the K-Correction to the standard relationship between absolute and apparent magnitude is required to correct for the redshift effect." This means that when the relative speed of the observer does not change, the measured redshift value of the celestial body will change as the intensity of light from the celestial body changes. In fact, a change in the red shift value will occur when using a color filter. As the filter weakens the light that the telescope captures, the redshift value changes. This phenomenon has nothing to do with changes in light frequency, but **only with**

changes in light intensity that a telescope captures. K-Correction is used to tinker with the application of theory. And our theory can satisfactorily explain this phenomenon, too.

We can also use this phenomenon to design changes in the redshift, implicating our point of view; that is, we can identify redshift values with telescopes focused on celestial objects in a universe that is not expanding.

4) The value of constant in Hubble's law, which describes the cosmic redshift, has caused long-term disputes in astronomical observations, with said values differentiated by up to 10 times. NASA has thus used the Hubble telescope to try to settle the controversy. Careful analysis, however, reveals that the controversy sustains. Is there an issue with the transmission of light waves when calculated by Doppler's law? Perhaps. The significant differences in the value of the Hubble constant obtained by different observers throughout history proves one thing: the frequency of **the received light has changed drastically with different instruments and/or observation methods**. In effect, the redshift is not entirely caused by the Doppler effect. And this can be explained by our theory of the change in speed of the observed image.

5) Over the decades, numerous scientists have believed that their observations are correct. Who then is wrong? For us, the answer is clear: no one. Rather **the theory explaining the observations they made and the conclusions they drew from their observations was wrong**! The correct theory should be able to encompass and explain all the different results obtained from different instruments – as our theory does.

6) All redshift values are generated by observation **instruments**. The Hubble constant has a certain relationship to the quality of the telescope. It varies with the sensitivity of different instruments and is a function of the sensitivity of the instrument. **To date, no theory has considered the instrumental factors when analyzing and calculating redshift**. But which astronomical data does not need to be obtained through observation instruments (telescope, the naked eye, etc.)? At the very least, scientists should consider how an observation instrument affects the Hubble constant. If the issue is simply instrumental noise and its elimination through technical refinement, observation results of redshift data should not exhibit the kind of deviations they did.

7) According to Hubble's theory, a celestial body at the "**proper distance**" — that is, within 30 million light years of Earth — should not exhibit a red shift. But virtually all close-range stars, including the Sun, exhibit redshifts; a phenomenon that Hubble's theory cannot explain on its own. So, scientists patched it up with an associated theory: the redshift caused by gravity.

8) According to the Hubble redshift theory, the farther a celestial body

is from the observer, the faster it recedes from the observer. Unfortunately, this presumption runs counter to normal classical physics when applied to explosions. In classical physics, the energy of an explosion expands in a spherical shape from its center. As it does so, the intensity of its energy decreases exponentially as the distance from the center of the explosion increases. The Hubble redshift theory, which contradicts classical physics in this regard, places no limit on the speed at which a cosmic object recedes from an observer.

9) According to a 2013 study by Bond et al, the star, HD140283, which is about 13 billion years old, is only 190 light-years away from the Sun. In a universe where celestial objects are ever receding from an observer, how is it that this star remains 190 light-years from the Sun? Does dark energy play a part; a part we do not yet understand? Who can say?

10) The proof of the previous geometric scale shows that if the Big Bang universe really exists, the observation results of Hubble's law should get different observation results in different directions. One formula (4-1) cannot be used to calculate all the observation results from different directions.

The above 10 points are the main reasons that prompted us to propose a new theory for calculating the redshift of celestial bodies based on the speed of celestial images.

We tried to use a new theory to explain the phenomenon of celestial body redshift from a new angle, and theoretically got better results, because our proposed explanation resolved all the contradictions in the original theory. All the different observation results obtained through the application of different observation instruments can be reasonably interpreted and be used, instead of being finalized by NASA's so-called most advanced satellite telescope, making the data obtained by other telescope observations meaningless, useless and needing to make various modifications or calibration to the data.

After studying the systemic effects consisting of the observed celestial objects, the propagation characteristics, distance of celestial light, and the sensitivity and response time or time integration of the observation instrument by classical physical methods, we need to abandon the hypothesis of expanding universe, and explain the main mechanism that produces the cosmic redshift from a new perspective, without considering the pseudo-theology.

2. A New method to reinterpret the redshift of celestial light

This new method revolves around an analysis of the time required for the entire process of celestial light wave propagation, that is: the time required for each link from when the celestial body emits starlight to when the computer outputs the astronomical observation image. When that time span is known, inclusive of the time required for the computer to resolve an image of the object, we can calculate the speed of the observed image in space.

The speed of light, the speed of celestial light and the speed of celestial images

Einstein defined the speed of light as the maximum speed of the universe.

So, how about the speed of celestial light?

Usually, people measure the speed of light in a laboratory on Earth, and confuse the light speed measured from a lab with the speed of starlight.

But in fact, on closer examination, the speed of celestial light and the speed of light we are now commonly using are very different.

In 2006, we published an article entitled "The Speed of Photons and the Observed Light Speed from a Star" in the journal *Matter Regularity*, the first article that mentioned that the overall statistical light speed emitted from celestial objects is not the same as the speed of a single photon.

Classical physics believes that the velocity of a single photon, about 300,000 kilometers per second or so, can be accurately measured in the laboratory. As such, it is what people commonly use for the speed of light and is the only value recognized by the scientific community.

After travelling great distances, starlight arrives in the vicinity of the relative observable limited radius of the telescope. (You can refer to the previous chapter on "Hidden Celestial Objects" to review this notion – along with content below on "different functional regions of photon batches"). Since the number of photons emitted by a celestial body – its light wave – at that distance is insufficient to maintain our view of the object extra, other batches of photons that both supplement and integrate with the light wave are necessary. This process makes the star's overall speed of light propagation relatively slower near the relative observable limited radius of the telescope. That is, the speed of light from far away stars becomes slower than the speed of light measured in the laboratory!

Does this cause problems with Einstein's assertion that the speed of light is constant? In fact, it does not. As noted by Lǚ Wu, it is not the speed of the photons emitted by the star that has changed, rather the observed image of the object has changed speed. It does not matter that Einstein is wrong or correct, but the calculation of the speed of the observed image of the object is!

In the summer of 2014, we took Lǚ Wu, who was just admitted to the University of California, Berkeley to Northern and Central Europe for two weeks to celebrate the end of his high-school life, that is, the end of his rebellious youth that gave us headaches, before he briskly headed off to the larger world.

One day while shopping around in Amsterdam, Netherlands, I spoke to him about the point of view from my paper "Photon Speed and the Speed of the Stars Observed" written 8 years ago. After listening, Johnny immediately replied: "It's not the star photon's speed that has changed, but the observed image that has changed speed. Einstein is not wrong, but the calculation of the speed of the observed image is!"

His words caused me some thought, so I was able to jump out of the trap of established thinking and into a new perspective to re-examine and think about this problem.

We were discussing this issue on a busy street in Amsterdam. In Amsterdam, exotic scenery becomes quick glimpse and casually superficial.

Johnny and I often discuss some of the issues I consider. The main idea is to take away his time and put the excess part of his adolescent energy toward research. He had worked with me on the subjects in this book for two consecutive summer vacations and helped me to deepen the ideas and write the research papers. Many of his views amazed me. The Hidden Celestial Object concept was completed with his help. He wrote the English paper.

I also discussed with him my views on the theory of Einstein's gravitational field, that if the gravitational field deflects lights emitted by celestial objects, then the stars should dance while we are watching them.

His comment was: Einstein's general theory of relativity should not be too involved in the details.

This comment was beyond expectation. I do not know how he could have figured that out with his young head. This should be a sixty or seventy-year-old man who is able to comprehend the secret.

This is my first time to ponder this point: Tao that can be described is not universal and eternal Tao. Lao Tzu's five thousand word <Tao Te Ching> hides like a mountain behind thick clouds and fog, abstract, profound, fuzzy, and high on hype and occult secrets. Because the Tao is highly abstract and does not rigidly stick to the detail of specific factors, thereby opening the door to all the wonderful, different perceptions of people, even leading to the door of eternal good fortune.

Einstein's theory in detail has left a variety of flaws, so those little guys, like me, can stand here and give their opinion on him.

But high on Lao Tzu, whose Tao has been on ever to comprehend, and from one state to another to comprehend a higher realm, where people have no way to see its flaws and give their opinion on the flaws.

Few people have ever tried to refute Lao Tzu, but only comprehend his theory with further insight, deeper and endless.

But Einstein's theory has never seen a lack of criticism, although mainstream science rejects such criticism, but the mainstream scientific community has rarely answered the criticism and questioning on relativity theory from non-mainstream scientific community. Can they answer my comment describing later chapters of this book?

In fact, this principle can also be extended to social life.

In a country, if the ruler cares only about the routine life of firewood, salt, and oil, and leaves the ideological spiritual thought to those in the religious sectors who only talk about the spirit, heaven, hell, afterlife, etc. - illusory, but everyone needs to face them in the days after death, and there is always, after death, the reward and punishment of heaven or hell - the country is always harmonious.

And when the ruler of a country not only manages temporal affairs but also those of the spirit, the social harmony in this country is not likely to improve.

If the management of the spirit also excludes the afterlife: heaven and hell, etc., that will lead to deletion of public behavior norms that come from the soul.

Using only the law, it is impossible to constrain people's selfish desires in every second.

Modern society is a society dominated by capital. The nature of capital is to stimulate people's selfish nature to do its utmost to pursue profit. Without the constraint of heaven and hell, the pursuit of profit in the heart of some little people can reach the point of madness, with profit like drug poisoning. More money, perhaps, does not make things better; it may even make things worse...

These are the questions my discussion with Johnny started on the street in Amsterdam.

The following is our discussion results on the celestial light propagation, now presented to you in this chapter.

The total time involved to output an image of a celestial object is counted from the time the celestial object emits its light to the moment when a computer resolves an image of the object.

In modern astronomical observations, information from extragalactic galaxies and distant celestial bodies must be received by various telescopes, and most of them must be processed by computers before observation results can be obtained. These results are basically presented in the form of images.

The **observed image** is the effective output information obtained by the telescope after receiving the information light from the celestial body provides, and its processing by computer. The time used for resolving the observed image is the total time the entire process takes.

Since we can perceive the photons emitted by distant celestial bodies only by using a telescope, we cannot directly measure how fast a celestial body is moving or, for that matter, the image of that body. So, although we can accurately measure the time from emission to reception of a single photon or a group of photons within a defined range (such as one day, one year, ten years, etc.) in the laboratory, but by using a telescope, a different situation prevails — where we can't accurately measure the total time between the light emitted by a celestial body and the output of the image by the computer. In brief, we have not measured the total time T it takes for the information we received and processed into a computer image. Be that as it may, this time T is closely related to the speed v in Hubble's equation (4 -1). If this time changes, the redshift value of the celestial object will change (Figure 4.3).

In Figure 4.3 above, T1 is the star light travel time. When we start the calculation, the photons are emitted by the celestial continuously as shown in Figure 4.4. We can roughly divide them into batches. The starting moment counts as the first batch of photons emitted from the celestial body, and travels at the speed of light until they reach the telescope. T2 is the time required for the telescope to receive subsequent batches of photons — if integration is required — plus the time required for the computer to receive, process and resolve, as output, the incident light information.

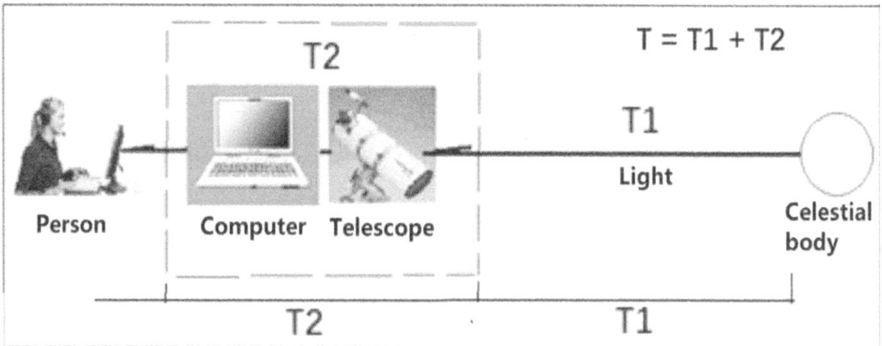

Figure 4.3 From right to left. The light from the celestial body C reaches the telescope after T1 time, as input information, and after T2 time processed by the telescope and the computer, the image is finally output at time T. But in fact, any celestial information must be transformed into an image that can be understood by humans before it can be truly "seen". The time T2 spent by the telescope and the computer framed by the dotted line is very important in the analysis of the light propagation speed of the celestial object.

In the following discussion, for the sake of simplicity, we simply mention that it takes total time T for the telescope to process the light, which includes the time used by the telescope and computer.

In astronomy, so far, when analyzing information related to the speed of light propagation of celestial bodies, the time T2 spent by the telescope has never been considered. But in fact, any celestial information must be truly "seen" after being transformed into an image that we can understand. The time T2 spent by the telescope and the computer framed by the dotted line is an important variable in analyzing the light propagation speed of a celestial body. Recall as well that only by using telescopes can we determine the value of the redshift that the celestial object exhibits.

In this way, light is emitted from the celestial body, processed by the telescope, and finally resolves as useful information or a visible image. The total time required can be simply calculated by the following formula:

$$T = T1 + T2 \qquad (4\text{-}2)$$

The total time T required for this transmission is not the same as the time T1 required for the photon to travel from the celestial body to the telescope.

T is the total time from when the light starts traveling to the moment the computer resolves, as output, a visible image. T1 is the time it takes for the photon to travel from the celestial object to the telescope. The total time also includes T2, which is the time spent by the telescope and computer in resolving useful information or a visible image.

We can describe this situation in another fashion. For example, a company that provides courier service sends out a parcel to a customer, and 24 hours later the customer receives the parcel. The delivery event has ended, and the courier company's work has been completed. But the client has not yet received in hand the desired product; it is still wrapped. This stage is equivalent to T2, the photon from the celestial object having arrived at the telescope, but the photon arrival information has not been extracted into an image or other useful information for storage or use. To calculate the full amount of time it takes for the process to conclude, one must also consider the time it takes to unwrap the package.

We can describe this situation in another fashion. For example, a company that provides courier service sends out a parcel to a customer, and 24 hours later the customer receives the parcel. The delivery event has ended, and the courier company's work has been completed. But the client has not yet received in hand the desired product; it is still wrapped. This stage is equivalent to T2, the photon from the celestial object having arrived at the telescope, but the photon arrival information has not been extracted into an image or other useful information for storage or use. To calculate the full amount of time it takes for the process to conclude, one must also consider the time it takes to unwrap the package.

If the customer immediately opens the package after receiving it, then the process consumes a relatively short period of time, perhaps only a few seconds. In this guise, the time it takes the customer to open the parcel is equivalent to the time it takes the observer to extract the information required from the photons received. This holds, of course, for any number of photons that reach the telescope, and that satisfies the minimum amount of energy required to process, resolve and output an image or images. Now the time it takes to process an image of a celestial object can be quite brief, but some time is involved.

Similarly, when processing telescope optical information from afar, the observer can encounter a delay, perhaps because the information was too weak for the telescope to respond to, requiring more time given to observation. In this respect, uncertainty enters the process — the time T2 it takes to obtain the image takes longer than expected.

The sensitivity of the telescope to light also greatly influences T2. The more sensitive to light a telescope is, the shorter the time it will take for the telescope to resolve an image of the celestial object.

For telescopes with the same sensitivity, the attenuation effect of celestial distance is more obvious. This does not affect the time T1(the time it takes for the light to travel to the telescope), but it does influence whether or not the telescope can adequately respond to the photons it captures in time; that is, T2.

Therefore, including T2 in the factors that affect the red shift value can explain the large variability in the Hubble constant we have been discussing (see Fig. 4.2).

A brief review here will help us. According to our discussion on hidden celestial objects, when approaching the limited observable radius of celestial bodies, due to the attenuation effect of distance on the intensity of light, the light emitted by a distant celestial body will decay below the minimum energy required for a telescope to resolve an image. As a result, the observation of the celestial object extenuates until the telescope can resolve an image of the celestial body, which the astronomer can see. The uncertainty involved is relative to the telescope used but can be calculated.

When we calculate the speed of celestial images, not only the travel time of photons, but also the time the observing telescope used. This time was not used by professional astronomers. Why? Because the telescope was not considered when calculating the speed of light! The objects considered are not only photons, but also the response time of telescopes that were not considered before when calculating the speed of light!

It is important to include in any final formula the time T2 it takes for resolving an image of a celestial body by instrumentation previously excluded from consideration in this regard. Doing so may bring unexpected results.

As shown in Figure 4.3, the total time it takes, from a star emitting light to its resolution, as an output image, by a telescope, is called the **Image Transmission Time T** of the celestial light. It consists of two parts: T1 and T2. T1 is the time it takes starlight to travel from the star to the telescope. T2 is the time it takes to process the image the telescope receives and the computer works up. This, the main object of our study, helps us to understand possible new causes of redshifts exhibited by celestial bodies. To obtain what we need here, we mainly consider the calculation of T2 below.

The calculation of the propagation time of this image takes into account the following conditions:

A) Within a limited time interval, only a limited number of photons can be emitted by the celestial bodies, so it can be considered that the celestial body emits light waves in batches;
B) During the propagation process, starlight weakens in intensity, according to the attenuation law, measured by the inverse square distance between the telescope and the star.
C) The telescope requires a certain number of photons that accumulate to a defined level to initiate a visible response.
D) When the photons entering the telescope are too weak for the telescope to respond to, astronomers commonly attenuate their observation time to suit their needs, with adequate response times varying with different telescopes.

Understood in this fashion, the light from a distant celestial body, which has the same light intensity everywhere on Earth and in near-Earth space, is affected by the sensitivity of the telescope to light. And this recognition is the basis for our suggestion that we need to reinterpret how the redshift is calculated, which also requires abandoning the theory of cosmic expansion.

Obviously, after the light of a celestial body reaches the telescope, the image transmission speed changes with the image processing time, and the image processing time changes with the sensitivity and response time of different telescopes. By examining the entire process of starlight from emitting to the resolution and output of images, we have obtained the original rough mathematical model of the new method of calculating the cosmic redshift. This new model contains Hubble's law and can explain phenomena that Hubble's law cannot.

There is no need to argue about the speed of light, because the speed of light is set. However, **neither the speed or light nor an alteration in the frequency** of light is of concern here. Rather it is an alteration **in the intensity** of light that causes variations in redshift values. This perspective obviously differs from Hubble's law, which is modeled after the Doppler effect. It is a new way for us to explain the phenomenon of cosmic redshift observations – and follows the same principle that red shift value changes are due to the weakening of the light intensity in the K-Correction that astronomers use as a matter of course.

In modern astronomical observations, the observation integration time is used arbitrarily, but the used integration time is not included in the processing results, and these are directly classified into redshift values. In fact, the farther the celestial body is, the weaker the light intensity transmitted to the telescope, the larger the observation integration time required by the telescope, and the larger the redshift value. Here, the changing of the redshift value is not caused by the receding movement of the celestial body C, but is caused by the variation of the distance between the telescope and C. **The variation of the distance causes the change of the input light intensity, thus changing the redshift value.**

Observations, such as those made by the Hubble telescope, were used to measure the redshift value of distant celestial bodies. Did NASA scientists consider in their calculations the time it took the telescope to process the photons received into visible images formed by computer? No.

Is there a possibility that the red shift is caused by the integration time required by the telescope and the output (an image) by computer? Of course there is. Perhaps the redshift exhibited by celestial objects is not caused by their recession from an observer in an ever-expanding universe, but by the telescope itself in its processing of weak information.

We will continue to discuss all the details related to this problem.

3. Derivation of a new model that replaces Hubble's law

We derive the new model according to A), B), C), and D) discussed above.

A) Revelations Given by K-Correction

In astronomical observations, people often use color filters for special effects when taking photos. When doing so, the color filter(s) will decrease the light intensity received by the instrument, thereby changing the redshift values observed.

Put another way, the change in the intensity of the light entering the observation instrument will also change the final redshift value given by the image of the celestial body output by the instrument.

K-correction is a common method used to correct potential changes in the redshift value provided by a telescope; the result of interference by certain factors in celestial observation. K-correction thus is closely related to, and involves, the energy of the light source as it enters the telescope per unit time per unit area.

A typical example follows: when using a color filter, the redshift value will change. K-correction will mediate that change. That is to say, K-correction tells us: not only will a change in frequency or wavelength cause a change in the redshift value, but also a change in the energy of the light wave will cause a change in the redshift value. Once again, this occurs because a color filter weakens the intensity of light entering the telescope. After this reduction of light intensity, K-correction enhances the response time T2 of the telescope, thus causing a change in redshift value. This phenomenon has nothing to do with changes in the frequency of light, but **with changes in the celestial body's light intensity** entering the telescope.

Of course, different telescopes with different sensitivities responded differently to changes in light intensity. Two factors and their effect are important to recall here: the change of light intensity and the response sensitivity of the telescope will cause the redshift value of the celestial body to change, as calculated by the observation. Be that as it may, in current astronomical observations, if a color filter is not used, these two factors are not taken into consideration when calculating the redshift value of the observation. And this gives us good reason to calculate the redshift value from a new perspective.

B) Analysis of Celestial object emitting batch quantities of light

When the celestial object emits light, it is clear that in a very short time, only a limited number of photons are emitted. Because even the whole celestial itself emits into photons, there would be only a relatively large batch of photons. This situation is illustrated in Figure 4.4. So, it is reasonable to divide the continuously emitting photons as batches in a continuous short time Δt intervals.

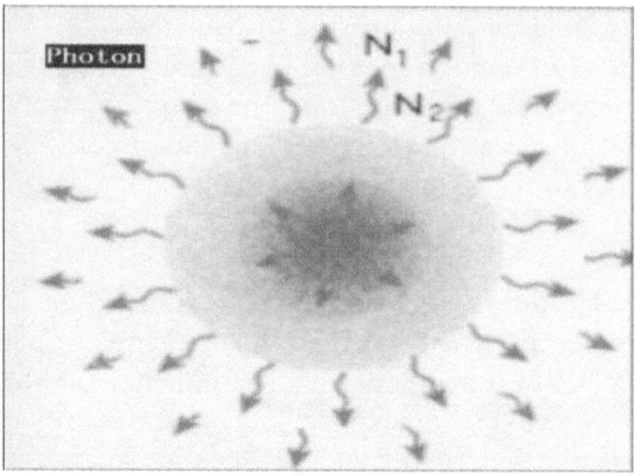

Figure 4.4 Schematic of star emitting batches of photons

In Figure 4.4, the celestial object continuously emits photons.

In order to facilitate research, we can artificially divide a continuous time into the same minimum time interval: Δt_i is used to represent any one such time interval. In the figure, the celestial object sends out N_i photons per unit area to the observation instruments for every tiny time interval Δt_i.

Since the photons have the same speed, if there is no loss during the transmission process, the number of photons in each batch will remain unchanged. Each batch of photons covers a sphere with a unique radius. Since each photon is constantly moving outwards, countless batches of photons form a photosphere that is composed of all these photons.

C) The light intensity received by the telescope varies with the change of the distance

However, as shown in Figure 4.5.

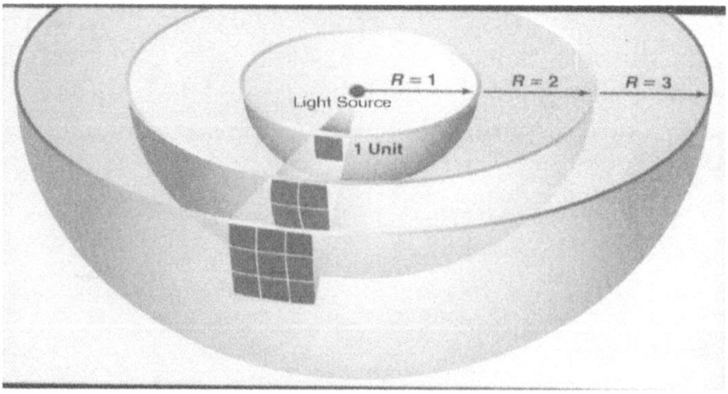

Figure 4.5 Schematic diagram of the attenuation effect of distance.

After the same batch of photons are emitted from the light source, as the distance from the light source increases, the surface of the photosphere covered by the batch of photons expands at a high speed. While the light intensity per unit area covered by the batch of photons shrinks rapidly in inverse proportion to the square of the distance.

Assuming that the light intensity emitted from the light source is L, when the radius R (distance) is 1 unit, the light intensity on the surface of the photosphere is $\frac{L}{1^2}$; when the radius R=2 units, the light intensity on the surface of the photosphere is $\frac{L}{2^2}$; when the radius R=3, the light intensity on the surface of the photosphere is $\frac{L}{3^2}$... That is to say, as the radius R (distance) increases, the photosphere expands, and the light intensity on the surface of the photosphere decreases according to the square of the distance $\frac{L}{R^2}$.

On one hand, the number of photons in the same batch remains the same. As the transmission distance increases, the photosphere becomes larger, and its covered total area increases. Therefore, the same number of photons need to cover a greatly increased area, so that the number of photons per unit area is reduced. There are fewer photons per unit area, and of course the light energy per unit area drops. As a result, the number of photons received by the telescope is reduced as the distance increases.

So if the telescope is located in the innermost position in the Fig. 4.5, the number of photons entering the telescope is so large that the telescope will respond immediately; and if the telescope is in the outermost position in the figure, because of the distance attenuation, the photons received by the telescope may be less than its lowest excitation response number (say 200 per unit area), and this batch of photons cannot stimulate the response of the telescope, so it needs to wait for the arrival of subsequent batches of photons to accumulate them. When the weak batches of photons accumulate on the telescope to the intensity required for telescope excitation, the telescope will send all these photons accumulated from multiple batches of light to the computer for processing. In this way, the time required for the photons entering the computer is not only the travel time T1 required for the photons to travel from the emitting celestial body to the telescope at the speed of light, but also the cumulative integration time required for each batch of photons to arrive later. We generalize this integration time into T2.

(If you require review at this point, please return to the content on hidden celestial objects in chapter 3. The same reasoning is used here, and we can understand the above paragraph with the same simple mathematics.)

All the basic principles used in this book are the simple five summarized below:

1) Telescopes, detectors, and all observation instruments are **receivers**.
2) The light emitted from the celestial body needs long enough time to travel through space before it reaches the telescope.
3) The light emitted by celestial bodies attenuates rapidly as the distance from the light source increases.
4) After the celestial body is destroyed, the light it has emitted still continues to spread;
5) The slow speed and short lifetime owned by humans restrict the understanding of the grand and long life of the universe.

What are these simple truths! There is no need for advanced knowledge and esoteric mathematics.

D) Analysis of the light intensity required to excite the response of a telescope

A telescope with a few meters in diameter is not, of course, as sensitive as sophisticated laboratory equipment that can respond immediately to the arrival of one photon. The human eye, which has even less sensitivity to light, requires a larger number of photons to enable sight. In addition, the number of photons that the eye can collect to enable sight must be greater than any interfering noise.

The same situation prevails with a telescope. Either it collects a sufficient number of photons to enable its response, and overcome interfering noise, or it cannot resolve an image of the observed celestial object or send that type of information to a computer to output a visible image for scientific analysis; after which the computer cleans up its storage capacity and awaits new information.

When the light is very strong, as soon as the light is irradiated onto the telescope, it reaches the number of photons required by the telescope within a short time period, so the telescope can react almost immediately to the light, processing it and then outputting the image. Although the completion of these steps is very quick and the time is short, it still takes some time. This time is named **Full Light Intensity Responding Time FLIRT(TEL)**. One specific example of this is when you focus the telescope at the Sun.

So, for any telescope, there is a minimum number of photons required in the input. Only the total incoming photon number is equal or bigger than the minimum photon number, the telescope will respond, the photons will be processed, and the computer will send its output, clean up the storage, and be ready to begin accumulating photons for a new image. This minimum light intensity to excite a specific telescope TEL is named Q (TEL).

This process can be explained as follows:

A modern telescope used to obtain celestial images has a storage buffer for arriving photons. The buffer is like a three-dimensional matrix in which the two-dimensional plane corresponds to the observation area of celestial objects, and each point on the plane corresponding to the third dimension is used to record the photons entering the point during the observation period. After the telescope receives an arriving photon, the photon will be stored in accordance with its corresponding storage location in the third dimension of the matrix corresponding to the recorder, the queue being in chronological order. After a certain time, the matrix is filled to a certain extent, the accumulated light intensity on the telescope is up to Q (TEL), and all the photons are output to a computer for processing.

If, in a very brief time, a large number of photons arrive (when the light emitted by the celestial object is relatively strong or the distance between it and the telescope not particularly great), then the counter matrix fills up quickly, with the information sent immediately to the computer.

But if the celestial object is located at a distance close to its Relative Limited Observable Radius (RLOR), the number of photons arriving will be very few, and the telescope will need sufficient time to receive and integrate enough photons to send to a computer for processing. The farther the distance is, the longer the waiting time.

Let's analyze the modern telescope equipped with a CCD camera as an example. It works as follows: during the exposure or integrated process, when the photons emitted from a celestial object arrive at the sensor, it produces a response of electrons. The electrons are stored into a potential well until the end of the integration or exposure process. Only then will the computer generate the corresponding image.

In fact, a paper can be written on accurately describing this narrative in mathematical form, and then abstracting the physical laws within it.

E) Analytical reasoning that combines all the above factors

In order to calculate the time required for the astronomical observing image, it is necessary to consider the influence of the distance attenuation effect on the statistical velocity of photons emitted by any batch of celestial bodies.

Therefore, we began to analyze the law of light intensity attenuation of different celestial bodies in the usual long propagation process, and together with the analysis of the observation ability of the telescope, in order to achieve our final research purpose.

The above is considered from the perspective of a telescope. We need to further consider, what if the light is very weak? Then the telescope can't react immediately. And what are the causes of the faint light of celestial bodies? Of course, they are the distance between the light source and the telescope, and the light intensity of the celestial body itself.

As shown in Figure 4.4, celestial body C continuously emits photons into the surrounding space.

Assume that C emits N_i photons in a very small-time interval Δt_i of the i-th time. If a telescope W is located close to the celestial body C at this time the intensity of the photons emitted by the celestial body C is bigger than Q(TEL), it can make a telescope immediately fill the potential energy well of the telescope.

When these N_i photons continue to diffuse away from the celestial body C to the surrounding space, as shown in Figure 4.4, as the distance increases, the sphere expressed by these photons (its photosphere) enlarges, and the light per unit area on this sphere (its light intensity) diminishes, equally so does number of photons uniformly distributed in each unit area of the photosphere diminish. Finally, this process will reach a **Critical Point P_{CRIT}** shown in Figure 4.6.

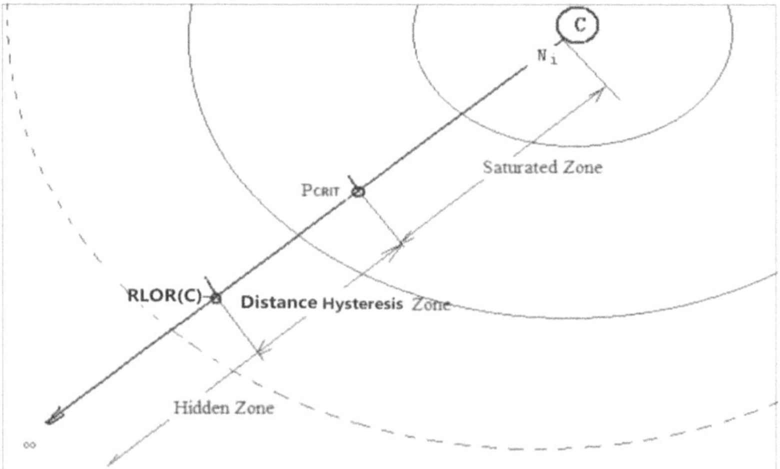

Figure 4.6 The change of the telescope's required response time in different distance intervals of celestial body C. When the telescope is between the celestial body C and the saturation response point P_{CRIT}, the response time of the telescope to the light of C is the full intensity response time FLIRT; When the telescope is in the distance lag region between P_{CRIT} and the Limited Observable Radius (LOR), the response time of the telescope to the light of C increases with the increase of the distance.

At a distance greater than this critical point P_{CRIT}, the light intensity of this layer of photosphere is less than the required minimum intensity Q (TEL), it can't make this telescope W respond immediately.

So, the distance of the light spreading route from C to P_{CRIT}, is named **Saturated Zone**.

Starting at point P_{CRIT} and going further away from C, the telescope will not respond to the first photon batch. It starts accumulating photons sequentially coming to the telescope, until the accumulated light intensity equals to Q(TEL), then it will respond.

As shown in Figure 4.6. This photon batch accumulation will last until the photon bache arrives the RLOR —Relative Limited Observable Radius defined in the chapter of discussing the HCO. Its value is decided by the noise and the sensitivity of the telescope. The distance of the light spreading route from P_{CRIT} to RLOR is named **Distance Hysteresis Zone**.

Beyond the RLOR is the **Hidden Zone**, in which no matter how to integrate, the light intensity from C cannot be accumulated to the level of Q(TEL).

Considering the different response times of the telescope to the light intensities of the celestial body when the telescope is at different distances from the celestial body gives us a different approach to solve the redshift observed from a different solution, different from Hubble's expanding universe. Thus discards the pseudo-God Big Bang.

The attenuation effect, fueled by distance, is an important factor that cannot be ignored for correctly analyzing the speed of an observed image.

4. The derivation result of the mathematical model of the new theory for the cause of the cosmic redshift

Based on the above analysis, we will conclude the mathematical model for calculating the different responding times of the telescope to the light of the celestial body at different distances, and present the results below. For readers not interested in the mathematics applied, We urge you to examine my other books: *Debate of Light and Dark (2016)* and *Fading Modern Cosmology (2020);*

What we analyzed above is the time it takes for the telescope to respond. The speed of response can be obtained by calculating the distance between the telescope and the celestial body being observed.

The Speed of Observed Image (SOI) after the telescope responds can be obtained as follows:

$$SOI\,(C,\ P_{TEL}) = \frac{D\,(C,\ P_{TEL})}{(T(C,\ PTEL) + FLIRT\,(TEL))}$$

$$= \frac{D\,(C,\ P_{TEL})}{[T\,(C,\ P_{TEL})\,(1 + FLIRT\,(TEL)\,/\,T\,(C,\ P_{TEL}))]}$$

When $0 < D(C,\ P_{TEL}) \leq D(C,\ P_{CRIT})$

The telescope TEL is located at P_{TEL}. Its distance to the celestial object C is D (C, P_{TEL}). The formula in different zones is shown in Table 4.1.

Table 4.1 Calculate the Speed of Observed Image SOI of a celestial body in different zones:

SOI =	The telescope is in Saturated Zone $\frac{V_{EMIT}}{(1 + Z_0)}$	when $Z_0 = \frac{FLIRT(TEL)}{T(C,\ P_{TEL})}$ $D(C,\ P_{CRIT}) \geq D(C,\ P_{TEL}) > 0$, $V_{EMIT} = \frac{D(C,\ P_{TEL})}{T(C,\ P_{TEL})}$	(4.1)
	The telescope is in Distance Hysteresis Zone $\frac{V_{EMIT}}{(1 + Z)}$	when $RLOR(C) > D(C, P_{TEL}) \geq D(C, P_{CRIT})$ $Z = Z_0 + Z_1 + Z_2$ $Z_0 = FLIRT(TEL) + T\,(C, P_{TE})$, $Z_1 = \sum_{i=0}^{j} \left(\frac{\Delta t_i}{T(C,\ P_{TEL})}\right)$ $Z_2 = f\,(\Delta\ D(C,\ P_{TEL}))$ $\sum_{i=0}^{j-1} \left(\frac{N_i}{4\pi(P_{TEL})^2}\right) < Q(TEL)$, $\sum_{i=0}^{j} \left(\frac{N_i}{4\pi(P_{TEL})^2}\right) < \geq Q(TEL)$	(4.2)
	The telescope is in Invisible Zone N/A	When $D\,(C,\ P_{TEL}) > RLOR(C)$	(4.3)

Note:

SOI	Speed of Observed Image
V_{EMIT}	Emitted photon speed, $V_{EMIT} = \frac{D(C,\ P_{TEL})}{T(C,\ P_{TEL})}$
Q (TEL)	The minimum required number of photons in excitation to cause a response of the telescope TEL
P_{CRIT}	Saturation response point, the telescope located at this point to meet $\frac{N_0}{4\pi\, D^2\,(C, P_{CRIT})} = Q\,(TEL)$
P_{TEL}	Telescope TEL located at this point
FLIRT(TEL)	Time of Full Light Intensity Responding of a telescope TEL when $I\,(C, N) \geq Q\,(TEL)$

N_i	Photon number of batch i the star emitted at Δt_i
RLOR(C)	Relative Limited Observable Radius of celestial C
Z	The Hysteresis Parameter $Z = Z_0 + Z_1 + Z_2$
Z_0	Full Light Intensity Hysteresis parameter, $Z_0 = \frac{FLIRT(TEL)}{T(C, P_{TEL})}$
Z_1	The Distance Hysteresis Parameter $Z_1 = \sum_{i=0}^{j} \left(\frac{\Delta t_i}{T(C, P_{TEL})}\right)$
Z_2	The distance displacement shift parameter $f(\Delta D(C, P_{TEL}))$. It is the function of distance change between C and P_{TEL}

In the table, $\mathbf{Z = Z_0 + Z_1 + Z_2}$ is consistent with the calculation form of the existing calculation formula for cosmic redshift theory.

From the result obtained, we can see it is basically consistent with the original old result of calculating the redshift. In other words, **the mathematical model obtained with the new theory can explain and apply the redshift results obtained from existing astronomical observations and calculations.** However, the derivation of the new theory does not assume that the redshift of celestial bodies is caused by the receding motion of celestial bodies. This discards the assumption that the universe is expanding and also **discards the assumption that the redshift of celestial bodies is mainly caused by the receding of celestial bodies away from the telescope.** But the observation and calculation results are retained.

Such calculations can also make use of all observation results of different factions that participated in the Hubble Constant debate.

We therefore now possess the possibility of abandoning the Big Bang theory, and the ever-accelerating speed of celestial bodies in an expanding universe. As the universe returns to its natural state of calm, we retain and apply all the original astronomical observation results related to redshift.

In other words, the new theory of explaining the observed redshift data of celestial bodies is completely different from the old, but the results of observation and calculation of redshift are basically consistent with the data.

Chapter Summary

The analyses made in this chapter are the result of research and discussion with Lü Wu on the phenomena of concern. Due to limited personal resources, the work is very rough. However, in this chapter we do grapple with a sufficient number of the issues that the pseudo-god Big Bang theory prompts and provide an alternative that does not rely on consensus opinions, however much they are supported by sophisticated space-based instrumentation and complex mathematical models. As ever, when advances in knowledge occur, they do so by marginalizing superstitions that may have the appeal of science but do not satisfy the rigors that the scientific method requires.

After all, our research provides a way out of the pseudo-god Big Bang model, so that the study of the universe does not have to be immersed in the smoke of a pseudo-God Big Bang exploded by atoms.

Since this is a new path, then there also may be other paths to explore in this realm that bring us closer to the truth. The key is to put aside the influence of superstition and Big Bang Pseudo-Theology, and it is possible to discover various paths closer to the truth.

How clean and beautiful the universe appears without the smoky Big Bang!

Part 1 Summary

Academia's Efforts to Prove the Big Bang Theory by Science All Have Failed

What a wonderful term science is.

In the name of science, the absurd becomes authentic, the unreasonable becomes rational.

Therefore, the attitude of the academic community towards exploring the universe is becoming more and more sloppy. Many research projects that cannot withstand scrutiny have used money to clear the way, and based on the most expensive instruments, they have continued to do things that cannot withstand query for many years. It's like using more and more advanced receivers, measuring (receiving) near the Earth for decades, and then drawing a map of the distribution of the cosmic microwave background in the whole universe.

Several major NASA official current "scientific" pillars of the Big Bang theory have been scientifically proven to be incorrect in this book from a scientific standpoint. This is due to the myth that the source of the Big Bang is an atom. Therefore, attempts to prove the science of this theory from a scientific point of view have no basis and are easy to prove false.

Science needs to be proven by repeatable experiments. And the repetition must be justified. Measuring microwaves year after year **near Earth** and pushing the results out into the **whole universe** is not repeatable scientific experiments, but purely waste taxpayer's money and unacceptable!

Attempts to use "science" to defend pseudo-theology have the consequence that, after being scientifically and irrefutably proven to be fundamentally wrong in this category of "scientific proof", true theology is also implicated and harmed.

Even the messengers of God should stand up to criticize pseudo theology and defend the true will of Heaven, right?

So, what is the right way to do it?

In our opinion, scientifically proving the whole and true face of the universe is unknowable to human beings, and thus scientifically planning how much of the universe can be known at what stage of scientific development is a responsible scientific attitude to human beings.

Those parts that humans can never reach and never know are left to "God knows"!

Part 2

Humans Will Never Be Able to Understand the Universe Fully and Correctly

In this part, we will prove from a **scientific** point of view that human beings are incapable of fully and correctly understanding the universe.

A comprehensive and correct understanding of the universe is the dream that human beings will never be able to achieve.

So, who can do this?

Who can know what is really going on in the universe?

Nobody knows!

Only Heaven knows!

Only **God** knows!

In modern times, the astronomical and cosmological communities' express great confidence in their work, and believe that they have, or with their methods can, fully understand the universe and how it works. More recently, with the investment of huge amounts of money in building and perfecting ever sophisticated instruments and interpretive software, they believe that human beings may be closing in on the ultimate laws of nature.

In this part, we begin by pointing out the simple but numerous mistakes that astronomers and cosmologists continue to make.

Then we will explain how those mistakes occur when astronomers and cosmologists focus their research on distant celestial objects.

In the end, the truth is clear: we human beings, try as we might, continue to confuse pseudo-science with the mysterious face of the Creator!

Tiny human beings can't see through the mysterious true face of the Creator!

Chapter 5

Why has NASA Been Issuing Obviously Wrong Observational Reports for Many Years?

The authors had different opinions in recent discussions after the chapter was completed. One point of view is that the narratives reported in this chapter are wrong, and the other point of view is that the difference between understanding and "self-evident" caused by different perspectives is not a big problem.

The former point of view is the narrative of this chapter and will not be repeated here. Readers will know while they continue reading. Here is a brief introduction to another different point of view.

This view holds that the telescope sees a newly formed black hole that is only 30 years old 50 million light-years away, or that the light information of this black hole travels 50 million light-years to reach the lens of the telescope when it is 30 years old.

There is some truth to such an explanation, but there are a few points worth considering:

1. How long have humans observed this black hole?
2. In human observation, is 50,000,000 light-years away, a short period of

30 years, within the error range of observation?

3. Why not just state that this 30-year-old black hole is actually 50,000,030 years old? You should know that this is not a scientific research report, but news for ordinary people without professional knowledge.

From the perspective of data processing, 30 years is really a small number in astronomy. In particular, the redshift theory is used to calculate ages and distances.

Figure 5.00 is a simple illustration of how one applies Hubble's law to calculate distance and age. Basically, it is calculated using statistical laws. In this algorithm, it is unconvincing to put 50000000 and 30 together for calculation.

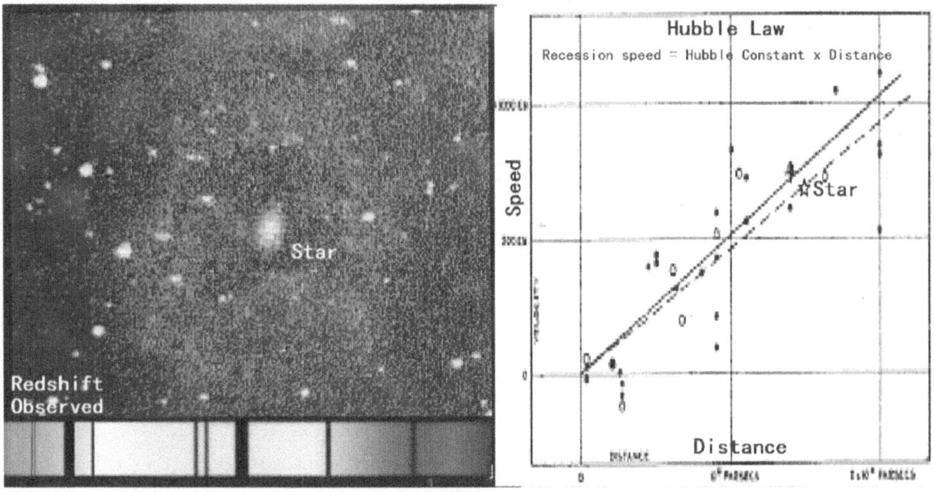

Figure 5.00 Observe Star in the middle of the left figure, and get the redshift value at the bottom of the left figure. Then in the right figure, use this redshift value to roughly estimate the relevant value of the Star according to the statistical method given by Hubble's law.

Below is our old view. Readers are asked to make more analysis and judgments in comparison with the two different opinionx, so we will not repeat them here.

Let's start with an old piece of NASA news released on November 15, 2010. Serial Number: 10-299.

https://science.nasa.gov/science-news/science-at-nasa/2010/14nov_babyblackhole

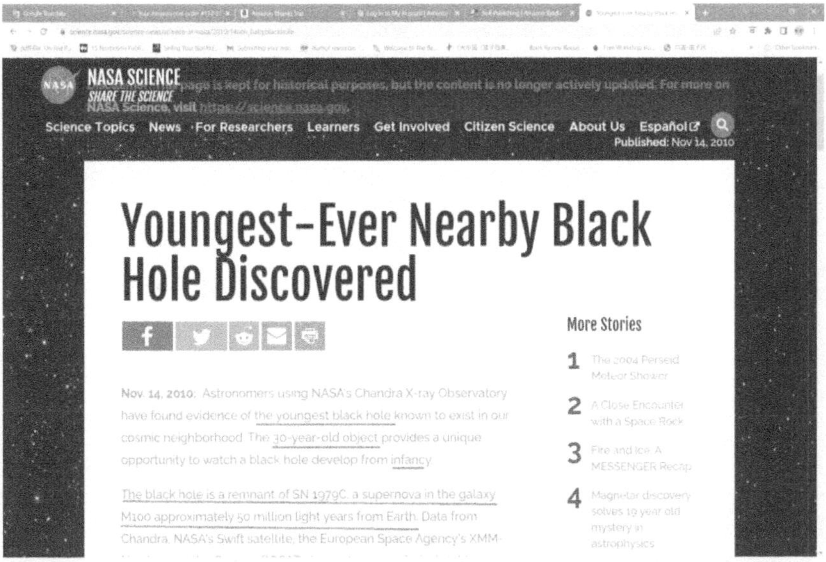

Figure 5.0 Retrieved again on 11/07/2022

Title：*Youngest-Ever Nearby Black Hole Discovered*

Nov. 14, 2010: Astronomers using NASA's Chandra X-ray Observatory have found evidence of the youngest black hole known to exist in our cosmic neighborhood. The ***30-year-old*** *object provides a unique opportunity to* ***watch a black hole develop from infancy****.*

The black hole is a remnant of SN 1979C, a supernova in the galaxy M100 approximately ***50 million light years from Earth****. Data from Chandra, NASA's Swift satellite, the European Space Agency's XMM-Newton and the German ROSAT observatory revealed a bright source of X-rays* ***that has remained steady during observation from 1995 to 2007.*** *This suggests the object is a black hole being fed either by material falling into it from the supernova or a binary companion.*

"...it is very difficult to detect this type of black hole ***birth****..."*

"...The idea of a black hole with an ***observed age of only about 30 years*** *is consistent with recent theoretical work..."*

Several issues arise immediately that require discussion, both concerning the news and the science behind it:

1. Is the news wrong? If so, why? (Yes, it is wrong!)
2. Is this one of many similar errors? Yes!
3. Do such errors occur systematically under the guidance of the famous color magnitude relation (CMR) theory? (Yes!)
4. Why do such mistakes occur? Whose fault is it?

5. Is this a typical example of the use of a mathematical formula, rather than whether or not its use is reasonable?
6. What lessons can we learn from these kinds of erroneous reports?

Let's discuss these issues one by one. As you read, consider carefully how to properly evaluate the relevant information against the alternative viewpoint presented at the beginning of this chapter.

1. Why is this news wrong? Is this report really wrong?

The news states quite simply that *NASA astronomers saw* ***a 30-year-old infant*** *black hole at about* **50 million light-years** *away from the Earth.*

This is an obvious mistake that a simple sentence can make clear: the propagation of light takes time. "Light-year" is the distance that light travels for one year.

Any information about the 30-year-old black hole can only propagate to as far as 30 light-years away. But the report said that the black hole is located about 50 million light-years away from the Earth.

Any 30 years old baby black hole can be discovered by the telescope on or around the Earth, this baby black hole has to be located near the Earth. The distance between the 30 years old black hole and the Earth has to be equal to or less than 30 light-years.

It is impossible for any information about this 30-year-old black hole to cross the 50 million light years within its current lifespan of 30 years to reach the Earth. This light information is still on its way to the Earth at a distance of 49, 999, 970 light-years from the earth.

In effect, this NASA report reads as a new tale with mythic elements.

Here we first do a simple knowledge popularization: we must have a clear understanding that **any** telescope, no matter whether it is a radio telescope, a space telescope, or space detector, all **observation instruments, can only passively receive information** transmitted from celestial bodies. Whether it is the light of a star, some information of a black hole, or related information of a celestial body explosion, it must be first issued out by the celestial body, and then transmitted through the space distance between the celestial body and the Earth, reaching the Earth or the space near the Earth, and being received by the instruments there. Only then can humans receive relevant information.

In other words, any human instrument for observing the universe can only be a **passive** receiver of information from celestial bodies, without exception!

Usually we raise the telescope, and what is in the lens immediately appears in front of our eyes. We raised the telescope to look at the moon, and the moon immediately appeared in the lens. But this is just an illusion. The reason for the illusion is that the objects the telescope is looking at are relatively small in distance, and star light can travel such a small distance almost instantaneously. But the fact that the lightning arrived before the thunder was enough to tell the story.

It must be emphasized that: because the light information emitted by celestial bodies is always being transmitted dynamically, what the telescope receives can only be the information at a certain moment. In general, there are the following **Principle of the Limited Receivable Time of Celestial Information**:

For a single celestial object, if the celestial object is D light-years away from the Earth, then only the information it sent at or before D years can be received by the telescope on the Earth. For example, earlier than D year, the information sent in (D + k, k > 0) year has reached the telescope k years ago, and the information has already shined on the telescope for k years. And the information sent later than D year, for example, the information sent in (D – m, m > 0) year, will take an additional m year to reach the telescope and be received.

For a cluster composed of multiple of single celestial objects, if the distance of each single celestial object from the Earth is D_i, ***i = 1, 2, ... light years, then the corresponding pictures of each object*** D_i ***in the telescope is corresponding to one pixel of the screen shot. The screen shot of the telescope is a composite of pixels of*** D_i ***years, each pixel corresponding to the different time of each single celestial object in the cluster. However, the cluster information picture does not contain any information sent at a time later than the nearest*** D_n ***light-year to the Earth, that is, there is no information less than*** D_n ***years in the image received by the telescope.***

Thus, we can be certain that NASA's interpretation of when this young black hole was born is completely wrong! Nor does this conclusion require the expertise of a scientist.

2. Is this one of a large number of common errors? (Yes!)

Do astronomers **occasionally** make similar errors in their research? If the answer is "yes", then there is no need for additional discussion: anyone, including astronomers, will occasionally make mistakes.

This is many errors that have been produced continuously for decades under the guidance of theory.

Table 5.1 below lists five such errors

Celestial Name	Distance between Observer& Celestial (**light-year)**	Reported Age of the Newborn Celestial **(year)**	Name of the Newborn Celestial	Report year
NGC 7318	*27 million*	*2 million*	Stephan's Quintet	2001
Abell754	*0.76 billion*	*0.3 billion*	Newborn galaxy	2004
NGC7320	40 million	10 million	Stephan's Quinte	2004
M100	*50 million*	*30*	Youngest Baby Black Hole	2010
NGC4485 …	*25 million* ….	*million* …	interacting with NGC 4490 …	2019 …

Unfortunately, this is not an accidental error, nor is it a rare error.

We can see this kind of bug report happening for many years in a row. Most reports are very sensational when they are published by the media.

There are dozens of related research articles listed in the references in this book and my previous books (such as *Fading Universe, Clarify Big Bang* and *Debate of Light and Dark)* in which similar errors occur. Other research articles that we have read, analyzing the research processes used, express the same type of errors, violating the principle of the Limited Receivable Time of Celestial Information. They all claimed that they found celestial bodies much younger than the least distance between them and Earth, as measured in light years. Please refer to the references in the appendix of the book.

3. The errors occur systematically when using the famous Color-Magnitude Relation (CMR) theory

After consulting the relevant materials, and carefully studying the source of the problem, we found that the calculations in these erroneous research reports came from the use of a mathematical model known as Color-Magnitude Relation (CMR). Experts are accustomed to doing their calculations using mathematical models, and publishing their results as such. But they don't seem to be considering basic issues, which includes determining whether the results of their calculations are reasonable or not. This kind of belief in certain mathematical models that have attained consensus approval — without first examining the rationality of the model from various angles — is a common problem in the scientific world. In this book, however, we note such errors as evidence of a belief, which if it goes unexamined can rise to the level of superstition.

All the examples listed in Table 5.1, as well as many research reports that are not explicitly listed, used the CMR mathematical method, that is, the Color-Magnitude Relation mathematical method to theoretically analyze the observation data from telescopes, from which their research results came.

For public consumption, of course, some of these research reports appear in lay form written by journalists, with various levels of depth that can, at times, help us understand the errors we have identified.

For example, in a news report titled " STAR CLUSTERS BORN IN THE WRECKAGE OF COSMIC COLLISIONS," a journalist wrote, as excerpted:

Studying the star clusters and dwarf galaxies in Stephan's Quintet provides insights into how galactic encounters may have driven galaxy evolution in the early universe. The ***quintet resides 270 million light-years from Earth*** *in the constellation Pegasus.*

"Hubble's superb resolution allowed us to discover and pin down the ages of these star clusters,"

*"**Analysis of the colors** of the clusters indicates several distinct epochs of star formation that appear to trace the complex history of dynamical interactions in this compact group.* ***As a cluster ages, the hottest, bluest stars die, and the color of the entire cluster becomes redder. Therefore, the redder the cluster is, the older it is. The amount of light emitted by a star cluster in different color bands can indicate its age.****"...*

The clusters in those regions span a range of ages, from ***about 2 million to more than 1 billion years old****...*

The positive aspect of this report is that it gives a simple and clear explanation of how astronomists use the CMR method to determine age based on color.

Although the CMR model provides scientists with one of their main methods for calculating and analyzing astronomical observation data, especially distant nebula collisions, it does not prevent errors from appearing in the results obtained, including obvious errors. Why were such errors overlooked? These wrong results were so plainly obvious, but either the authority experts or masses ignored the errors.

We have a small question about the distant nebula collisions that are often reported in astronomical observations: Why do such collisions always occur in distant places billions of light years away? Why have they never been closer to us? According to the "universal principles" used by experts in studying the cosmic microwave background and the Big Bang model (Figure

0.1), the chance of a nebula collision is equal anywhere in the universe.

This question has prompted us to consider it in greater depth. The basis of the CMR model is the redshift theory applied to analyze the observing results of celestial bodies, as commonly used in astronomy. This theory is also the basis of our belief in the Big Bang and for how we use advanced telescopes when observing distant celestial bodies. If something goes wrong with the method, imagine the panic that would erupt in the world of astronomy and cosmology?

We have discussed the causes of the redshift of the universe and other related issues before, and the discussion here can be regarded as its supplement in a certain respect.

4. Identifying the reason why such mistakes occur? – Is it the fault of the math involved?

Cosmologists have consistently made the same mistakes in their observations of nebulae in collisions that produce new stars.

Now we can't think cosmologists are so shallow that they don't know that what happened 300 million years ago at a location 800 million light years away would be invisible to us. We need to do a more detailed study and analysis of this strange thing: t why so many cosmologists make similar mistakes and formulate rules to eliminate them.

The question is: why are there so many weird bugs in general?

We absolutely believe that every cosmologist or astronomer understands that the light emitted by celestial bodies takes time to cross the distance in space.

But why do they forget such basic scientific common sense when they are doing their research?

We have considered this issue for several years. But we didn't want to spend too much time on an in-depth study of astrophysics theory. We believe that, in general, without using advanced mathematics but instead ordinary philosophy, general logic, and simple mathematics we can basically infer the correctness of an event or a theory. However, in order to discover the truth we sought in the facts available to us, we were forced to read original research papers from which the reports we reviewed derived.

And what we found will interest you: It's the math that caused the errors we encountered repeatedly.

The Beauty of Mathematics Blinds the Experts

Mathematics is the diamond in the crown of science. The beauty of mathematics lies in its abstractness and generality. However, when abused, mainly as we found by mathematicians and theoretical physicists, its beauty quickly fades.

Take for example the famous theoretical physicist and cosmologist Stephen Hawking, specifically in his book we noted earlier: *A Brief History of Time*. Masters of their science, like Hawking, despite expressing their esoteric thoughts in plain, accessible language, also like to hide behind strange symbols or advanced instruments. Hawking's book deserves respect, if not necessarily for its use of sophisticated research and results, but rather for his willingness to reveal his thoughts to everyone. His book enables us to understand a great deal about time and how we conceive it.

There is, however, this revealing passage on singularities in his book:

During the next few years, I developed new mathematical techniques to remove this and other technical conditions from the theorems that proved that singularities must occur. The final result was a joint paper by Penrose and myself in 1970, which at last proved that there must have been a Big Bang singularity provided only that general relativity is correct and the universe contains as much matter as we observe. There was a lot of opposition to our work, partly from the Russians because of their Marxist belief in scientific determinism, and partly from people who felt that the whole idea of singularities was repugnant and spoiled the beauty of Einstein's theory. However, ***one cannot really argue with a mathematical theorem. So in the end our work became generally accepted and nowadays nearly everyone assumes that the universe started with a Big Bang singularity. It is perhaps ironic that, having changed my mind, I am now trying to convince other physicists that there was in fact no singularity at the beginning of the universe****—as we shall see later, it can disappear once quantum effects are considered.*

Notice how his conception of the formation of the universe changed when he considered or re-considered the mathematics behind it. In fact, when considering Einstein's concept of the universe, what evidence can prove its truth? Until now, such proof has yet to appear, including from its author: Einstein – who failed to find a solution to the cosmic formula he presented. Did Einstein's concept of the universe automatically gain acceptance because it came from Einstein? Can the universe be described by ONE mathematical formula? As it seems, theoretical physicists commonly identify the universe with whatever their mathematical models describe and predict.

Hawking, of course, generally followed master Einstein's path. When the master says it is a lamp, he will add oil, and when the master says it is a temple, he will kowtow.

And his success, which inspired others to imitate him, or the example he provided, foreclosed perhaps too much on going their own way.

Aren't current astronomical and cosmic studies repeating these old "successful" paths?

Actually, there is a mathematical formula for a singularly: $\frac{x}{0}$. On its own this formula is generally meaningless or infinitely expansive. Notice again the process Hawking went through expressed in the excerpt from his book given above. Hawking turned this formula into a "scientific" proof for the Big Bang theory, boasting as he did so that he also developed "new mathematical techniques." Later he changed his mind, denying the existence of a singularity. But because, as he put it, "one cannot really argue with a mathematical theorem," however complex it is, the initial work was generally accepted. Thereafter, when faced with Hawking's reversal – "there was ... no singularity at the beginning of the universe" – his "other physicists" continued to support the theories he had abandoned. What a mess it turned out to be.

We don't know how the equation for a singularity, $\frac{x}{0}$,was accepted or how it was derived from Einstein's theory to support the Big Bang, nor do we have any interest in learning about it. We only know that $\frac{x}{0}$ is a **meaningless** thing in mathematical **application**, and no feat of technology can change such a basic fact. Whether it came from Einstein or Hawking's use of pure mathematics, the support given to the Big Bang theory does not militate against its mythic character. We only need to recognize it, and put an end to playing with the universe with mathematics.

When confronting our skepticism, sharpened by our research, people often tell us that humans have landed on the moon and space shuttles are flying in space. How can you not trust scientists?

Others say: Are people all over the world wrong, only you are right?

Indeed, we are also wondering: How come such simple errors were published by NASA, and no one raised any questions about them?

Why?

Thinking more deeply about this matter, doubts will arise about the capacity of astronomy and cosmology to discover and explain how the universe came into being and how it works. Is our research on the cosmos advancing correctly or has it wandered off into a fool's paradise illuminated by fancy mathematical models?

We don't know how astronomers feel when they look back at this piece of news today? This kind of common-sense mistake should not be made by any professional astrophysicist!

So many experts have not questioned this most naive error.

Ask again once more, why?

Thinking deeply, you will have doubts about the prospects of the seemingly profound and lofty cosmic astronomy: Is contemporary cosmic astronomy advancing on the right path? If even such simple mistakes will occur in large numbers and systematically for a long time under the guidance of the theory, then don't we have to ask a series of why?

Should we firmly oppose such pseudo-theology preached in the name of "science" or take a more measured approach?

Our purpose in this book is to sound an alarm by prompting readers – scientists included — to ask why too many low-level mistakes have occurred as we have identified them. In this way, perhaps the research communities involved in exploring the cosmos will ask the same question, inspiring the kind of insight that will enable us to better understand the universe. For one, we believe that when the Big Bang theory is finally proved erroneous, and that error gains the status of scientific consensus, looking back on its evolution will enable young scientists to determine a better way ahead.

I got the Regents Fellowship in 1987, and received my PhD from the School of Engineering of University of California Berkeley, in 1990. Some years earlier, in 1987, I was awarded a Regents Fellowship. I wrote many serious papers on astrophysical phenomena, but none of them were published in the astrophysical literature media. In fact, I submitted most of the content in this book to journals dealing with physics, astronomy, or nature. All of my papers were rejected. Now that I am retired, all of this no longer bothers me.

Today, during a time when science is of paramount importance, not only for our understanding of the universe but also for our continued existence on Earth, the lay public tends to believe in the teachings of celebrated scientists. Other, more common scientists usually follow the rules established by celebrated scientists in order to survive in the fields they work and teach in. Perhaps they should ask themselves this question as well: Do celebrated scientists use their advanced mathematics and "superior" astronomical telescopes to forcefully sell their wonderful theories to other scientists and to the world at large?

What then happens when a theory is proved wrong by way of a large number of simple and intuitive errors made when using that theory. Hawking's experience is cautionary in this regard.

When this situation occurs, we need to know whether or not these errors are the result of accidents in the research done, lack of oversight of that research, or errors that have occurred continuously, sometimes for decades under the theoretical guidance and mathematical models so widely used. If the latter holds true, our research communities have much to answer for.

In 2005, we asked this question: "Is math old?" In 2016, we answered: "The math is old!"

Fortunately, we were not the only ones who asked and answered this question. There are many independent thinkers who are fighting in their own way. In 2018, a young mainstream theoretical physicist, Sabin Hossenfelder, came forward to explore it all in her book, *Lost in Math*. As she put it:

How [is it that] this modern preoccupation with beauty blinds us from seeing the natural world as it really is. Led by their aesthetic criteria, physicists have concocted mindboggling new theories, invented dozens of new particles, and declared that distant places across space are connected by wormholes. But observation has been unable to confirm almost all of these ideas — in fact, many of them are actually untestable. In spite of these contradictions, theoreticians are convinced that their graceful equations and elegant formulas contain higher truths about nature. And these 'too good to not be true' theories have left the field stuck in a cul-de-sac.

We are very encouraged by this book. Suddenly, here is a rebellious text written by a young scientist who seeks clarity over complexity, and confirmation by accurate observation. Is the thick, hard ice of consensus finally showing signs of melting? Is a new truth on the horizon?

We look forward to the blooming spring of flowers to bring back the colors in the universe that cosmic science has faded!

5. A typical example of formula usage, without considering whether such usage is reasonable or not

Hubble Space Telescope's official website presents such a topic: Star Clusters Born in the Wreckage of Cosmic Collisions. Let's analyze this news carefully to see if we can find useful clues to help us find a correct way to process the information obtained.

Our analysis on this wonderful news is based on the following two points:

First, all cosmic observers and data analysts are scientifically educated, and they definitely know the following basic astronomical knowledge: it is impossible to see any celestial body younger than X years at X light-years away.

Secondly, there must be some reason for their errors, because while they focused on conducting data analysis, they ignored the previous rule.

What is the reason for the latter?

Let's take a closer look at some of the information that inspired us in the report below: https://hubblesite.org/contents/news-releases/2001/news-

2001-22.html

NGC 7318 is a pair of colliding galaxies about ***300 million light-years from Earth.*** The team is particularly interested in the ages of the clusters in the northern starburst region and in NGC 7318B's elongated spiral arm. The astronomers found evidence of a series of *starbursts beginning at the time of a collision* ***20 million years ago*** *and* ***continuing over millions of years***. The youngest clusters are relative newborns in astronomical terms, *about* ***2 million years old****...*

"*It has been* ***20 million years*** *since the encounter, and yet* ***we're seeing star clusters that were born 2 million years ago,***" says team member Jane Charlton of Pennsylvania State University. "Previous studies suggested that the ages of *these young clusters were* ***10 million to 20 million years old***. But with the Hubble telescope we could determine that the ages were even younger. The range of ages of these young clusters indicates that the star formation in that region may be 'self-propagating."

We can see the errors in the report from the above description. NGC7318 is at 300 million light years from the observer. The light emitted 2 or 20 million light-years ago is still far away from the observer, and it will not arrive until almost 300 million years later. At present, the observer cannot see it.

Where did the error come from? Let's look at the following passage in the same report:

"Analysis of the **colors** of the clusters indicates several distinct epochs of star formation that appear to trace the complex history of dynamical interactions in this compact group. ***As a cluster ages, the hottest, bluest stars die, and the color of the entire cluster becomes redder***. *Therefore,* ***the redder the cluster is, the older it is.*** *The amount of light emitted by a star cluster in different* ***color bands can indicate its age***. *"*

So, we know that after the telescope obtained the requisite data, experts based their analysis on relatively new color-related theories: that is, CMR along with computational theory, and eventually cosmic redshift theory. From this kind of synthesis, the analysis proved are wrong in different aspects, even perhaps ridiculous. The basis of this CMR theory is the cosmic redshift theory.

Therefore, **re-examining the cause of redshift theory should be an urgent task** for astronomical and cosmic physics.

Figure 5.1 is one of the many research articles retrieved by random search. We extracted a brief excerpt from it about a galaxy cluster in the constellation Hydra, which reveals some relevant information.

The research article, published in 2016, was titled, "Search for a Non-equilibrium Plasma in the Merging Galaxy Cluster Abell 754" and authored by more than 12 scientists. It identifies Abell 754 as 800 million light-years away from the Earth. The merge happened 0.3 Gyr (300 million years) ago.

2 Abell 754

Abell 754 is a well-studied major merger cluster on the plane of the sky at $z = 0.0542$ (Struble & Rood 1999). Its surface brightness is elongated along the southeast-northwest direction. Previous works studied its galaxy distribution (e.g. Fabricant et al. 1986, Zabludoff & Zaritsky 1995), the profile of the radio emission (e.g. Kassim et al. 2001, Bacchi et al. 2003, Kale & Dwarakanath 2009, Macario et al. 2011), the X-ray morphology and the temperature structure in the ICM (e.g. Markevitch et al. 2003, Henry et al. 2004, Macario et al. 2011). These works reveal the followings:

- Neither of the two major clumps in the galaxy distribution corresponds to the X-ray brightness peaks.
- Spatial variations of the temperature are found with highest temperature kT > 10 keV.
- The 1.4 GHz radio diffuse emission consists of two large components (Bacchi et al. 2003), roughly correspond to the two large main galaxy concentrations.

These features indicate Abell 754 is undergoing a recent major merger. Hydrodynamic simulations carried out by Roettiger et al. (1998) reproduce the observed X-ray features by considering an off-axis merger between two subclusters with a mass ratio of 2.5: 1. **They also imply the merger is a very recent one (<0.3Gyr).** Markevitch et al. (2003) suggests that a more complex picture for the merger in which the third subcluster or a cloud of a cool gas is sloshing independently from its former host subcluster. Macario et al. (2011) first confirmed the shock front with the Mach number of $M = 1.57^{+0.16}$ at the southeast part of the cluster by measuring the temperature jump with Chandra.

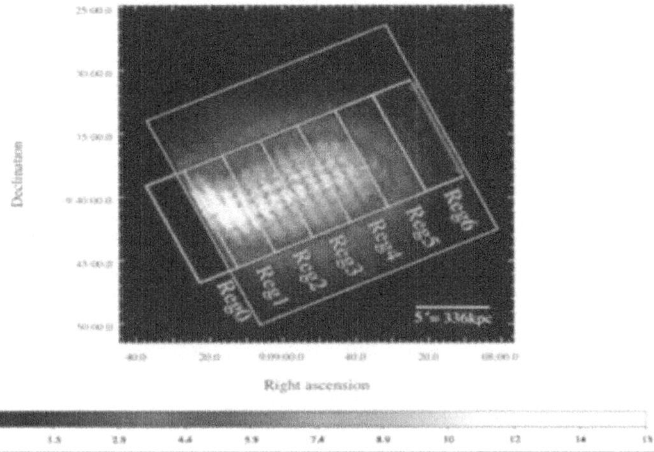

Figure 5.1 Another error example

Actually, the merge can't be observed from the Earth.

First of all, names of other researchers who work on related phenomena appear in the article. All use relatively consistent methods. Secondly, calculations are based on the CMR theory, and the color in the picture is related to the red shift. Then notice this statement: “the merger occurred recently (<0.3 Gyr)” — that is, the experts agree: 800 million light-years away, they have seen the Nebula merger that happened less than 300 million years ago.

Count how many experts are involved in the paper above: six authors, six single scientists, and seven teams of scientists who collectively publish research reports.

The seemingly consistent research results they presented were all wrong!

6. What lessons can we learn from these kinds of erroneous reports?

The first lesson is the enormous cost incurred by these simple mistakes, including the time taken by these experts to prove their point and the associated funding.

In 2005, in my Chinese book, *Who Should Talk about Cosmos*, I questioned NASA's report: that Abell 754, 800 million light-years away from the Earth, saw new galaxies born 300 million years ago. The report was authored jointly by more than 20 prominent cosmologists. They published their findings after concluding observations with the world's most advanced telescopes on the top of high mountains for more than two years. In other words, these experts wasted two years of their lives, with a lot of money spent and valuable time used on their equipment.

Then, the world's media focused heavily on NASA's press release, and no one objected to the apparent misreporting. Today (May 7, 2021), a Google search with the "Abell 754" keyword retrieved approximately 184,000 results in 0.40 seconds! How many resources are required to do this?

In addition, there are many research projects and reports that follow the same erroneous path.

Almost all of the research papers listed in the references in this book are resource intensive.

For example, the research paper, "The Colour-magnitude Relations of ClJ1226.9+3332, A Massive Cluster of Galaxies at z = 0.89" was jointly completed and published by six authors, six scientists and a team of seven scientists.

How much time, money and expertise does this obvious, low-level error cost? This waste should have been avoided, since the error has been pointed out at least by 2005.

Sadly, until 2021, there are still studies making the same mistake.

Are there any good solutions to prevent these kinds of errors from continuing to occur?

Why has the astronomical community ignored such obvious mistakes for so long, or just reluctantly agreed to discuss them? Our articles listed in the reference never had a chance to be published in a main stream academic journal!

Chapter Summary

The ancients asked the sky while toasting to the Moon about affairs that only the Moon lady might know about. Modern people look up at the stars, but think about events that are billions of kilometers away.

From our contemporary perspective, because the moon is so close to us, it seems perhaps a trivial task to explore it and discover what we can. The astronauts who set foot on the moon years ago found a desolate planet. There they found no Moon Lady, white rabbit, or Guanghan Palace. The ancient myth was left with only a vanishing shadow.

How does modern astronomical and cosmological science respond as it seeks to enhance our vision of the universe? Should we be surprised that the explanations the experts are giving to us are becoming more and more incomprehensible?

Indeed, while aerospace science allows mankind to set foot on the moon, explore Mars and Mercury, and from time to time provide achievements that inspire us, the same cannot be said for astronomy and cosmology. In countless scientific papers new myths arise, from wormholes and black holes to dark matter, dark energy, and more. What kind of telescope can you use to see new galaxies and young black holes newly born from distant celestial bodies? What kind of detector can you use to detect weak microwaves evenly distributed throughout the universe?

We try hard to explore the world we live in, but only feel our own smallness, and the greatness of our Creator.

Huge distances overshadow our eager eyes as we stare at the universe.

No matter how good a telescope is, it is only a receiver and can only see celestial bodies within a limited distance.

No matter how imaginary thoughts can be, we can't see through the veil-of-distance that the Creator of the universe hangs in front of us!

Mankind has no way to fully understand the universe.

Simple mistakes on such a large scale over a long period of time tell us that contemporary astronomy needs to return to basic and fundamental scientific rules, abandon the Big Bang pseudo-theology that an atomic explosion created the universe, and start a new journey of exploration.

Don't be frightened by the world's most advanced telescopes on Earth or in near-Earth orbit.

Don't be intimidated by celebrated scientists.

Let us make our best effort to poke through the fog and see the unknowable universe more clearly and correctly!

Chapter 6

Are the Celestial Objects on the Telescope Lens Real or Fake?

Some poems can be explained, some can't, and for some explanations are not necessary, It's like a flower in a mirror, or a moon on water. There is no need to be limited to their traces.

—— <Zen Xie. The Poet Talks Poem Intuitively>

The answer to this question in the title to this chapter is: we can never be sure that a distant celestial body is an entity or a fake light. It is impossible to be certain whether the distant celestial bodies seen by a telescope are real. None of the deep-space photos that NASA has shown us over the years can give us the certainty we need when considering the real faces that the universe turns to us.

Astronomers and cosmologists spare no effort to develop new telescopes. In their opinion, the more advanced the telescope, the greater distances can be seen, along with their truer colors. The more beautiful and fantastical the pictures they present, the easier it is to convince the people who view them.

But this is just an illusion!

It can be said that **none** of the images of distant celestial bodies that appear in any telescope lens is a real celestial body.

As the saying goes, we can't discuss ice with summer insects!

Let us explain to you this unpleasant truth from five aspects.

1. "What we see may not be true" is a law for observing the universe.

What You See Is What You Get is the most important means that cosmologists possess to expand or deepen their knowledge of the universe.

We always believe by seeing. We use our eyes to understand and explore the world.

The universe, however, is obscured by a most mysterious veil composed of great distances and endless time. Before this veil, which blurs our sight, we see phantoms, or find from a photograph a distorted image born from light without a source. Reality is flying far away from us as illusion plays its game of hide and seek.

We once heard a Nobel Prize winner's speech on the radio saying, "What we see is what we get." But actually, this sentence, in terms of understanding a universe calculated in light years, can best be described as wishful thinking with eyes open.

Everyone thinks that what we see is real. And while we certainly hope so, we also must ask if it's possible when considering the universe.

It can be said that what is seen on the Earth and in the space near the Earth is basically real rather than false.

In places far distant from the Earth, such as stars in the Milky Way and farther, what you see may not necessarily be a true entity.

When we look at the ancient universe with childish eyes, we must understand what the universe shows us, and what we have at hand. We must also understand the potential and actual flaws in the instruments we use, and we must use wisdom to explore the secrets of the universe if those secrets are explorable. So far, cosmic science has not included these values in its rush to study and understand the universe.

What you see, in this respect, is not what you get when observing the universe.

Only by fully understanding this point can we embark on a journey to grasp those mysteries of the universe we are able to grasp.

Before we can find the correct way to see through the blindfold that the universe has laid on humans, shouldn't our cosmic experts also leave a little room for the imagination for ordinary people who are not experts?

In this chapter, we paint a big question mark over the cosmic images taken by our space telescopes, and the image of the universe that they help to form in our brains. For every picture of the universe that we have or will see in the future, we must have a clear understanding of just how this image comes to be, and how we understand it.

Of course, the question mark we have used can also touch the mysterious veil of the universe, illuminating it briefly. But it does not have the power to lift it; we cannot see the true face of the universe behind this mysterious veil.

How can we solve the mysterious questions that the universe poses to us, how can we lift the veil that obscures the universe, and how can we respond to serious questions about the universe that trouble us.

Let us work hard together to see a real universe!

2. Seeing May Not to Be True 1: Is It a Real Star or Residual Light from a Destroyed Star?

Looking up to the sky, countless stars come into focus. From generation to generation, people have studied the stars, fascinated by them and the dreams they inspire.

Have you ever wondered whether these stars, perhaps your favorite ones, or those you are most familiar with, may not be real stars? Perhaps it's only their residual lights we can see?

Sounds a bit mysterious, but it is the truth.

Mr. Hawking has told us in "A Brief History of Time" that should the Sun go out, it will take 8 minutes for people on Earth to know it. Because the Sun is 8 light minutes away from the Earth, from the moment the Sun goes out and does not shine it takes 8 minutes for the light from the Sun to reach Earth before it spreads farther into space. In other words, after the Sun goes out, the sunlight that has been emitted will continue to shine on us for 8 minutes before finally disappearing.

Let's refer to Figure 6.1 to further understand and expand it:

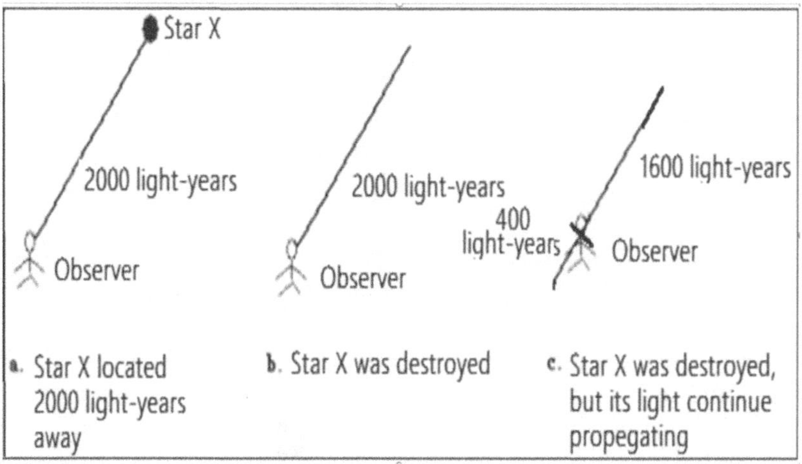

Fig. 6.1 shows the light transmission before and after the death of the star X

In Fig.6.1:

Frame a: Suppose that after humans invented the telescope, an X-star was observed at 2,000 light-years away from us, as shown in part a of the figure. The black circle in the figure represents the star, the straight line shows the distance from the star to the observer, and here we assume it is 2000 light-years.

Frame b: Suddenly one day, the star was destroyed for some reason as shown in portion b of the figure. However, the light it has emitted was not destroyed but continues on its path to be seen by an observer as shown in part b of the figure. Since star X no longer exists, the corresponding circle representing it is no longer drawn.

Frame c: Suppose we have been observing X for 400 years after the destruction of the star to the present, which part c shows in the figure. Well, however frustrating it might be, this light wave, which represents the destroyed star, can continue to be observed from Earth for 1600 more years!

This star X, which was destroyed 2000 years ago, is the same star that we have been observing for 400 years. We have received a large amount of data from it. But the star, having been destroyed, has not existed during the centuries we have been observing it. We have been observing it all this time! And we will continue to observe it for more than a thousand years!

Stars that no longer exist we can still observe for many generations! Astronomically, the 2000 light-year distance between star X and us is a very small value. Most stars are farther away from us. Even if those stars were destroyed, they would still exist for a very long time in our observations.

Although 2000 years is a figure that we use to make our point, in fact, it can be any number. Any stars far away from us are likely to cause this weird situation: over a long period of observations, the star we have been observing may no longer exist. And if that is the case, the starlight we can see presents an illusion of its source.

How long has human astronomical knowledge been recorded? Thousands of years? In other words, if any star on a human recorded astronomical map is more than a few thousand light-years away from us, we never know whether it was destroyed before our observation, or it has been destroyed during our observation period.

Of those stars with distances less than the length of time humans have been observing them, we know that they existed when we began observing them. This is the meaning of the comparison of visible distance between the naked eye and the Hubble telescope. The data collected by human eyes can determine the true existence of a star for thousands of years, a much longer period than this relatively new telescope can determine. But were these stars, which were determined to have existed at the beginning of the observation, possibly destroyed in more recent years? We cannot know. In other words, at this moment we are facing the starry sky, we cannot even be sure that any star is a real entity or not.

And we haven't been able to test it by actual observation. Because the history of human observation is too short, and the life of human beings is too short.

What we see is real! Non-existence is also true! What we see is illusory at the same time! Can we trust our eyes? Can we trust the space telescope? Can we believe the dazzling pictures that NASA has shown us?

The universe is too big! Human life is too short! The speed that humans can control is too slow! Humans can't see the authenticity of the universe!

I only caught a glimpse of nothingness! Oh, not a group, it is a group of billions!

Gosh! We lead people to doubt almost all the stars they see. It is really sinful!

Stars, please forgive me. How dare we shake you off the mathematical framework of the universe!

Illusory truth

During the Mid-Autumn Festival in 2000, while climbing a mountain with colleagues, impromptu I wrote a short poem, excerpts are as follows:

To the Moon Sister

Why only you can use your slender hands,
Casually teasing and stroking everyone;
Why I cannot gently give a kiss
On your white jade cheeks ?

Send a meteor to me ah,
Let me climb up on Qionglou with its long tail.
Bring to your cold Guanghan Palace
A series of modern laughs.

Maybe I'll set up a website there,
Let you feel the passionate heart of the net-friends;
Maybe I'll put on a few Rock and Roll music,
Sing and dance with you
under the sweet-scented Osmanthus tree......

Why do you keep silent
Stay alone with a gentle smile?
Your streamers fill heaven and Earth,

I only hold a handful of light illusory.....

I only hold a handful of light illusions! Today even though the dream came true, I am really holding a group of light illusions! The visual light illusory, the light illusory for the stars of the sky! The light is illusory for the whole universe!

The universe is really like a Master of Art, free to reconcile the colors of reality and nihility at will.

People, little people, how lovely of your ridiculous arrogance!

This annoying scene can be extended from two aspects: from the quantitative point of view and from the perspective of distance.

The distance of illusory

From the perspective of distance, even if the distance between the celestial objects we're observing and the Earth is dozens or hundreds of light years, we cannot be certain that those objects exist other than the light they have emitted. We cannot identify the authenticity or phantasy of our seeing! The Hubble telescope in space also doesn't help us here either. The light-year measures distance in cosmic, not human, terms. It is too great for us humans to comprehend. We use kilometer or mile as our units to measure distance by; an utterly insignificant measurement when used for cosmic objects.

The amount of illusory

From a quantitative point of view, we can doubt whether it is a phantom that existed for more than a few hundred years of light observed by every human being. When we look up at the stars of the sky, they naughty blink at you, as if asking: Do you know what I am? Am I just a group of beautiful light dreams, or a distant reality?

More badly, we cannot prove, and we cannot confirm! This is the suspicion of human thinking! This is the true sorrow of mankind!

Eyes really met the images in front of us, but are they the real existence?

The universe is so ambitious. Human life is too short. The speed man can control is too slow! Man cannot see the authenticity of the universe!

I just hold a group of illusions! Not a group, but millions of groups.

God! We guide people to suspect almost all stars in their existence, are we sinful?

Stars! Please forgive me. We dare to shake you off from the mathematical framework of the universe!

3. Seeing May Not to Be True 2: The Principle of Uncertainty in The Location of The Celestial Objects

This is another situation where we cannot see the true face of the universe.

In the following schematic diagram 6.2, which is the principle of uncertainty of the position of cosmic objects, the circle represents the cosmic object, and the cosmic object A (assumed to be a star) is N light-years away from the observer. The light emitted by A at position 'a' received by the observer "now" is emitted by A at N years ago. Since all objects in the universe are in motion, the star A itself has moved to the position 'b' during the N years that this piece of light propagates from 'a' to the observer. It can be seen from this that the picture of the universe that the observer really "sees" is not the real existence of the universe at the moment when the observer observes!

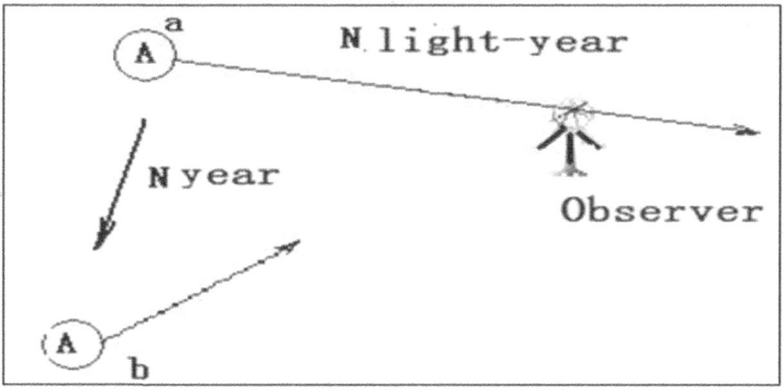

Figure 6.2 Uncertainty of the position of the universe object.

Unfortunately, the observer doesn't know A's motion patterns, so where A will be in the universe after N years, the observer can't be sure. The larger the N, the larger the variable, and the more uncertain the orientation of 'b'.

The worse thing is during the N-year movement, if A's distance between the observer and 'A' exceeds the so-called "escape distance," the information sent by the star A can no longer reach the observer.

What is even worse is that when we look at the stars as a whole, as the stars are so far away from us, the uncertain values of their positions relative to the Earth are also very small, making it difficult for us to go from a holistic and integrated point of view to grasp a real starry sky.

The stars are exactly what the observer observed, but it is really not actually within the images of the observer. The scene of the universe in front of the observer is a fantasy image! What they see is not what exists! Seeing is not believing, the old saying once again is broken!

Discussion: If the stars are basically relatively static? Then the principle of uncertainty in this position does not hold. But if so, then the Big Bang theory will be automatically broken.

4. Seeing May Not to Be True 3: Different times on one same image taken by the telescope

The Hubble Space Telescope has taken countless photos. People try to understand the universe by these photos, to infer the law of development of things in the universe.

Many important concepts about the universe, such as supernovae, are mainly derived from the images obtained from the universe.

However, do these photos that we rely so heavily on really reveal the true face of the universe? No! The **different points on the same picture** of the distant celestial bodies seen by the telescope represent **different times**, and these time points may differ by thousands or tens of thousands of years. This image with huge time differences between points cannot reflect the correct status of the area shown in the image.

In order to facilitate the discussion, we need to divide the pictures of the universe obtained by human beings into several roughly categories.

Pictures of the universe can be divided according to distance into close to the Earth, and away from the solar system images. Human research space within the solar system and constellations is quite detailed. Since the distance is relatively close, these images, rough or fine, are very realistic. But this is not all. Remember in the last decade it was also reported that members of the planets of the solar system are subject to change. Even the nearby 1 light-year size solar system were ambiguous, so how can we talk about the edge of the universe 13.8 billion light-years away?

For spatial images away from the solar system, we can roughly divide them into two categories: one is very diagrammatic and very distant. The measurement of their distance is a problem, which must be approximated with the aid of the parallax angles at the poles of the Earth's orbit around the Sun. The data thus obtained is a kind of rough perception of enormous cosmic objects.

Another is pictures of distant space taken with the help of advanced instruments such as the Hubble Space Telescope. These photos are very clear, and researchers have drawn a lot of conclusions from the analysis of these photographs. Because of this, advanced telescopes have become an important tool for people engaged in astronomical research. Major achievements in astronomy are discovered through telescopes.

Due to the powerful capabilities of the telescope, people are increasingly striving to produce ever more advanced telescopes. These telescopes with limited practical capabilities gave the right to speak on the theme of the universe to a small group of people who have mastered the most advanced telescopes.

However, are advanced telescopes capable of playing a decisive role in furthering human understanding of the universe?

NASA is doing so. People are thinking so. But this is completely wrong! What people see are pictures of the elusive fantasy beautiful scene. People never can glimpse the real face of the deep cosmos!

Think about it for a while, and you will find that the answer that is taken for granted after a quick thinking, is not able to be so sure to be granted.

For example, in Figure 6.3, assume it is a cloud in the sky.

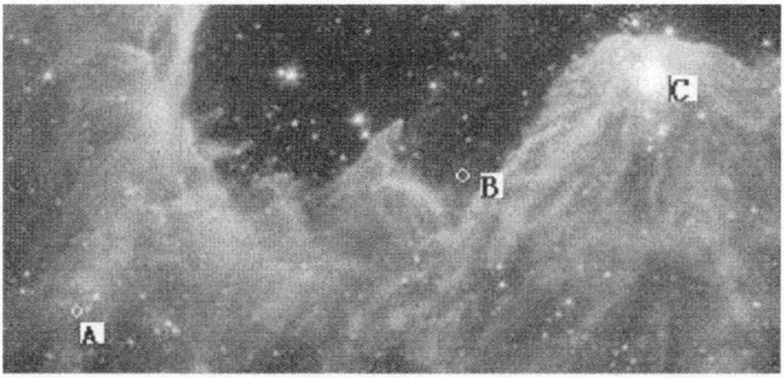

Figure 6.3 Cloud — everyday common cloud, or cosmic nebula cloud

If the distance between point A and point B is tens or hundreds of kilometers, it is very easy at one moment to grasp the whole picture, its three-dimensional structure and hierarchy, and use the instrument to measure its size clearly.

Figure 6.3 is a nebula of our universe. The distance between the telescope and the nearest point A of the nebula shown in the picture is 10 million light-years. This cosmic nebula is so big that observing from the telescope the distance between point A and point B is 300 million light-years. Under this assumption, how would our understanding change? Does the image reflect the true status of this nebula?

First, a simple fact: Suppose each of point A and point B at the same time emitted a particular group of lights, when the lights emitted from point A, say, 10 million years to travel to our eyes; the lights at the same time issued from point B would be 300 million years later to arrive our eyes! When the light emitted from point B reaches our eyes, at the same time we see the light from point A 300 million years earlier than the light emitted by point B! It is like saying that point B conveys the information of primitive society, while point

A conveys the modern information. Note that this is the information at different points in the same photo taken by the Hubble Telescope. The information at different points has different times in the same image.

This may be somewhat difficult to understand, so do the following more detailed description.

Figure 6.4 is a schematic view drawn from the observer's point of view.

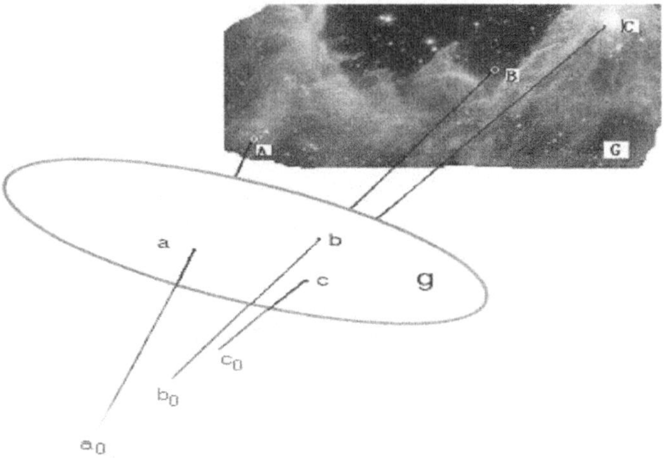

Figure 6.4 Schematic drawn of the nebula from the observer's point of view

As the figure 'g' shown, assume the position of the observer plane which may be our retina of the eye, or also be photos taken by the Hubble Space Telescope. To facilitate this, the observation image of the nebula is called the overall observation area G.

Take any three points from the nebula observed, assuming they are point A, B and C points drawn in the figure. We know the distance between the observer and point A is 'aA', the distance between the observer and point B is 'bB', and the distance between the observer and point C is 'cC'. The three distances are not equal. Since it is assumed that nebulae are millions of light years across, and assuming that the observer mission from the nebula is the nearest point A, and this is 10 million light-years from the nearest, and the farthest point from the point B to observer is 13 million light years, it is assumed that the position of C is at a distance of about 12.5 million light-years away from the observer.

Consider first that this nebula emits light toward the viewer at time T. When the light is emitted from point A at time T and arrives at the observer 'g' plane, lights sent at time T from point B and point C both are not visible yet. Point B light emitted at time T needed three million light-years more to reach the 'g' plane, and light from point C emitted at time T needed 2.5 million more light-years to reach the 'g' plane. When the first ray of light in this nebula emit at time T arrives at plane 'g', from the observer at 'g' to observe the nebula region G, can only see this one-point A.

When the light emitted from point B at time T reaches the observer 'g' plane, 3 million years has elapsed since the light was emitted at time T from point A, but point A continues to emit light, and the light emitted from A at 3 million years after time T will simultaneously reach the observer. In figure 6.4, 'b' is the point B at time T light emitted, and 'a' is the light emitted from point A as early as 3 million years before light emitted from B. In other words, if the 'b' spot in G is light from the nebula after it arrives at B to our eyes, then 'a' is the point A 3 million years earlier than 'b' at the scene. We have no way of knowing if anything has changed at point A over millions of years.

The plane is shown in Figure 6.5. It is a two-dimensional plane from a 2.4 oval circle of a sphere. The three points are shown in that picture.

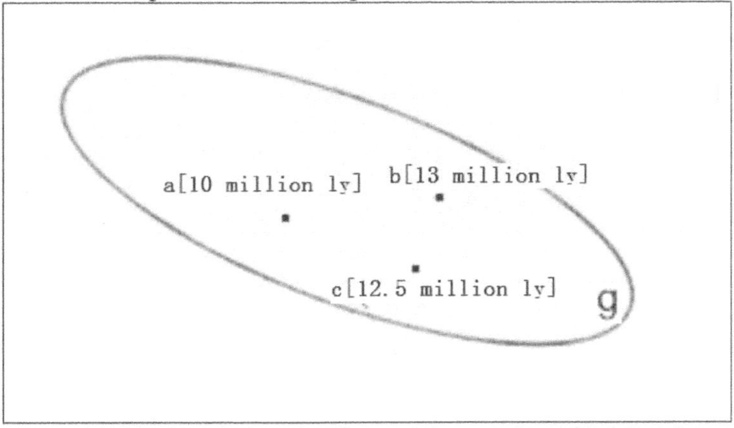

Figure 6.5. Taking the 'g' plane in Figure 6.4 out and marking the three points on this plane with their respective times, we get this graph.

And this is a big area of the image that people see through a telescope, which is composed of image points with different times, which reflects the situation of the region in different periods with a difference of millions of years in different points, rather than the real existence of the region at a certain moment.

From the Figure it can be seen that in the view of this telescope lens, the three points in this observation image that we deliberately study, each represent a different time, and the time differences between them are calculated in thousands of years. People who look at the picture don't know what has changed between the points during such long time periods.

Figure 6.5 is a star sky picture of the telescope. Figure 6.5 differs from the erroneously dated image discussed earlier in that point A is the point of the celestial object closest to the telescope. Telescopes will be able to see it when it is older than 10 million years. When the age of point B is greater than 13 million years, all three points A, B, and C can be seen. When the telescope has just seen point B, the telescope has seen point A for 3 million years and point C for 0.5 million years. Such images composed of time points in different periods certainly cannot reflect the true face of the space shown in the 'g' figure.

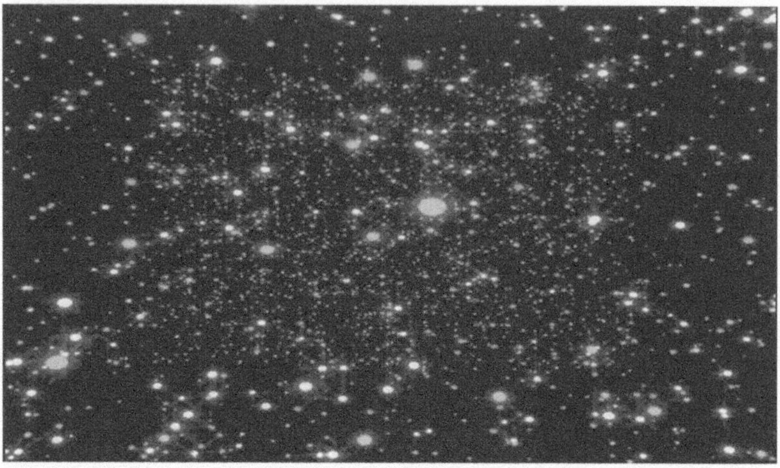

Figure 6.6 In the star map people photographed "now", the young may be less than a thousand years old, and the old may be more than 100 million years old. Their different ages of light report to the telescope 'now' at the same time. What are their states at the present time? Do they still exist? Did they explode? Where are they located? This is a section of the face left to us by the universe that humans can see, but it is not the real face of the universe at all. The true full picture of the universe will never be seen by humans.

Therefore, when from NASA we see photos that are sized in millions of light-years, those beautiful scenes of the photos are actually just illusory unreal scenes composed of different points of ages in units of ten thousand years. But don't take such serious calculations and deductions from it as a series of cosmic stories!

Seeing may not be believing after all.

The pictures of the universe we see are basically untrue.

The distortion of time and space coupled with our lack of complete understanding of what we're looking at can sometimes lead down the wrong path and bring us to false conclusions. People must have the courage to admit that they can never see the complete true face of the universe. The most important thing is to return to the basic spirit of science, based on facts, to have the courage to engage in critical thinking, but also to allow different views to be published.

The academic world should strive to find the part of its true face of the universe that mankind has the ability to know and should try to know the ultimate limitation for humans to know the universe. Rather than being dazzled by the beautiful graphics output from the computer like now.

The first condition to change this status quo is to return to the basic spirit of science, to speak based on facts, to have the courage to think critically, and to allow the expression of different opinions. Only in this way can we get rid of the false and keep the truth and move towards a smooth path to studying the universe.

5. No Matter How Good the Telescope Is, We Can't See the Whole Picture of the Universe

Let's take a look at Figure 6.7 below, it is (May 9, 2017) cut from the Zhihu website. The URL is http://www.zhihu.com/question /47072818

The question raised in Figure 6.7 is such: "As long as the telescope is good enough, can it see the celestial object located even at infinity?"

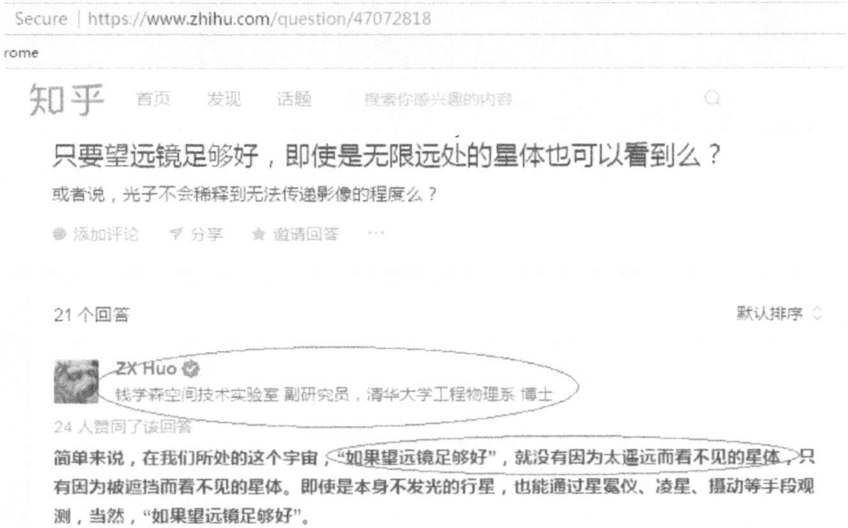

Figure 6.7 screenshot of the discussion page from Knowing Net web, the discussing titled "if the telescope is good enough, then no matter how far away the celestial body, it can be seen"

Here is an associate professor from "Qian Xuesen Space Technology Laboratory, Department of Engineering Physics, Tsinghua University" who gave the standard answer. He said: "If the telescope is good enough, there will be no celestial objects that are too far away to be seen."

How do you want to answer the question in Figure 6.7?

Among the people discussing this issue on the discussion page, there are cosmic major doctoral students, and astronomy enthusiasts. And their reply, unfortunately, was basically wrong.

Before answering why they are wrong, first make a statement about what it really means if they are all wrong.

Does it mean that there is a problem with the basic knowledge of cosmology? Does it mean that the information we got about the universe deviates greatly from knowing the universe correctly?

Should not people - professionals or amateurs - read this book?

Is it not that ...?

Let us first explain why "if the telescope is good enough, then no matter how far the cosmic object is, it can be seen by this telescope" is a **completely wrong assertion**!

There are 3 main reasons why this assertion is wrong. In fact, we have discussed them all before, here is to review again:

1) Any telescope can only passively wait for the arrival of celestial light information.

In other words, the space telescope is a pure receiver, like a radio. The Space Telescope doesn't have a shred of initiative to detect things like a radar! From a functional point of view, the space telescope is just a photoelectric identifier or recognizer. It is not a radar; it does not have that ability to proactively detect objects. It is impossible for people to launch a detecting pulse and wait it back like a radar - simply consider the time consuming for back and forth - we can't afford that.

The word "telescope" gives the impression that when we are using it, we can instantly see the objects in the distance more clearly. However, when the distance scale is so large that it is counted in light years, using the telescope does not necessarily immediately see the object we want to see! We have to wait for its light wave to reach the telescope, and then distinguish it from the various light waves among many other uninvited celestial objects.

2) The distance increases, the light intensity decreases. The celestial light intensity decays by the square of the distance between the light source and the telescope.

No matter how powerful a celestial object emits light, after it has traveled far enough distance, its light information will be attenuated to a very weak, difficulty to discernible. A distance of 1 light-year can attenuate sunlight to $\frac{1}{4\ billion}$! And usually this decay is very rapid, mathematically speaking, the limit of $\frac{1}{x}$ is 0, and the limit of $\frac{1}{x^2}$ tends to 0 at a faster rate. Think carefully about the distance units usually used by cosmic experts when talking about the universe!

3) The noise in the universe can cause big interference to the telescope's recognition ability

If the light information of a celestial object is so faint that it is below the noise of the environment where the telescope is located, the information of this celestial object cannot be discerned by the telescope from the noises. In other words, the telescope "turned a blind eye to the celestial object"!

The above three points tell us that the factors that determines whether a telescope can see a celestial object lie not only in the telescope, but also in the brightness and distance of the observed celestial object. If the photons emitted by a celestial object do not reach the telescope, the telescope cannot see the celestial object. If only a small number of light waves that less than the noisy waves are arriving at the telescope, the telescope will not be able to transmit the information from the celestial object to us even the telescope receives this weak information. The number of photons reaching the telescope is determined by the brightness of the celestial body and the distance from the telescope. And if the distance between the telescope and the star is too big that the weakened light intensity arrives on the telescope below a certain level, the star is not able to be seen by the telescope.

We classify the noise in the universe into two generalized categories. One is **controllable noise** such as telescope-related sensitivity, resolution, field of view, etendue, etc.; and those unrelated to telescopes like atmospheric interference, and so on. Another category is those impossible to change through human efforts**, uncontrollable noise**, such as the universe background light and so on. Through our own efforts, mankind can reduce the controllable noise to a very low level, but cannot do anything about uncontrollable noise.

The maximum distance that a celestial body can be seen is not determined solely by how good the telescope is. When the development of the telescope reaches a certain accuracy, how far it can see is not decided by the telescope. At this point, it is pointless to continue to improve the power or accuracy of the telescope.

For our eyes, we can see four photons if we exercise our eyes, but the uncontrollable noise level that comes into our eyes is eight photons, then we cannot see the information of four photons, and at most we can identify only eight photons or more.

A simple example of a telescope that cannot see a celestial object is that if there is a star like the Sun but at more than a million light-years away from the Earth, the Earth's telescope cannot see the star. Because according to the data given by NASA it can be calculated that when the sunlight reaches about 1.01 million light-years away, only one photon per square centimeter per second reaches that place, while the space telescope may need 200 photons per second per square centimeter to be able to distinguish a celestial object. That is to say, celestial objects' light intensities like the Sun's or weaker, if they were located a million light years away or farther, their light cannot be seen on the Earth.

In mathematical language, those objects are at distances greater than the observable radius, so the telescope cannot see them. Such a celestial object we named it a **Hidden Celestial Object**. There are so many of them that you would be surprised: the number of Hidden Celestial Objects is much more than the celestial objects we can see with the best telescopes!

When discussing dark energy, we used simple mathematics to define the **Absolute Observable Radius** and the **Relative Observable Radius** of a celestial object. Here I'll briefly review them.

The three objects involved in **Limited Observable Radius** are the observation instruments (eyes or telescopes, etc.), the observing target (Halley's Comet, etc.), and the distance between these two.

For the **Relative Limited Observable Radius (RLOR),** changing the nature of these three objects affects the length of the observable radius. For the **Absolute Limited Observable Radius (ALOR)**, the telescope itself is not important; more important is the part that cannot be controlled by humans: Uncontrollable noise.

However, when the celestial body is away from us more than the Absolute Observable Radius of the celestial object, we can hardly do anything to increase our vision of the celestial body. There are, of course, a very small number of such hidden objects that can be inferred by the observation of the behavior of other nearby observable celestial objects. But most of them cannot be seen. In particular, the celestial bodies outside the scope of the so-called Big Bang universe are all hidden objects. We cannot estimate how many of them there are.

The absolute observable radius is unaffected by human behavior. Including the improvement of telescopes. If the distance between the celestial object and the Earth is beyond the absolute observable radius, the celestial object will never be seen!

At this time in such a situation, the wise men rely on the basic scientific principles and the correct scientific reasoning to "see" the presence of these never-seen objects in his eyes. Of course, there are people who only believe in "what we see is what we get," refusing such reasoning and imagination! After all, it's something that we can never see, that exists outside of our sight. It needs imaginative and scientific quality capacity to "look"! People without imagination, even if they really have an authoritative title, can't see it.

This invisible celestial object, we call it a **Hidden Celestial Object (HCO)**. We defined ten categories of Hidden Celestial Objects using simple mathematics. The main prop of the theory of the Big Bang, the dark matter that used to explain the critical explosion, the contemporary cosmologists are desperately looking for, is just one of them, and it is a kind of normal celestial object.

My bet with NASA in my book "Debate of Light and Dark" is that, given NASA for a hundred years, they can't find more dark matter than hidden celestial objects.

These are after words, wait for the next discussion.

Roughly speaking, the way mankind explores universal mysteries can be summed up in four words:

"***Passively waiting, daring speculation.***"

These are very sad characters.

"Passively waiting" means that mankind can't fly across a distance of even one light year without using many thousand years! So, mankind is forced to stay on Earth and to study two kinds of objects coming from outer space. The first consists of small amounts of meteorite-like objects dropping on the Earth. The other is starlight, which consists of all kinds of light. According to their different frequencies, they can be called microwaves, electromagnetic waves, ultraviolet light, infrared light...

Meteorite-like objects can bring us some information, but their quantity is too little, they mostly represent a too small distance, and we often cannot determine their source. They cannot depict the face of the universe.

Light is the primary target for humans to study the universe.

It is impossible for humans to actively emit radar to explore the universe.

Humans can only wait passively for the arrival of light or electromagnetic waves from outer space. It is not that people do not want to take the initiative, it's that no initiative is possible. People don't have the slightest ability to proactively detect distant luminous objects.

Coupled with the slow speed of human travel, the distance that human beings need to cross to reach other celestial bodies becomes unthinkably huge. Even the places cosmologists have called "the backyard garden of Earth," located just a few light years away, we are not able to reach. Humans can only hover in place in the vast space of the elusive star road.

Chapter Summary

From the discussion in this chapter, we can see that because of trying to support the non-scientific pseudo-theological Big Bang theory, the astronomy and cosmology community has fallen into the trap of replacing science with myth and cannot extricate itself. All kinds of "scientific" myths and stories have been made up, and the entire mainstream astronomical and cosmology community has fallen into the trap of not talking about real science and not thinking innovatively.

Simple mistakes like "saw a 30-year-old black hole 50 million light-years away" casts doubt on the basic scientific training level of NASA's experts. Otherwise, how can such simple mistakes be repeated over and over again? The only explanation is that the myths are told too much and the basic scientific principles are forgotten.

No matter how advanced a telescope is, it is limited by its absolute observable radius, and it is impossible to see the whole picture of the universe. The huge distance blurs the faces of distant celestial bodies, and people will never see their true colors. Thus, the limited speed and the narrow vision scope of human beings limit the ability of human beings to observe and understand the vast universe to a very small scope. Humanity has not yet realized this.

Human beings need to realize this, so that they will not waste a lot of taxpayer money on meaningless so-called scientific research in the name of science.

Humans must realize this in order to truly understand the greatness of the Creator!

Only the Creator can know the true face of the universe!

Part 2 Summary

Humans do not have the ability to see a real universe

Sitting on the banks of the Milky Way, watching the ups and downs of stars, the sightseeing changes. Former philosopher, witty discussing stars, intelligent ideals pierced through space. There are also confused nebulae, and chaos in the eyes and ears. In recent years, artificial smoke screens, diffuse sky, scholar professors, running after the void, uncertain fogs pervade the space, and it is difficult to see the road returning home.

Even as strong as NASA, in the face of simple truth, has to be submission. The academic mansion of cosmic science, although no longer has free space to compete, and basically give itself to the telescope, other scientific instruments, and authority, but the spark of thought, although weakening, will not easily extinct; basic scientific principles and norms, still, are the basis for our understanding of the world.

The illustrations in this part are enough to show that although it seems that who has the best weapon - the best telescope in hand - has the utmost power to speak, this erroneous thinking ultimately leads to absurdity that the general public can understand, and it is in conflict against the basic scientific principle. Since these kinds of facts unfold before our eyes, we also have to turn away our heartful belief in cosmological scientists and in the space telescope we have all the time and turn our respectful gaze for the space scientists to the exploring eyesight of thinking.

The thinking of modern cosmologists is rather peculiar. They believe that everything is in their hands, and they can know everything. They believe they are close to the end of exploring the ultimate law of nature.

In his 1988 book <A Brief History of Time> Hawking said:

"On the basis of cautious optimism, we remain convinced that we may be nearing the end of the search for the ultimate laws of nature."

"The end!" The ultimate laws of nature!"

Look at the tiny human beings, so arrogant like cute!

We would also like to use this book to whisper a few words to these respected NASA and contemporary cosmological authorities: If you blindly follow the established goal of proving the pseudo-theological Big Bang universe exploded by an atom, and reject other objections, it will ruin cosmology science and advance nature, and will encourage more non-principled imagination and scientific fiction-based reasoning in rigorous science, and will go further and further away from the truth.

In our opinion, it is never possible for humankind to scientifically penetrate the true face of the whole universe.

Humans can neither see the true colors of distant celestial bodies through telescopes, nor can they travel outside the solar system. While they neither can see clearly nor can reach them, how can they talk about understanding the true colors of the universe?

We, the insignificant human beings, should recognize our limit of understanding of the universe. Then we will not use the one atom Big Bang mystic story, make unnecessary exploration and waste taxpayer money.

Instead of applying the myth of an atom, holding up a space telescope showing dazzling unreal pictures and shouting science, we surrender our reverence to the Almighty Creator by admitting our ignorance.

We feel like we are that child who has torn off the emperor's new clothes, who has not only offended more than just the emperor, but all the guards, accompanying, booed, and even partially onlookers.

We need a bit of motivation to create.

We would like to use my feeble cry to pry open the Big Bang space closed by NASA with the space telescope and inspire the young generation to move forward with courage, imagination, pursuit of truth and potential creativity.

Let us work together to shake this cosmic mansion built by math and blackened by the Big Bang smog, telescope and fantasy!

Part 3

Understanding the Theory of Relativity from the Most Basic Modeling Principles

In recent years, relativity, mainly general relativity, has become one of the main scientific pillars of the Big Bang theory. This makes any discussion of the universe inseparable from the theory of relativity.

Indeed, the theory of relativity seems a near miracle in the development of human science and technology. A half century ago, during the 1970s, books on physics portrayed general relativity as a theory that required further study. Today, mainstream science and technology simply bow to its great authority, and find in its various utilization values. That voices criticizing the theory of relativity have been marginalized or suppressed, both in the scientific literature and popular media, seems not to matter. It is enough that contemporary astronomical and cosmological circles regard it as a theoretical pillar.

Opinions different from relativity have largely disappeared from the mainstream magazine media. There has even been repression through various official institutions, beyond the scope of academic debate.

At the same time, the methods used by critics of the theory of relativity are also inappropriate. Basically, they are entangled in some mathematical derivation or experimental verification of one or other aspects of the theory. In effect, they are local disputes that experts can't judge thoroughly and ordinary people can't understand. Coupled with the authority of those in academia who praise Einstein's theory, and official institutions who rely on it to survive, it is nearly impossible to criticize the theory of relativity mathematically or experimentally in any professional or public space.

In this part of the book, then, we will reinterpret the theory of relativity, beginning with its original models and definitions. And we will find that the theory is indeed more like a pseudo-theology with a whimsical flavor than a science.

Special Relativity Viewed from the Relationship between the Physical Model and Mathematical Model

The foundation of the Big Bang is an atom. The foundation of special relativity is a ray and a rod. They are also things that only God can operate. It's just that the universe bursting out of an atom is too amazing, and the results from a rod and a ray are less immediately sensational. These we discuss in detail in the following chapter.

Here we focus on whether or not, a correct and comprehensive physical model is the most decisive and necessary first condition to ensure that the mathematical model derived or deduced from it is also correct and comprehensive! If the physical model is incorrect, the mathematical model derived from it can never be correct, and the resulting application of the mathematical model will cause irresolvable errors. Some of these errors may be intuitive, some may be quite subtle, but all need careful consideration.

Let's make a simple illustration by designing a railway bridge.

If we want to build a railway bridge for high-speed trains, we need to examine the various operating condition parameters involved, such as the speed of the train and its maximum load. If the same data parameters obtained by examining a car on a highway is used to design a high-speed railway bridge, you can imagine what will happen.

And if the design is not well thought out, then various errors will occur after the bridge is built, for example, larger ships cannot pass under the bridge, or the bridge is broken by the heavy load of the vehicle, etc. This design is of course wrong.

In this light, we'll discuss special relativity from the following two points: the comprehensiveness of the model and the rationality of the model. These two points will be discussed in detail in Chapter 7:

1. Due to incomplete cognition, the special theory of relativity has a fatal error: Although Einstein discussed the situation of the crow flying over the train, he did not discuss the situation of the crow walking in the train. This allowed him to see only relative simultaneity and firmly deny absolute simultaneity; hence the special theory of relativity.

2. Look at the model used by Einstein to derive the special theory of relativity, which is similar to the model in our discussion above: using a car model to design a railway bridge. Einstein used a ray of light to affect a rigid rod that was completely out of contact with the light. Observe the relative motion between the two, so that the light ray can change the length of the rod, and also change the time of a clock on the rod. A mathematical model of special relativity is derived from a physical model that is completely disconnected from reality. The conclusion thus obtained is a **pseudo-theological inference.**

A similar problem exists in general relativity. For example, light travels along the geodesic line: the light of a celestial body never propagates in a single line but propagates 360° from the source to the surrounding space as a spherical wavefront. When a huge photosphere encounters a relatively very small celestial body like the Sun, which piece or part of the photosphere can travel along which geodesic line? Refer to Figure 0.2 below.

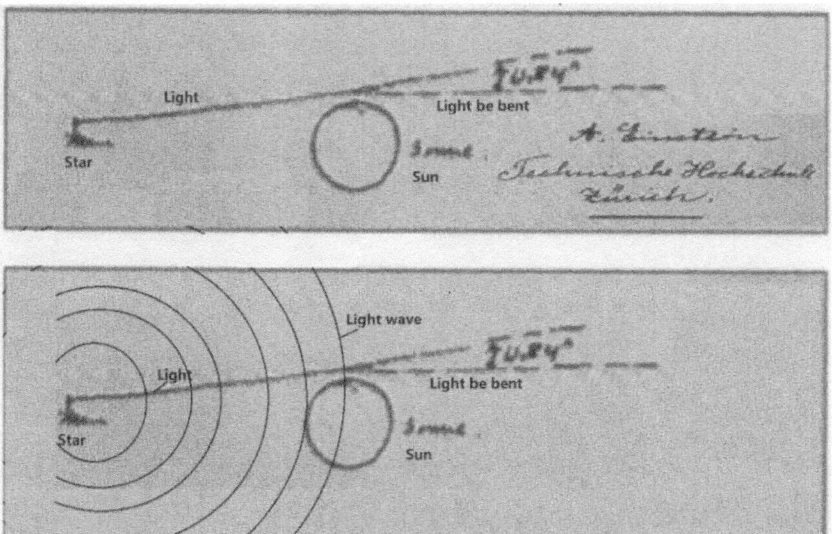

Figure 0.2 How does the star light propagate? Top figure: Einstein original handwriting displays the star light as ONE ray that is bent by the Sun. Bottom figure: The light waves the star was sending out were added to the figure. Now, how the spherical waves be bent by the Sun?

The General Theory of Relativity Considered from the Basic Definition of Space and Time

Although we will discuss these two points in greater detail in Chapter 9 and Chapter 10, it's important to note several aspects about them now:

What is space? At its most basic, space is empty, with nothing at all in it.

If you agree with this statement, then how, according to general relativity, can space with nothing at all in it be curved? Bend what? Bending "nothing"?

Please carefully distinguish empty space from the matter contained in the space. Anything like geodesics is something that space accommodates, not space itself.

What is the time? Can time go to the past? Is there time in the empty space?

When we think about all of this carefully, we will find that Einstein's concept of space and time is more wishful than scientific. It is the kind of scientific pseudo-god thinking we have been criticizing throughout this book.

Old knowledge, in addition to new concepts, are both important when considering the universe. Critical thinking here, however, is essential. And when we recognize errors, as much in theory as in direct observations, it's best to return to origins.

Many principles that inspire excellent research, particularly on models, are often simple; this may be one of them.

Chapter 7

From the Mythical Error of the Physical—Math Model to the Special Theory of Relativity

Regarding the special theory of relativity proposed by Mr. Einstein we have the following 5 concerns:

1. A light ray moving relative to a rigid rod. Can the light ray have any effect on the rigid rod?

2. In sections VI to X of [2], Einstein gave a lot of discussion on relativity with a system that does not meet the conditions of a relative system.

3. Einstein overturned the rule of synchronized transfer defined by himself.

4. Conflicts caused by being passive relatives.

5. Why did Einstein repeatedly deny the existence of "relative system that maintains absolute simultaneity?" There are actually countless relative systems that maintain absolute simultaneity in the world. Why can't we find them before?

There is a **rod** and a **ray**.

Please think of a way **to make the "ray" affect the length and the time flow on the rod.**

Please think about it, think again. Think freely, think wildly, think eccentrically, think imaginatively; come up with a way. You can set up the ray above the top of the rod, or below, left, right, whatever side of the rod you wish; the ray is 0.001 mm from the rod, or one meter from the rod, or as far as a light year; the rod follows a straight line, or up and down. In short, the ray moves freely; the rod is also free in shape and movement, as long as the set up achieves its purpose: To make the "ray" affect the length of the rod and the time flow on the "rod" by the relative movement between the ray and the rod.

Think, think, think ...

Can't think of any good way? Of course, a ray cannot be in contact with the rod because the ray has vanished as soon as it touches the rod. So, the ray and the rod must be completely independent from each other! How can two totally independent things affect each other?

If you can't come up with a method, it shows that you are not a master, not a professor, nor among the few select in the world who understand the issue! Certainly, you are not God!

If you come up with a way, then you can complete the following tasks, however fantastical they may seem:

If you put a clock on the rod, the movement of this ray should make the clock on the rod go slower. This situation is called "time dilation" in special relativity theory, or "the moving clock moves slower."

If you put a meter ruler on the rod, the movement of this ray should make the ruler shorter. This situation is called the "moving meter becomes shorter" in the theory of special relativity.

The former reminds us of time travel and long life, even immortality, while the latter frightens me: if, for example, a spacecraft is moving too quickly, the spacecraft shortens along its direction of flight, the engine deforms, and finally the spacecraft disintegrates.

Isn't it amazing, fantastical, mythic?

Let's look at the casting and flaws of this myth from the following two perspectives:

1. Because of Einstein's incomplete consideration of the issues related to the physical model, he based the theory of special relativity on an erroneous mathematical model.

2. Two objects that are physically independent of each other cannot change the physical properties of the other.

1. Bad physical models lead to Concerns about the rationality of the theory of special relativity

See Figure 7.1 below:

§ 1. Definition of Simultaneity

Let us take a system of co-ordinates in which the equations of Newtonian mechanics hold good.[2] In order to render our presentation more precise and to distinguish this system of co-ordinates verbally from others which will be introduced hereafter, we call it the "stationary system."

If a material point is at rest relatively to this system of co-ordinates, its position can be defined relatively thereto by the employment of rigid standards of measurement and the methods of Euclidean geometry, and can be expressed in Cartesian co-ordinates.

If we wish to describe the *motion* of a material point, we give the values of its co-ordinates as functions of the time. Now we must bear carefully in mind that a mathematical description of this kind has no physical meaning unless we are quite clear as to what we understand by "time." We have to take into account that all our judgments in which time plays a part are always judgments of *simultaneous events*. If, for instance, I say, "That train arrives here at 7 o'clock," I mean something like this: "The pointing of the small hand of my watch to 7 and the arrival of the train are simultaneous events."[3]

It might appear possible to overcome all the difficulties attending the definition of "time" by substituting "the position of the small hand of my watch" for "time." And in fact such a definition is satisfactory when we are concerned with defining a time exclusively for the place where the watch is located; but it is no longer satisfactory when we have to connect in time series of events occurring at different places, or—what comes to the same thing—to evaluate the times of events occurring at places remote from the watch.

We might, of course, content ourselves with time values determined by an observer stationed together with the watch at the origin of the co-ordinates, and co-ordinating the corresponding positions of the hands with light signals, given out by every event to be timed, and reaching him through empty space. But this co-ordination has the disadvantage that it is not independent of the standpoint of the observer with the watch or clock, as we know from experience. We arrive at a much more practical determination along the following line of thought.

If at the point A of space there is a clock, an observer at A can determine the time values of events in the immediate proximity of A by finding the positions of the hands which are simultaneous with these events. If there is at the point B of space another clock in all respects resembling the one at A, it is possible for an observer at B to determine the time values of events in the immediate neighbourhood of B. But it is not possible without further assumption to compare, in respect of time, an event at A with an event at B. We have so far defined only an "A time" and a "B time." We have not defined a common "time" for A and B, for the latter cannot be defined at all unless we establish *by definition* that the "time" required by light to travel from A to B equals the "time" it requires to travel from B to A. Let a ray of light start at the "A time" t_A from A towards B, let it at the "B time" t_B be reflected at B in the direction of A, and arrive again at A at the "A time" t'_A.

In accordance with definition the two clocks synchronize if

$$t_B - t_A = t'_A - t_B.$$

We assume that this definition of synchronism is free from contradictions, and possible for any number of points; and that the following relations are universally valid:—

1. If the clock at B synchronizes with the clock at A, the clock at A synchronizes with the clock at B.
2. If the clock at A synchronizes with the clock at B and also with the clock at C, the clocks at B and C also synchronize with each other.

Thus with the help of certain imaginary physical experiments we have settled what is to be understood by synchronous stationary clocks located at different places, and have evidently obtained a definition of [redacted] "synchronous," and of "time." The "time" of an event is that which is given simultaneously with the event by a stationary clock located at the place of the event, this clock being synchronous, and indeed synchronous for all time determinations, with a specified stationary clock.

In agreement with experience we further assume the quantity

$$\frac{2AB}{t'_A - t_A} = c,$$

to be a universal constant—the velocity of light in empty space.

It is essential to have time defined by means of stationary clocks in the stationary system, and the time now defined being appropriate to the stationary system we call it "the time of the stationary system."

Figure 7.1. From "§1 Definition of Simultaneity" in "On the Electrodynamics of Moving Bodies."

Figure 7.1 is taken from the English version but with two critical words deleted to exactly follow the original German paper of Einstein's "Zur Elektrodynamik bewegter Körper". (See the tracking story in a later part)

The foundational publications on special relativity are Einstein's paper, "On the Electrodynamics of Moving Bodies" published in 1905, and his monograph "Relativity: The Special and General Theory" published in 1920. The latter is an extension and supplement to the former. The two do not substantially change the issue we are discussing here.

The pitfalls of "simultaneity" and "synchronization"

As the name of §1 suggests, this section is used to define the concept of "simultaneity". This also tells us that Einstein believed that the original meaning of the word "simultaneity" did not express exactly what he needed when he used the word in his theory of special relativity, so he redefined it.

So, what does "simultaneity" look like as Einstein redefined it? Please read §1 carefully and answer the following question: "What is the exact definition of simultaneity in Einstein's theory of relativity?"

The title of §1, "Definition of Simultaneity," which is less than 1100 words. Therefore, our question should be easily answered after reading the text in the figure. But in fact, reading it ten times cannot answer this simple question with confidence and accuracy.

As can be seen from §1, Einstein did not give a clear and unambiguous definition of "simultaneity" as the title of the subsection suggested. Instead, the mathematical definition of "synchronization" in the thesis was given as below:

> Let a ray of light start at the "A time" t_A from A towards B, let it at the "B time" t_B be reflected at B in the direction of A, and arrive again at A at the "A time" t'_A.
> In accordance with definition the two clocks synchronize if
> $$t_B - t_A = t'_A - t_B \qquad (1)$$
> Thus, with the help of certain imaginary physical experiments we have settled what is to be understood by synchronous stationary clocks located at different places, and have evidently **obtained a definition of "simultaneous," or "synchronous," and of "time."**

The formula (1) and the number and boldface of the formula in §1 are all added by us for the convenience of description.

In §1, "simultaneously" is used five times and "synchronization" seven times. Regarding the definition of "simultaneity", there is only a relevant sentence shown in the text box above "and thus obviously also obtains the definition of 'simultaneous'."

But we don't understand this "obvious." Do you? What definition is obtained? What is the definition of "simultaneous"?

So, Einstein set a trap for the reader in §1: Although its title was "§1 Definition of simultaneity", he did not give the definition explicitly, instead he gave a precise mathematical definition of "synchronous". Normally, "simultaneous" is not equal to "synchronous".

This has trapped many people, and it is also certain that there are many so-called "masters" who do not really understand the theory of relativity. It was not until **18 years later that** those two words were quietly added to "On the Electrodynamics of Moving Bodies" in the book *The Principle of Relativity* that Einstein personally arranged to reprint, making it clear that **"simultaneous" and "synchronous" have the same definition**!

We spent a lot of time tracking down this answer, please refer to how we did it later in this chapter.

Through the above discussion, we have made it clear that the precise definition of Einstein's "simultaneous" or "synchronous" is the formula (1). In relative theory, if two objects in relative motion conform to formula (1), they are "simultaneous" or "synchronous".

Then, can we set the conditions for the distinction between relative simultaneity and absolute simultaneity? With the help of formula (1), of course it is possible. But this needs to be done little bit later in analysis of §2, because at this moment there is no such requirement yet.

The theoretical basis of special relativity

Figure 7.2 is "§2 On the Relativity of Lengths and Times" of Einstein's paper "On the Electrodynamics of Moving Bodies."

See how a ray changes the length and time flow of a rod – virtually changing the world.

§ 2. On the Relativity of Lengths and Times

The following reflexions are based on the principle of relativity and on the principle of the constancy of the velocity of light. These two principles we define as follows: —

1. The laws by which the states of physical systems undergo change are not affected, whether these changes of state be referred to the one or the other of two systems of co-ordinates in uniform translatory motion.
2. Any ray of light moves in the "stationary" system of co-ordinates with the determined velocity c, whether the ray be emitted by a stationary or by a moving body. Hence

$$velocity = \frac{light\ path}{time\ interval}$$

where time interval is to be taken in the sense of the definition in §1.

Let there be given a stationary rigid rod; and let its length be l as measured by a measuring-rod which is also stationary. We now imagine the axis of the rod lying along the axis of x of the stationary system of co-ordinates, and that a uniform motion of parallel translation with velocity v along the axis of x in the direction of increasing x is then imparted to the rod. We now inquire as to the length of the moving rod, and imagine its length to be ascertained by the following two operations: —

(a) The observer moves together with the given measuring-rod and the rod to be measured, and measures the length of the rod directly by superposing the measuring-rod, in just the same way as if all three were at rest.

(b) By means of stationary clocks set up in the stationary system and synchronizing in accordance with § 1, the observer ascertains at what points of the stationary system the two ends of the rod to be measured are located at a definite time. The distance between these two points, measured by the measuring-rod already employed, which in this case is at rest, is also a length which may be designated "the length of the rod."

In accordance with the principle of relativity the length to be discovered by the operation (*a*)—we will call it "the length of the rod in the moving system"—must be equal to the length l of the stationary rod.

The length to be discovered by the operation (*b*) we will call "the length of the (moving) rod in the stationary system." This we shall determine on the basis of our two principles, and we shall find that it differs from l.

Current kinematics tacitly assumes that the lengths determined by these two operations are precisely equal, or in other words, that a moving rigid body at the epoch t may in geometrical respects be perfectly represented by *the same* body *at rest* in a definite position.

We imagine further that at the two ends A and B of the rod, clocks are placed which synchronize with the clocks of the stationary system, that is to say that their indications correspond at any instant to the "time of the stationary system" at the places where they happen to be. These clocks are therefore "synchronous in the stationary system."

We imagine further that with each clock there is a moving observer, and that these observers apply to both clocks the criterion established in § 1 for the synchronization of two clocks. Let a ray of light depart from A at the time t_A, let it be reflected at B at the time t_B, and reach A again at the time t'_A. Taking into consideration the principle of the constancy of the velocity of light we find that

$$t_B - t_A = \frac{r_{AB}}{c - v} \quad and \quad t'_A - t_B = \frac{r_{AB}}{c + v}$$

where r_{AB} denotes the length of the moving rod—measured in the stationary system. Observers moving with the moving rod would thus find that the two clocks were not synchronous, while observers in the stationary system would declare the clocks to be synchronous.

So we see that we cannot attach any *absolute* signification to the concept of simultaneity, but that two events which, viewed from a system of co-ordinates, are simultaneous, can no longer be looked upon as simultaneous events when envisaged from a system which is in motion relatively to that system.

Figure 7.2. Copied from "On the Electrodynamics of Moving Bodies." §2

§2 gives the following model:

> Let a ray of light depart from A at the time t_A, let it be reflected at B at the time t_B, and reach A again at the time t'_A. Taking into consideration the principle of the constancy of the velocity of light we find that:
>
> $$t_B - t_A = \frac{r_{AB}}{c-v} \text{ and } t'_A - t_B = \frac{r_{AB}}{c+v} \quad (2)$$
>
> Where r_{AB} denotes the length of the moving rod—measured in the stationary system. Observers moving with the moving rod would thus find that the two clocks were not synchronous, while observers in the stationary system would declare the clocks to be synchronous.

In the text box above, the reference bodies in this relative system are a ray and a rod. Equation (2) means that when the rod is moving, the light ray and the moving rod are not synchronized. But when the rod is stationary, the light ray and the stationary rod are synchronized, as described by (1) in §1.

From here, Einstein draws the following conclusions, which are the basis for the establishment of special relativity.

> So, we see that we cannot attach any absolute signification to the concept of simultaneity, but that two events which, viewed from a system of co-ordinates, are simultaneous, can no longer be looked upon as simultaneous events when envisaged from a system which is in motion relatively to that system.

In the relativity system, Einstein thought the static system is synchronous; and the dynamic system is not synchronous. Therefore, absolute simultaneity does not exist, so there is only relative simultaneity. This is the most basic theoretical basis for the establishment of the special theory of relativity.

But we will prove below that in a qualified relative system that conforms to Einstein's rules, **absolute simultaneity exists completely**, so the most **basic theoretical basis for the establishment of the special theory of relativity is not solid.**

Why does absolute simultaneity exist?

We discuss this issue on four parts:

- Regarding the conditions that qualify Einstein's system as a relative system.
- Regarding absolute and relative simultaneous or synchronous events.
- Some examples of relative systems used by Einstein.
- There are many relative systems in the world that maintain absolute simultaneity.

1) A relative system that meets the requirements of Einstein — a qualified relative system

In the foundational paper "On the Electrodynamics of Moving Bodies" published in 1905 and the monograph "Relativity: The Special and General Theory" published 15 years later, Einstein perfected various conditions for relative systems and added many examples of relative systems.

According to Einstein's exposition, the composition conditions of a qualified relative system that meet his requirements are roughly as follows:

A relative system consists of two reference bodies that do inertial motion (i.e., uniform linear non-rotational motion). One of the reference bodies must be moving. The other reference body, if it is moving, defines this relative system as dynamic; if it is stationary, then this relative system is static. We won't clearly specify all other required conditions here, but implicitly include all other required conditions in our discussion. This means we will be discussing a qualified relative system that meets all of extant Einstein's requirements. All "relative systems" we use in the following discussion refers to such a qualified relative system.

2) Absolute and relative of simultaneous or synchronous

According to the above discussion, if a relative system is to maintain absolute synchronization or simultaneity, the condition that needs to be satisfied is: when the relative system is a dynamic system, the motion result of the dynamic system can always make the equation (7-1) in §1 (transcription below) is established, namely:

$$t_B - t_A = t'_A - t_B \qquad (7\text{-}1)$$

In Einstein's cognition, only the static system satisfies the equation (7-1), while the dynamic system does not. That is to say, Einstein believed that in a relative system, only the static system can maintain absolute simultaneity, while the dynamic system cannot maintain absolute simultaneity. Einstein was adamantly against the existence of a dynamic system that maintains absolute simultaneity, because if it existed, the theory of relativity would be untenable.

More than a hundred years ago, when Einstein built the castle of relativity, he buried absolute simultaneity under his relativity castle as the foundation and fertilizer of his theory. For a long time after the publication of the theory of relativity, he had been talking about the non-existence of absolute simultaneity.

In **1905**, Einstein defined the concept of simultaneity in his foundation theory of relativity related in "On the Electrodynamics of Moving Bodies" and, for the first time, put forward the conclusion in §2 that "**we cannot attach any absolute significance to the concept of simultaneity.**"

In **1916**, Einstein published Relativity: Special and General Theories. In the book, Einstein used a lot of space and different roles to illustrate the special theory of relativity, and once again, he explained the significance of relative simultaneity. For example, in the book's Section IX, "Simultaneous Relativity," he used two lightning strikes sent out from its front and back ends of the running train at the same time, with each lightning speeding to the midpoint of the train, and the train also started to move at the same time. The example that lightning cannot meet at the midpoint of the train explains the relativity concept.

In **1920**, in a letter to the Russian philosopher and mathematician, Solovine, Einstein emphasized again: " That is why the theory of relativity **rejects the concept of absolute simultaneity, absolute speed, absolute acceleration**, etc., they can have no unequivocal link with experiences."

Since then, due to Einstein's influence, not only in relativity, but also in physical theory, there has been no absolute simultaneity. Einstein established the theory of relativity by negating absolute simultaneity.

Why did Einstein repeatedly deny the existence of absolute simultaneity?

Because the existence of absolute simultaneity illustrates the utopian nature of the pseudo-god of special relativity.

Everything in nature has an inherent harmonious beauty. This beauty is contained in the relative balance of everything: where there is yin, there is yang; when there is beauty, there is ugliness; dynamic-static, cold-heat, permanent and temporary, relative, absolute, limited and so on. If something is missing, for example, if there is no heat and only cold, or only ugliness and no beauty, is this world still harmonious??

In philosophy, it is generally considered that relative and absolute are two indispensable aspects of all existence.

But in fact, according to Einstein's definition of absolute and relative simultaneity, we can identify many relative systems that maintain an absolute simultaneous or absolute synchronous character in reality, such as a raven walks back and forth on the running train, or a sprinter running on the Earth ……

Of course, there is no instance of such a relative system that maintains absolute simultaneity in Einstein's work. However, we can analyze the example of maintaining relative relativity in Einstein's work, and we can find what we need

3) Some examples of relative systems used by Einstein

In 1905, in "On the Electrodynamics of Moving Bodies," Einstein used a very long, rigid rod as Reference-body-B and a light ray moving back and forth at both ends of the long rod as the Reference-body-A. This system of a ray and the moving rigid rod is called a dynamic system. Obviously, this dynamic system is a relative system with relative simultaneity. When the ray moves back and forth over a stationary rigid rod, the relative system of this ray and a rigid rod is called a static system, which has absolute simultaneity.

In 1916, Einstein published *<Relativity: The Special and General Theory>*. In the book, Einstein used a lot of pages and different roles to illustrate the application samples of special relativity. Below we tried to collect **all** of the roles of the examples he used in the book.

In section V: A **raven** flies at a constant speed along the railway embankment. If we observe it from the railway embankment, its motion is uniform and straight. If we observe it from the moving carriage, we will find that it maintains a uniform linear motion, but has different speeds in different directions.

In section VI: The **railway carriage** has a velocity v; a **man** traverses in the same direction of the train carriage with a velocity w. According to classic kinematics, the combined velocity W that the man advances relative to the **embankment** during the process is W = v + w.

In section VII: Replace the man above with a light ray whose velocity is c. w is the speed of light relative to the carriage, that is w = c - v.

In section VIII: There are train carriages running at a constant speed, the railway embankment, lightning from two points A and B opposite each other, and an observer at midpoint M of AB. Here, the definition of simultaneity is: A~TM =B ~TM; that is, if the observer sees lightning from A and B at the same time, then they are said to be simultaneous.

In section IX: There is the train carriage moving at speed v, the railway embankment, lightning from two points A and B, and people moving in the carriage. Two events (such as two lightning strikes, A and B) are simultaneous with respect to the railway embankment, but are they simultaneous with respect to a moving train? The answer is negative.

Section X: The **railway carriage** has a velocity v; a **man** traverses in the opposite direction of the train carriage with a velocity w. According to classic kinematics, the combined velocity W that the man advances relative to the **embankment** during the process is W = v - w.

4) In the world there are many relative systems that maintain absolute simultaneity

By slightly changing the examples of relative system that Einstein used above, the following relative systems can be obtained:

Changing "A ray flies along the rod" to "light wheel rolls on the rod";

Changing "the raven flies back and forth over the train" to "the raven walks back and forth in the train".

Changing "man walks in the train carriage relative to the railway embankment" to " man walks in the train carriage relative to the carriage".

Such modified systems, then, are still qualified relative systems, without violating any of Einstein's rules for a qualified relative system. In fact, an instance of Einstein's relative system is that of a man walking in a train carriage. A raven walking in the train carriage follows the same pattern.

Now, please answer the following elementary-level arithmetic problems:

How long does it take for a person to walk a long train carriage in the same direction along the train's movement? What is the time it takes for a person to walk in the opposite direction to the motion of the train?

How much time does it take for the raven to walk a long train carriage in the same direction along the train's movement? What is the time it takes for the raven to walk in the opposite direction to the motion of the train?

The same questions about the light wheel to roll forward or backward on the moving rod.

The answer is that in all the above-mentioned paired relative system motions, the time used to go back and forth is the same, and they all satisfy formula (7-1), that is, they are all systems that maintain absolute simultaneity.

Einstein repeatedly denied the existence of relative systems that maintain absolute simultaneity, but in fact there are countless examples of relative systems that maintain absolute simultaneity,

It is straightforward to prove that a qualified relative system is one that preserves absolute simultaneity, which only need to prove that the relative motion in the system always satisfies (7-1).

Let's explain this with the following example. We call it **Example-A**.

A relative system consists of a train carriage and an attendant walking back and forth from both ends of the carriage between A to B, the speed of the train carriage is $V_{carriage}$, and of the attendant is V_{man}.

When the train attendant moves from end A to end B of the carriage, he moves in the same direction as the carriage. The speed of the attendant is $V_{man} + V_{carriage}$, but at the same time the carriage leaves him at the speed of $V_{carriage}$. So, his final speed is $(V_{man} + V_{carriage}) - V_{carriage} = V_{man}$.

When the attendant moves from end B to end A of the carriage, he moves in the opposite direction to the carriage moves. The speed of the attendant this is $V_{man} - V_{carriage}$, but at the same time the carriage also carries him at the speed of $V_{carriage}$. So, his final speed is $(V_{man} - V_{carriage}) + V_{carriage} = V_{man}$.

In this relative system, Reference-body-A = train attendant, Reference-body-B = train carriage.

In other words, the speed of the carriage does not affect the speed of the attendant at all. Therefore, even if an ant walks slowly, whether it is from the A end to the B end, or from the B end to the A end, it can finally reach its destination.

In this way, the time it takes for an attendant to go from the A end to the B end of the carriage is always equal to the time from the B end to the A end of the carriage. In other words, Example-A is a relative system that maintains absolute simultaneity.

Moreover, even if the system is observed from outside the carriage and train, even if from Mars, it is a relative system that maintains absolute simultaneity.

The calculation results tell us:

The time it takes for a raven to walk through a train carriage in the direction of the train uses the same amount of time as it takes for the raven to walk in the opposite direction of the train!

It takes the same amount of time for a person to move back and forth in a moving carriage!

It takes the same amount of time for the light wheel to roll back and forth on the rod!

That is to say, they are all relative systems that **always conform to formula (7-1)**, that is, they are all **relative systems that maintain absolute simultaneity**!

Of course, we can cite many examples of relative systems that maintain absolute simultaneity, which pervade our lives.

Please consider the following small quiz:

Why do sprinters not need to consider the direction of rotation of the Earth when in competition?

Simply considered, the Earth turns from west to east, so when following the direction of the Earth's rotation, the speed should be faster than the reverse because the speed of running along the direction of the Earth's movement is the superposition of the speeds of the athlete and that of the Earth, and the reverse is the two speeds subtracted. Therefore, sprinters should run faster in the direction the Earth is moving than when running in the opposite direction. So, which direction should have a greater impact on world record times for sprinters? Is this correct? Why do races occur wherever they do, without considering the direction athletes run? Shouldn't the world record be as infinitely precise as possible?

An ant crawls at a constant speed on the wall of an elevator. Whether the ant climbs from the bottom of the elevator to the top or from top to bottom doesn't matter. If the ant crawls with the same speed, then the time required for it to go from bottom to top or from top to bottom is the same, no matter whether the elevator is running or not.

People walking back and forth on the deck of a moving ship maintain absolute simultaneity. Likewise, a stewardess walking back and forth on a high-altitude aircraft maintains absolute simultaneity.

……

All of this means that systems that maintain absolute simultaneity exist in abundance.

So, what makes Einstein ignore and deny the existence of all these systems that maintain absolute simultaneity, and force on us the idea that only systems that maintain relative simultaneity exist? Of course, the requirement imposed by special relativity itself is a significant reason. Otherwise, the mathematical derivation in §3 of Einstein's thesis in, "On the Electrodynamics of Moving Bodies," cannot proceed.

Another reason is that the physical models collected and used by Einstein were not complete, and his understanding of relative systems that maintain absolute simultaneity was flawed. He had always insisted that there was no relative system that maintained absolute simultaneity.

Of course, there are certain conditions for the existence of "absolute." Some things do not have absolute attributes. For example, weather forecasting with our current technology is not absolutely accurate. But arithmetic does have absolute properties. For example, in elementary school arithmetic, 1+1=2 is always correct. In the relative system, if a relative system that always satisfies formula (7-1), then it is the relative system that maintains absolute simultaneity or absolute synchronization. This is determined by the definition (7-1) given by Einstein! That's why we insist on making it clear what "simultaneity" is.

Then why can't we find relative systems that maintain absolute simultaneity before?

To Recognize the existence of relative systems that maintain absolute simultaneity begins with answering the question posed at the beginning of this section:

Then what is the precise definition of "simultaneity"?

To clearly determine that equation (7-1) is not only a formula for defining synchronization, but also a formula for defining simultaneity, is a tough job. It is impossible to get the idea by reading the original German, Spanish, and Chinese versions of Einstein's paper "On the Electrodynamics of Moving

Bodies". Only after reading the English version paper quietly revised by Einstein in 1923, 18 years later after the quiet modification made in the English version, added to the original German version published in 1905. Then **finally it can be determined that equation (7-1) is the key**.

Applying equation (7-1), we can find many relative systems that maintain absolute simultaneity.

Summary of Conditions of the Two Different Forms of Simultaneity in a Relative System

Equation (7-1) is Einstein's precise definition of the relative system's condition of being in synchronism or simultaneity with mathematics. According to this condition, we can deduce the condition that a relative system maintains absolute simultaneity or absolute synchronization, that is, the system should always satisfy equation (7-1).

What is the fundamental difference between a relative system that maintains relative simultaneity or maintains absolute simultaneity? Summary of the above discussion, we can conclude that the conditions for a relative system to maintain absolute simultaneity or only relative simultaneity are as follows:

In a qualified **dynamic** relative system, if two reference bodies are **independent of each other**, this system is a system that ***maintains relative simultaneity***; if two reference bodies are **interdependent on each other**, this system is a system that ***maintains absolute simultaneity***.

A qualified **static** relative system is a relative system that **maintains absolute simultaneity.**

Due to the simplicity and specialty of static systems, we will not discuss them further. If it is not specified, the relative systems discussed in this book later refer to dynamic systems.

Generally speaking, in the examples of relative systems that Einstein used, the majority involve light rays. As the Reference-body-A in the relative system, the light ray must be completely separated from the Reference-body-B, and therefore, they are **completely independent** of each other. The light ray disappears as soon as it meets the Reference-body-B.

Einstein also refers to several Reference-body-As that are not light rays. But they also are independent of the Reference-body-Bs.

Earlier, we mentioned the example of the flying raven that Einstein used. It is a lone ranger flying in the air. It has no direct connection with the other reference bodies and is an independent existence. But if the raven walks or runs on Reference-body-B, the two reference bodies are not independent of each other.

In short, looking at all the examples of various reference bodies in the relative system that Einstein used to aid him in proving the theory of relativity, we find this: The two reference bodies in all examples of the relative systems Einstein used, are completely independent of each other!

In his book "Relativity: Special and General Theory" Einstein said: "it became evident that there is not the least incompatibility between the principle of relativity and the law of propagation of light, and that by systematically holding fast to both these laws a logically rigid theory could be arrived at. This theory has been called the special theory of relativity." Of course, a light ray must be independent of any other object to exist!

Indeed, there is no incompatibility between Einstein's relativity system and the law of light propagation because light, as Reference-body-A, can only constitute a relative system that maintains relative simultaneity; in other words, this light only exists in a system of Einstein's relativity world. However, in a relative system in which Einstein denies its existence of relative simultaneity while maintaining absolute simultaneity, light cannot exist. Of course, this has nothing to do with the two being compatible or incompatible.

When Einstein extended his theory of relativity from light-related laws to the general system of matter in motion, irreconcilable contradictions arose.

Conditions consistent with Einstein's Two Relative Systems: System Maintaining Absolute Simultaneity and Maintaining Relative Simultaneity:

In a qualified relative system consisting of moving Reference-body-A and Reference-body-B:

- If Reference-body-B is stationary, then the relative system is called a static system, and the static system always maintains absolute simultaneity.
- If the Reference-body-B is moving, the relative system is called a dynamic system.
- In the dynamic system, if the Reference-body-A is **independent** of the Reference-body-B, the relative system maintains **relative simultaneity**; this is the relative system that Einstein strongly endorsed.
- In the dynamic system, if the Reference-body-A **inter depends** on the movement of the Reference-body-B, then the relative system maintains **absolute simultaneity**. Einstein strongly denied the existence of this relative system. In such normal relative systems, the two reference bodies that make up the system are interdependent, and the movement of one reference body is necessarily affected by the other one.

Einstein mistook a system that did not meet the relative condition as a relative system

In Sections VI to X of "Relativity: Special and General Theory", Einstein discussed the theory of relativity extensively with a system that does not satisfy the conditions of the theory of relativity.

> **TextBox** Excerpt from "Relativity: Special and General Theory", Page 15
>
> **VI THE THEOREM OF THE ADDITION OF VELOCITIES EMPLOYED IN CLASSICAL MECHANICS**
>
> LET us suppose our old friend the railway carriage to be travelling along the rails with a constant velocity v, and that a man traverses the length of the carriage in the direction of travel with a velocity w. How quickly, or, in other words, with what velocity W does the man advance relative to the embankment during the process?

The text in the text box above is the short excerpt we took from it. As can be seen from the text box, Einstein formed a relative system with the railway embankment, the train carriage and the man moving back and forth in the carriage. In five subsections VI to X of the book, Einstein used this model to discuss the theory of relativity.

Einstein composed the train carriage and man together as reference A, the railway embankment as another reference B, and let the synthesized reference A to be relative to reference B, to form a relative system.

This system looked like many other examples of relativity systems that Einstein used, so Einstein called it "old friend". However, this old friend is not a qualified relativity system as discussed by Einstein. It does not belong to the system of relativity, and cannot be used to discuss the theory of relativity.

Let's answer the question in the last sentence of the TextBox 1 above: what is the speed W of the man relative to the embankment?

The speed of the combined motion of the man and the carriage is the speed of the man relative to the embankment.

When the man and the carriage move in the same direction, the combined speed of the man and the carriage relative to the embankment is: $W = v + w$

When the man and the carriage move in the opposite direction, the combined speed of the man and the carriage relative to the embankment is: $W = v - w$

However, the most **basic requirement** of a relative system satisfying Einstein's requirements is that each **reference body must maintain a uniform motion**. When the moving speed of the object changes, the object cannot be used as the reference body in the relativity system.

Now, W is the combined speed of the man and the carriage, and W does not keep the same speed all the time. So, the system composed of the man-carriage synthetic reference body A relative to the embankment reference body B is not a relativity system that meets Einstein's requirements.

Einstein used an unqualified relativity system to discuss his theory of relativity a lot. We have not looked closely at whether his use of the unqualified system in Sections VI-X to study theory of relativity has raised points worthy of concern. Interested readers can continue to conduct research about this.

Here Einstein confuses himself with a composite system （A reference body was composed by 2 or more objects, such as person-train above. ）

Sure, Einstein also confused the world. He overturned his own rule about the relative system, leaving an embarrassment to those “who know relativity!”

Questions brought about by Mr. Einstein's example of a system that does not meet the relative conditions

In connection with the previous discussion on the relative system that maintains absolute simultaneity, Einstein’s wrong example cannot but lead us to the following doubts:

Why did Einstein create a person-carriage synthetic reference body to be relative to the railway embankment? Aren't the two relative systems composed of either person relative to carriage, or carriage relative to embankment, both simple, clear and qualified relative systems?

The key point is that the relative systems composed of either by person-carriage or by carriage-embankment are systems that maintain absolute simultaneity. If Einstein used the person-carriage system, then the assertion that Einstein has repeatedly emphasized for many years that "the absolute relative system does not exist" will completely become a wrong view.

What is depressing is: in my memory, although the calculation of the same direction and reverse movement is discussed in the mathematics textbooks of elementary and middle schools, there is no such pattern as the conductor walking in the moving carriage.

Perhaps Mr. Einstein also saw this, so he only used the flying raven to be relative to the carriage, and the lightning flashing in the air to be relative to the carriage, etc. but never discussed his relative system with the simplest system such as a train attendant being relative to the carriage!

Einstein also overturned the rules of synchronization transmission defined by himself

This "person-train-railway embankment" composite system used by Einstein also overturns the **synchronization transmission rule** defined by himself. In §1, after providing the details necessary for a synchronous condition of formula (7-1), Einstein defined the rule as below:

> 2. If the clock at A synchronizes with the clock at B and also with the clock at C, the clocks at B and C also synchronize with each other.

However, according to formula (7-1), the train carriage is synchronized with the railway embankment, and the train carriage is also synchronized with the people, but the people and the railway embankment are not synchronized! Thus, in the example of this relative system used by Einstein, overturned Rule 2 of his own definition in above text box

It can also be seen from this that not only the example given by Einstein is wrong, but also the assertion 2 in Einstein's paper is wrong from this wrong example, and their mistakes reinforce each other to demonstrate the other's mistakes.

2. It is pseudo-theology that a light ray makes a rigid rod shorter and the time on the rod slower

We have been **studying** the theory of relativity and testing its equations for more than ten years (see the relevant parts of our previous books and papers). But imagine our shock when we uncovered the true character of the theory: as **a device to obscure God with pseudo-miracles and complex mathematics.**

A pair of completely independent and irrelevant things — like a light ray and a rigid rod, like a raven that only flies over the carriage of a moving train, or like lightning that travels along the carriage and the moving train — how can they affect each other?

Ordinary and ordinary us, of course, can't play such big tricks with mathematics. But the master can! The pseudo-God of Lemaître can!

Einstein did it more than a hundred years ago.

How did he do it? Use math!

What a great magical and mythical medium mathematics can be! Given the definition like this, and making the calculation like that, then the "light" can change the length and the time flow rate on the "rod"!

Of course, you and we can't do it, only the great supernatural Einstein can do it! How did he do it? Use the methods outlined in Figure 7.1 and 7.2.

Should readers wish to review the two figures now, if only to reassess our conclusions, we suggest the following method: quickly browse the text in each figure from beginning to the end as if reading a novel. There is no need to understand everything, no need to understand the logical relationship, but only to roughly understand what the protagonists in this section are? Aren't they just **a ray and a rod**? From a ray and a rod moving back and forth in a wild manner, Einstein extrapolated the **basic concepts of his special theory of relativity.** And then two miracles surfaced: the "**moving ruler is shorter**" and the "**moving clock is slower.**"

Please also carefully read §1 in Figure 7.1 and §2 in Figure 7.2 in "On the Electrodynamics of Moving Bodies", they are the whole reasons for the establishment of the special theory of relativity. **All the mathematical derivations in all subsequent chapters of the paper are based on these §1 and §2, including the famous Lorenz transform** that can be used to calculate the shortened meter and slower clock.

What we want to discuss here is: the basis for the derivation of the mathematical model of the special theory of relativity is a physical system composed of a light ray and a rod that are always completely independent of each other. Why did Einstein put them together when comparing their motions? Is it possible a light ray magically makes a rod shorter and the time on the rod slower? If you listen to the messy mathematical explanations of the experts, you'll probably get just as dizzy as us did.

When a light ray moves around a rod, can the rod be changed? Show me!

Wait, wait ... isn't math just **brain** work? How can it be used to modify the motion state, geometry, and time flow rate of a **physical** rod? If it can, then the simplest way to gain eternal life is to allow a ray to shuttle back and forth beside us. Remember to position the watch you wear relative to the ray. The watch, which the ray would slow down, would also slow down the time involved with this parallel result: the time elapsed in our life would also slow down. Wonderful, isn't it? Long live knowledge of the eternal! Or is it just a myth?

Consider as well how it's possible to make the ray be relative to the rod? For example, if there is a ray and five similar rods, we want this ray to be "relative" to the middle rod. How can we do it?

Just think about how they can be relative. Using imagine to make them be "relative" to each other is the only way, but is it right? Isn't it fantastic to think that they are related to each other? This is the only way, right? Is there possibly any other way that makes them relative? How about using a rope to tie them together or a screw to screw them together or glue to stick them together? Or imagine another way to make them be relative to each other physically.

Consider making the light ray next to us relative to the car rushing down the road. When they are relative to each other, will the car be shorter and slower, and will anyone inside the car live longer?

What would happen if this light beam were relative to the flagpole on the moon that the American astronaut had set up there?

To put it bluntly, the light ray and the rod in Einstein's paper shown in Fig.8.2 have no strict scientific definition at all, there are no set conditions attached, and the paper reads as if it weren't science in the strict sense we understand and expect.

A little magic, right? Think again?

It's mysterious!

But **it is this ridiculous magical theory that has ruled human scientific thought for more than a century!**

For more than a century! For all human beings! Awesome? Funny? Ridiculous? ... What other words are available? Perhaps following describes one ten thousandth of it:

"There is no way you can do this, unless you do so arbitrarily, by an act of will or imagination."

Take a look at NASA's experiment: comparing a clock on the ground to a clock placed in a spacecraft. What did this experiment find? We just want to ask: If a plane flying in the sky is also consciously (some people imagined on the plane) or unconsciously (no one imagined) to relative to this ground clock, if some running trains are also relative to this clock, if another reverse-flying spaceship would also be relative to this heavy-duty clock... So, we would like to ask:

What can you do to make the clock in the experiment relative only to the clock on the spacecraft?

There is no way!

If there is no way, then if this clock is facing thousands of flying machines with different speed and flying direction, then:

Which one of these flying machines should this clock be relative to?

How is the time on this clock calculated?

Do you think the whole thing is absurd? Isn't this a "scientific" myth that NASA has played out in earnest?

It's a pity that more than ten precious years of our lives have been wasted in this ridiculous dismantling of this elementary mathematical game.

More generally, we believe that our thinking has suffered, perhaps from too much reading and learning, but more from the patterns and expectations that modern science has imposed on us, standardizing how we interpret phenomena. Whenever a problem comes up, whatever it is, we first turn to mathematics and drum up a theory to explain it (see "Chapter 13 Math is Old" for more detailed discussion of this issue).

Of course, with reading too few books, we felt timid and dared not express our objections, thinking that we don't understand the theory well enough. We need only listen to the boastful explanations provided by experts who, we believe, understand more than we do. After reading too many books, our thinking, trained as it was, became standardized. When seeking to understand a phenomenon, we first turned to a mathematical model, considering whether or not the formula it provides is reasonable. But we forget the most basic truth. **In applied mathematics rather than pure mathematics, to establish a correct mathematical model, the physical model must be correct first.** So, the first thing to consider here is whether the physical model is correct or not. The movement of any ray can never have any effect on any factor associated with a rod, neither on its speed of time flow, or its structure or length, etc.! We simply can't use mathematics to achieve such results! It's just like when a person is hungry. We cannot do mathematical calculations to make him full. No matter how we calculate, his stomach is still empty! If you can solve the hunger problem with pure mathematics, we'll bow down.

It's pretty clear then that special relativity is a magical mathematical game played between a rod and a light beam that can't communicate with each other at all. As much as it is a pseudo-god miracle for an atom to create a universe by exploding, the same holds a light ray changing in any manner a rigid rod.

Although the discussion that follows to the conclusion of this chapter is a bit redundant, it also provides readers with supplemental various ways to understand how the special theory of relativity is wrong headed. From a scientific point of view, people can further understand the wrong nature of the special theory of relativity.

Taking §1 in Fig. 7.1 and §2 in Fig. 7.2 as the basic model, Einstein deduces a similar Lorenz motion according to the relative motion of a rod and a light ray in the later sections of "On the Electrodynamics of Moving Bodies." In a dizzying array of definitions, settings, and formulas, let's see how a light ray causes the clock on the rod to change the length of the rod and the time flow rate on the rod.

We already saw from §2 "On the Relativity of Lengths and Times" of Einstein's "On the Electrodynamics of Moving Body" in Figure 7.2, in the dazzling definitions, formula and settings, how a "ray" changes the time of the "rod" and changes the length of it. This is the process of establishing all the relative concepts and theoretical foundations of "moving ruler shorter" and "moving clock slower" time dilation in the theory of relativity. Starting from §3 and all rest part of the thesis, were played with mathematics on the physical basis of §1 and §2.

Looking at most of Einstein's works, we can confirm that Einstein really established the theoretical basis of the special theory of relativity based on the physical model of a light ray and a rod.

Recall that it is impossible for two independent reference bodies to affect each other. In a relative system, then, that maintains the absoluteness of simultaneity, the results obtained – that "moving clocks becoming slower" and "moving rulers becoming shorter" exemplify how mathematical thinking can turn fantastical.

People who study the theory of relativity tend not to pay much attention to the concept, as established in sections §1 and §2, previous to section §3. The latter, of course, initiates the mathematical derivations involved, and it is this section that people most attended to. But this is absurd: Without a correct physical model, an applied mathematical model that correctly describes the world cannot be derived. The wrong physical model leads to the collapse of the corresponding applied mathematical model.

Confusion and Computational Conflict Caused by Passive Relativity from Fantasy Relativity

Now we have made it clear that in a relative system composed of two completely independent reference bodies, there will be no mutual influence between the reference bodies due to their mutual motion. And the relative systems in Einstein's special theory of relativity are all independent of each other, so it is impossible to produce the various phenomena he described.

Below, we **assume** that the phenomenon of Einstein's "moving ruler becomes shorter" and "moving clock becomes slower" exist. Then extending the lines of Einstein's thinking, let's suppose Einstein's independent reference bodies really interact with each other as he described. If they did, what kind of magical consequences would arise?

An object or system is also often unconsciously referred to as a reference body and is relative to another object or system. We define this situation as **Passively Relative**.

For example, when using an atomic clock on a spaceship and an atomic clock placed on the ground to do a relativity time dilation test, experts only focus on whether or not the data gained from the ground clock and the clock on the spaceship are different. In reality, however, these two clocks are simultaneously **passively** relative to many different reference bodies, such as white clouds, airplanes, and even birds flying at a constant speed in the air, the rotation of the Earth, countless cars and trains moving at a constant speed, the African male lion running at a constant speed, the whales in the sea … They all can be passively relative to these two clocks, and there is no way to stop them from being so.

Of course, anyone can imagine one object as only being relative to another specific object, but in reality, there is no way to specify such an exclusively relative object, let alone prevent an object from being passively relative to any other moving objects.

There are many moving objects with different speeds that are passively relative to the atomic clocks on the spaceship and on the ground. According to Einstein's calculations in theory of relativity, they each will have different effects on the time changes of the two clocks. In fact, for Einstein's system, which maintains relative simultaneity, since the participating reference bodies are completely independent of each other, there will be no connection or link between them. Therefore, in actual application calculations, when a reference body is set to be relative to many other moving reference bodies with different speeds at the same time, the relative calculation will produce contradictory results. Following is an example. (For readers not versed in the mathematics used here, please skip this section.)

Conflict results of simultaneous relative calculations on one clock paired to each of multiple clocks that have different speeds

We use the Kesan Online Calculator provided by CASC COMPUTER CO, LTD (Figure 7.3) for the calculations below.

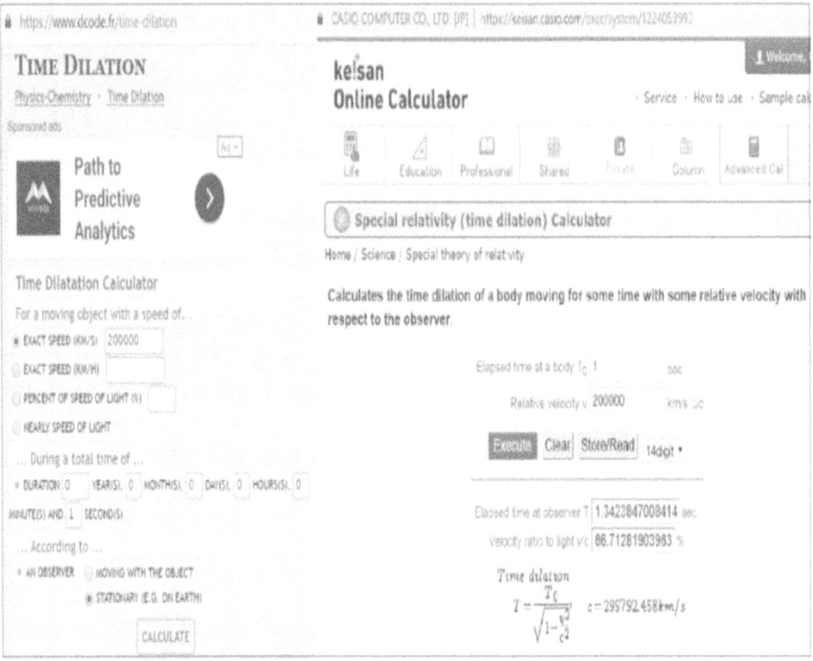

Figure 7.3 Special relativity time dilation calculator.

We construct a relative system for the calculation of the Lorentz transformation **in mind**. An atomic clock on the ground and eight atomic clocks in different spacecraft traveling at different speeds respectively constitute eight relative systems. The eight relative systems are divided into two groups. Each group starts at the same time, lasts for the same duration, and calculates its experimental results by Lorentz transformation, the duration of the first group is 1 second, and the duration of the second group is 30 minutes. The calculator in Figure 7.3 is used to perform Lorentz transformations: determining the time dilation of each related system. The results are shown in table 7.1 below.

Table 7.1 Calculation results for eight spacecraft with different speeds.

Experimen time (second)	spacecraft peed (k/second)	Ground clock Experiment time (second)	Spacecraft indicated Experiment time (second)
1	200000	0.74494293578673	1.3423847008414
1	150000	0.86582546589248	1.15496718379552

1	100000	0.94272742316888	1.0607520004442
1	10000	0.99944352013705	1.00055567897052
1800	200000	59.3589994339	0.64122891504257
1800	150000	58.48583860647	38.940930831342
1800	100000	16.909361703985	49.353600799567
1800	10000	58.998336246698	1.0022214693684

As can be seen in Table 7.1, as the duration increases, the time dilation value also increases significantly. Therefore, if this time dilation is true, it should be easy to verify in experiments.

Test start time: 12:00 noon. The calculation results obtained are summarized in Table 7.1 above.

Question about the results in Table 7.1: At one second after 12, what time should the ground clock point to?

And again, at 12:30, what time should the ground clock point to?

Einstein also provided us with an implicit conclusion: according to the change of latitude from the equator to the poles, the clock's movements will change slightly.

In other words, for clocks located at different latitudes, their clock hands will have slightly different positions. Clocks on the equator go fastest, and clocks at the poles of the Earth go slowest. Of course, these changes are extremely minor.

But minor changes are also changing. Even minor changes can accumulate into observable changes over a long period of time! According to Einstein's statement and algorithm, if identical clocks are placed along different latitudes, after one month, one year, ten years, one hundred years … all of them will point to different times!

In other words, Einstein proved that the times indicated by clocks on different latitudes in the world are different!

Dear physicists and/or cosmic scientists, why don't you do this experiment? After the same clocks are installed at different latitudes around the world, after one year, ten years, or more, let the world test the truth of Einstein's theory. Perhaps using an atomic clock will be more accurate.

The Disaster of "Moving Meters Are Shorter" caused by Lorenz transform calculation

According to the Lorentz transformation theory, "a moving meter stick is shorter" means this: in a high-speed flying system moving along an X-axis, the length of the flight, the distance traversed, will only change in the X-axis direction, but the lengths or distances in other dimensions remain unchanged.

Thus, there is a relativistic motion catastrophe: if a spaceship is flying at high speed along the X-axis, then, according to "a moving meter stick is shorter," all components on the spacecraft, including the spacecraft itself, will become shorter along the X-axis. This would result in the spacecraft not being able to fly and maybe even disintegrating.

How big is the change on the X-axis? Table 7.2 gives the answer.

Table 7.2 Calculating the length change of the spacecraft at different speeds according to the Lorentz transformation:

Spacecraft speed (Light-speed)	observed length (meter)	observed height (meter)	observed width (meter)
0	200	40	60
10 %	199	40	60
86.5 %	100	40	60
99 %	28	40	60
99.99 %	3	40	60

The data in Table 7.2 tells us that if the length of the spacecraft is 200 meters when it is stationary, then when the spacecraft is flying at 10% of the speed of light, the length of the spacecraft is shortened to 199 meters. If the speed of the spacecraft is 14% of the speed of light, the length in the X direction will be shortened by about 1%. Einstein believes that this is an **actual systematic contraction** along the X-axis, so it has nothing to do with the material of the spacecraft, and there is no way to prevent it.

Then, when the spacecraft is flying at 0.14 speed of light, its engine will be shortened by 1% along the X-axis, while the dimensions in the other directions will remain the same, so the round engine becomes oval. At this time, this engine cannot work properly, and the spacecraft itself cannot fly. In this way, a catastrophe for space navigation with a shortened relativistic moving ruler appeared!

The possible practical usage of "moving rulers getting shorter"

This aspect, while not studied thoroughly, is only proposed as an assumption.

Einstein's theory of relativity posits that a moving object in a relative system will exhibit a "moving ruler shorter" phenomenon in the direction of its motion. The phenomenon manifests physically; that is, the moving ruler would shrink.

We think that if the actual physical shortening is corrected to apparent shortening, it may be applied to astronomical observations.

The distance between two points in space does not change due to the motion of some object. Otherwise, the distance between these two points in space, as we measure it, would respond to any object near enough to affect it; for example, multiple aircraft flying in different directions.

However, under certain circumstances, such as the observation of the motion of the same celestial body by a telescope on a mountain top and a space telescope in near-Earth orbit, it should be possible to identify an apparent error that can be calculated by the Lorenz transformation.

The apparent error we are talking about here is the observed error, which does not affect the celestial body.

This section is purely causal thinking.

The author's opinion on Einstein's formula and other mathematical models

To help him express his unique understanding of the universe, Einstein devoted himself to learn the mathematical formulas of Riemann's geometry. Then Einstein devised his formula that depicts the universe. In truth, we were not interested in studying how Einstein derived his formula, but we are sure the universe cannot be described by one mathematical formula, even with a century's worth of research behind it! There is so much about the universe that we still do not understand that focusing exclusively on Einstein's formula is ridiculous. And we include here dark energy and dark matter, along with the so-called uniformly distributed microwave background that cosmologists spend so much time analyzing.

What do we learn from numerical methods? How about using a computer to solve problems that cannot be solved by given formulas? In real life, the problems solved by computer methods involve greater complexity than the problems solved by the mathematical formula. Neural networks in popular artificial intelligence may use tens of thousands of partial differential equation groups to identify a simple graph, whereas depicting the universe requires only one equation? It's all too funny, right?

Because Hawking added clever, new (but doubtful) solutions to Einstein's equations, he has gained immense authority when considering the universe, including his work on black holes.

However, even the perfect interpretation of a mathematical formula confirmed by accurate observation comprises only a very small part of the world. Indeed, the world's complexity presents issues that equations cannot resolve. As for immediate real problems we are faced with, these can normally be solved by approximate solutions that computer numerical methods produce.

Yes, for very many equations, their practical use is limited, and often valid only in a certain area. In contrast, an equation describing the universe is ever more difficult to perfect, with every detail correct, and every relationship to physical phenomenon comprehensively revealed.

Modern cosmologists, playing with Einstein's cosmic equation, make a fuss about singularities that he himself did not explain. The universe is bent for a while, then closed for a while, and then opened for another while.

So, is the universe, including time and space, limited to Einstein's mathematical model? Do you agree with cosmologists who derive their theories and proofs from Einstein's model? This is a question that our experts should think about.

At this time, we wanted to relax our heavy hearts, and suddenly remembered that when 2018 came, I saw a very simple trifle arithmetic problem on WeChat, but some people were blinded. Of course, it did not have the same effect that the mysterious proof of Einstein's did, but it quickly made it clear that even when working through simple mathematical problems certain rules applied.

Verify: 2023 = 2022

Prove: 2023

$$
\begin{aligned}
&= 2018 + 5 - 9/2 + 9/2 \\
&= 2018 + \sqrt{(5-9/2)^2} + 9/2 \\
&= 2018 + \sqrt{(25-45+(9/2)^2)} + 9/2 \\
&= 2018 + \sqrt{((16+9)-(36+9)+(9/2)^2)} + 9/2 \\
&= 2018 + \sqrt{(16-36+(9/2)^2)} + 9/2 \\
&= 2018 + \sqrt{(4^2-2\times 4\times\ (9/2)+(9/2)^2)} + 9/2 \\
&= 2018 + \sqrt{(4-9/2)^2} + 9/2 \\
&= 2018 + 4 - 9/2 + 9/2 \\
&= 2022
\end{aligned}
$$

How about it? Of course, it is wrong. But where is the fault? If you can determine why it is wrong in a few seconds after scanning it, or as soon as you read the first few lines, you get the answer, then you can also easily see through Einstein's mistakes.

The grandeur of the universe makes people feel their own insignificance. When we tried our best to steal a few secrets from God we found that many were falsely rendered; pretty, even beautiful equations, for example, that proved no more and no less than the lengths we would go to believe in them.

We strive to explore, but only feel the insignificance of ourselves and the greatness of the Creator!

Proximity to the ultimate truth of the universe is not something that can be claimed casually. As down-to-Earth scientists, we well know that before we can truly see the authentic content that the universe has exposed to us, we must work hard to traverse the road built by basic scientific principles.

A story that traces the correct meaning of the "Definition of Simultaneity" in Einstein's paper

One Chinese guy copied the "Relativity of Simultaneity" from my paper into his own paper, and applied the scientific and technological achievements. However, this analysis of the theory of relativity from the perspective of "relativity of simultaneity" and "absoluteness of simultaneity" is a formulation without precedent in the world before me. It took me a long process of more than ten years to realize the importance of criticizing the special theory of relativity from this perspective, and how difficult it was.

In 2017, at the invitation of "Regularity" journal, I published several papers of 25 pages in the "Wu Yuxiang Forum" in the fourth issue, discussing the problem of relativity, and initially touched on this topic. On the first page of the first issue of the journal of 2019, the author published "The Dilemma of Special and General Relativity from a Historical Perspective" also discussed this topic. The first paragraph of this paper is "1. The Special Theory of Relativity from the System of Maintaining the Absolute Simultaneity", which is the first paper in the world analyzing the errors of relativity theory from "there exists absolute relative systems that maintain simultaneity." In books such as "Questioning the Big Bang - Exploring the New Appearance of the Universe" published by the author in 2017, "Clarify Big Bang" published in 2018, and "Fading Modern Cosmology" published in 2020, the author is gradually deepening and improving the understanding of the universe from this perspective and criticizing the Relativity.

In brief, analyzing special relativity from the perspective of "simultaneity" incurs a great deal of thought over a long period of time.

Now when I showed the manuscript of my book, *Fading Modern Cosmology* to my son (then was a doctoral supervisor at Columbia University), he asked a question: "What is the exact meaning of Simultaneity?" He asked this question after reading Einstein's paper.

This question caused us to consider it carefully.

We noted earlier that the method Einstein used to write his thesis was strange. The text and formula are shown in Figure 7.1. While the concepts and mathematics involved in §1 and §2 are very simple, the entire section reads obscurely as if trying to catch some elusive something in the fog. We doubt whether Eddington, who is said to be the first to understand relativity, can answer why Einstein wrote in the obscure manner he did.

The title of this section is "§1. Definition of Simultaneity." Though it has only about 1100 words, after reading it again and again, it is still impossible to clearly answer the question: "What is the exact definition of simultaneity?" It cannot be expressed concisely

We noticed that in "§. Definition of Simultaneity," there are numerous descriptions of "synchronous" and even a mathematical definition of the term. So, what is the relationship between "synchronous" and "simultaneity" in Einstein's world? If we can define "simultaneity" with concise mathematics, like Einstein defined "synchronization," then we can also discuss "relative simultaneity" and "absolute simultaneity" through mathematics!

Again, what is the relationship between "simultaneous" and "synchronous"? There is a sentence in Einstein's text that speaks to this issue. It speaks to the stipulation that stationary clocks in different places are synchronized. Obviously, then, the definitions of "simultaneous" and "time" become clearer. However, how were the definitions of "simultaneous" and "time" obtained? Does "simultaneous" mean "synchronous"? It looks a bit like that. But why use two totally different words again and again? It's really confusing.

Some key points need to be confirmed against Einstein's original text. For example, with respect to the "principle of relativity" and "principle of constant speed of light," we need to determine (in the original paper written by Einstein in German) whether he proposed them as "definitions" or "postulates." From the "definitions" used in the original text of the German version, Einstein "defined" these two principles without explanation or proof. But descendant scientists regard them as scientific laws!

Looking at §1 in Einstein's relativity theory and concept in Figure 7.1

above, there is neither advanced mathematics (only the formula for calculating the superposition of linear motion in elementary school arithmetic) nor deep and difficult concepts. Nonetheless, he wrote the paper in a style that "only a few people in the world can understand." Is this the writing style of a great master?

We can only seek answers from the literature on Einstein, which is not too much (Figure 7.4).

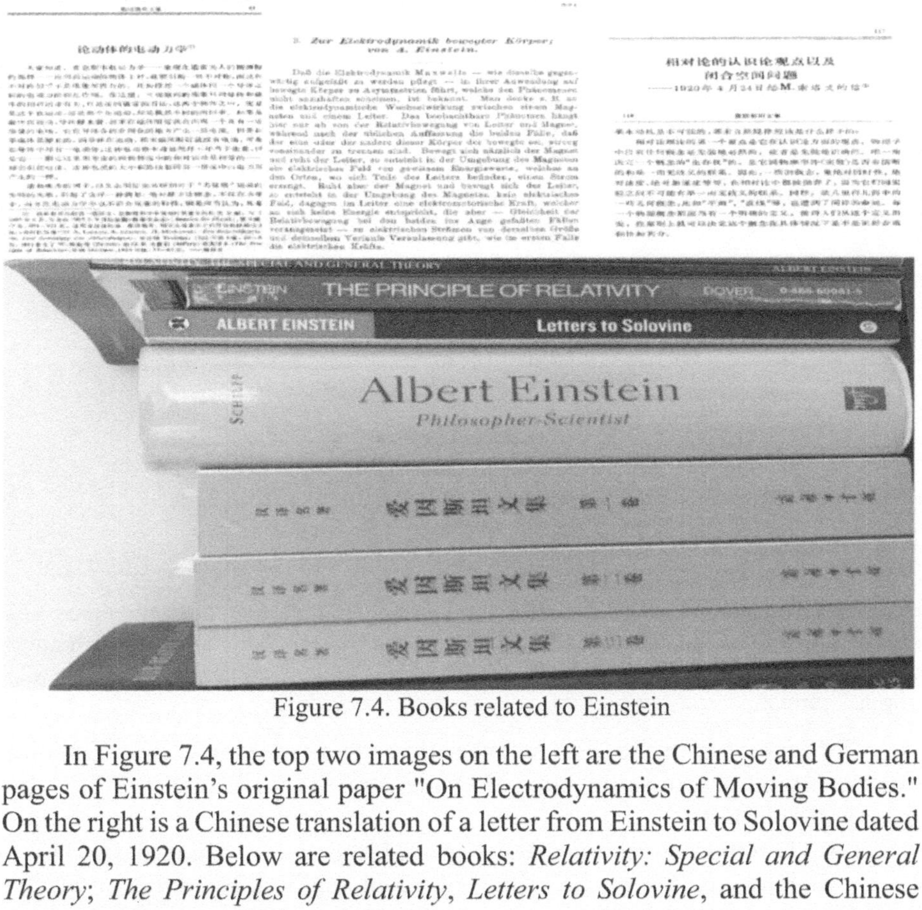

Figure 7.4. Books related to Einstein

In Figure 7.4, the top two images on the left are the Chinese and German pages of Einstein's original paper "On Electrodynamics of Moving Bodies." On the right is a Chinese translation of a letter from Einstein to Solovine dated April 20, 1920. Below are related books: *Relativity: Special and General Theory*; *The Principles of Relativity*, *Letters to Solovine*, and the Chinese version of *Einstein Collected Works*...

As noted, some key points need to be checked against Einstein's original German text. For example, did Einstein propose the "principle of relativity" and the "principle of the constant speed of light" as "definitions" or as "axioms." From the "*definitions*" used in the original text of the German paper, it is known that Einstein "defined" these two principles without explanation or proof.

The translation of the critical sentence in various languages is shown in Figure 7.5 below.

Original German is this : Wir haben so unter Zuhilfenahme gewisser (gedachter) physikalischer Erfahrungen festgelegt, was unter synchron laufenden, an verschiedenen Orten befindlichen, ruhenden Uhren zu verstehen ist und damit offenbar eine Definition von „gleichzeitig" und „Zeit" gewonnen.

Original Chinese is this 这样，我们借助于某些（假想的）经验，对于静止在不同地方的各只钟，规定了什么叫做它们是同步的，从而显然也就获得了"同时"和"时间"的定义。

Original Spanish s this : De esta manera con ayuda de ciertos experimentos f´ısicos (imaginarios), hemos establecido lo que se debe entender bajo relojes sincronizados que se encuentran en reposo en diferentes lugares y, por ende, obviamente hemos obtenido una definici´on de "simult´aneo" y de "tiempo".

Original English translation by Meghnad Saha is Thus with the help of certain physical experiences, we have established what we understand when we speak of clocks at rest at different stations, and synchronous with one another ; and thereby we have arrived at a definition of synchronism and time.

The translated English version in 1923 by W.Peerrett and G.B. Jeffery is: Thus with the help of certain imaginary physical experiments we have settled what is to be understood by synchronous stationary clocks located at different places, and have evidently obtained a definition of "simultaneous," or "synchronous," and of "time."

Figure 7.5 Translated texts in different languages of an important, but ambiguous, key phrase in German, Chinese, and English in Einstein's thesis "On the Electrodynamics of Moving Objects." Einstein quietly corrected his vague text expression in the 1923 English version.

In Figure 7.5, in the last sentence, **there are two words that Einstein quietly added a few years later that are critical for understanding the theory of relativity.**

Based on the English translation of 1923, the correct Chinese translation of this sentence, is that the definition of simultaneity is synchronization. Its complete correct Chinese translation should be:

*這樣，我們借助於某些（假想的）經驗，對於靜止在不同地方的各只鐘，規定了什麼叫做它們是同步的，從而顯然也就獲得了**"同時"，或曰"同步"**，和"時間"的定義。*

The most popular and most used English translation text was translated by W. Perrett and G.B. Jeffery in 1923. This is a very outstanding translation, with. no other, newer version found. The translation in this version answers the question I'm struggling with!

Notice in the last sentence in Figure 7.5, the phrase in bold red. It tells us that Einstein later considered "*simultaneous*" as equivalent to "*synchronous*." This translation was also approved by Einstein. Notice in Figure 7.6 below, the statement in English attesting to this approval: "**the version was reprinted under special arrangement with Einstein**."

Figure 7.6 below, the statement in English attesting to this approval: "**the version was reprinted under special arrangement with Einstein**."

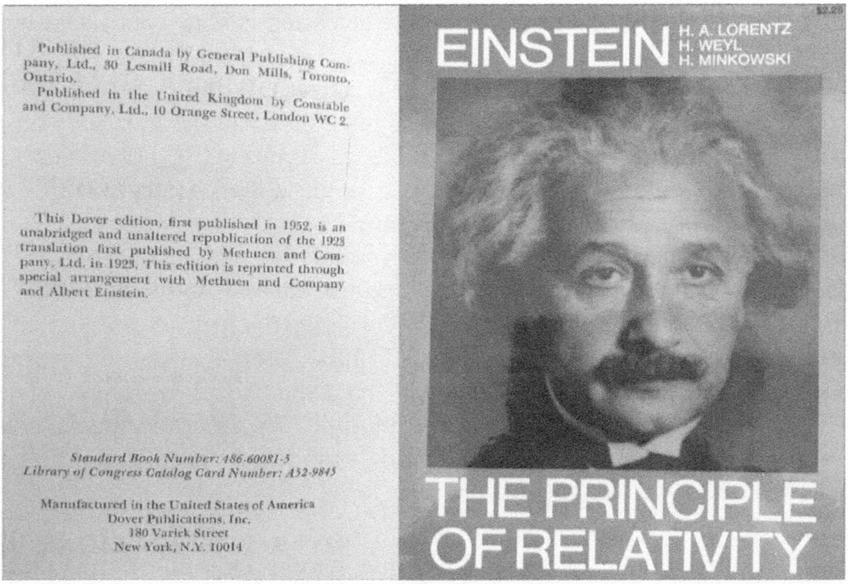
Published in Canada by General Publishing Company, Ltd., 30 Lesmill Road, Don Mills, Toronto, Ontario.

Published in the United Kingdom by Constable and Company, Ltd., 10 Orange Street, London WC 2.

This Dover edition, first published in 1952, is an unabridged and unaltered republication of the 1923 translation first published by Methuen and Company, Ltd. in 1923. This edition is reprinted through special arrangement with Methuen and Company and Albert Einstein.

Standard Book Number: 486-60081-5
Library of Congress Catalog Card Number: A52-9845

Manufactured in the United States of America
Dover Publications, Inc.
180 Varick Street
New York, N.Y. 10014

Figure 7.6. W. Perrett and G.B. Jeffery's translation of The Principle of Relativity. "On the Electrodynamics of Moving Bodies" is on pages 35–65. The last sentence in the middle of the left side of the picture says: "**This version was reprinted by special arrangement between Methuen and Albert Einstein**."

The English version of this sentence that removes confusion is the last English sentence in Figure 7.5, which shows that in Einstein's thesis, "simultaneous" is equivalent to "synchronous." This version comes from Figure 7.6 below, which specifically states that "**the version was reprinted under special arrangement with Einstein**."

In the middle paragraph of the left side of Figure 7.6, it is explained that the version of "On the Electrodynamics of Moving Bodies" collected in this book is an English translation without addition and deletion, published in 1923. The last sentence of this paragraph said that this 1952 reprinted book was published through special arrangement with Albert Einstein.

This to say, the meaning of "simultaneity" and mathematical expressions of it can be correctly understood by reading "On the Electrodynamics of Moving Bodies" in this English translation. For those who read the original German, or Chinese translations, or other translations, it is difficult to correctly understand the meaning of "simultaneity," and it is more difficult to express it mathematically. This is not about the level of knowledge, scholar or translation, but a problem made by Einstein's misuse of " synchronous" and "simultaneity".

In terms of content, contextual coherence, and references to Einstein's other works, W. Perrett and GB Jeffery's translation is very good, a lot better than other versions.

Now we know how to correctly and accurately understand Einstein's concept of "simultaneity," as expressed in his paper "On the Electrodynamics of Moving Bodies." That is, in Einstein's dissertation, "synchronous" and "simultaneous" are equivalent, which can be described by the formula that defines "synchronization," as seen in Figure 7.1.

Figure 7. 6 also tells us that Einstein found out why his thesis was only understood by a few people in the world, so amendments to it were made. But he didn't discuss this publicly but quietly corrected his fuzzy expression about “simultaneity” and “synchronous” in his English version. Then, for example, all German or Chinese readers who only read their local-language versions still do not clearly understand special relativity, because the only Chinese translation is based on the original German version, and in the German version the error is not corrected but sustained.

Interestingly, in the Chinese translation there is a footnote noting that the translators referred to W. Perrett and G.B. Jeffery’s English version as they worked. However, the most critical words **(or, "synchronize")** were not translated in the Chinese version or used in the German version. Because of this difference, readers of the Chinese or German versions can't understand the exact meaning of simultaneity. Without understanding its precise definition, it is impossible to understand Einstein 's special theory of relativity. That is to say, readers cannot fully understand the theory of relativity itself!

So, in this way, if you only read the German or Chinese version of "On the Electrodynamics of Moving Bodies," it is impossible to fully correctly understand the special theory of relativity. There are many such masters of relativity in the world.

As far as we are concerned, we would not read the English version if we did not want to delve into the issue of simultaneity. The translation of the Chinese version is very good, and it is in line with Einstein's original German thesis. It is a pity that neither the original German text nor its Chinese translation show that simultaneous and synchronous have the same meaning.

As a result, special relativity has been transformed into a heavenly idea that few people in the world really understand. Why did Einstein quietly add the words "simultaneous, or" to the translation of the English version published under his special arrangement in 1952, more than 40 years after the publication of "On the Electrodynamics of Moving Bodies"? Was it because he could clarify the concept of simultaneous, which was obscure in his original thesis? It certainly seems so.

And though Einstein quietly corrected his original phrasing, readers from all countries who have not been able to read the approved English version, where the correction sits, including German readers who only read the original German text and Chinese readers who only read the Chinese translation, remain at a disadvantage. They are unclear as to the meaning of "relativity" and, thus, the theory of relativity.

The master is a master. The mathematics used in the first section of Einstein's thesis are only elementary school mathematics, but they can be presented in such a way that people cannot understand them. This is not nonsense talk. It really shocks us!

But before we continue discussing, there is one more interesting thing. In the 1920 English translation of "On the Electrodynamics of Moving Bodies" by Meghnad Saha, he replaced the word "simultaneous" with "synchronous." The title of this translation for §1 is "Definition of Synchronization." That is, Meghnad believed that in Einstein's dissertation, "simultaneous" and "synchronous" mean the same thing and there is no need to use two different English words to represent the same concept and cause unnecessary confusion.

This translation solves the last doubt about the precise definition of simultaneity and enables people to understand the paper. This also shows that our current understanding is correct from another angle, that is, the formula in Figure 7.1 ($t_B - t_A = t'_A - t_B$), not only precisely defines "synchronize" with mathematics, but also precisely defines "simultaneity".

On the other hand, while Meghnad's translation replaced all use of the word "simultaneous" with "synchronous," it did not, at the same time, provide a definition of "simultaneous." Nonetheless, Einstein used "simultaneous" in his other works. So Meghnad's translation can cause problems. Perhaps this is the reason why Meghnad's version was not approved by Einstein.

The discussion of this issue can also lead to our further doubts: When Einstein was writing and publishing his paper, did he really think that synchronous = simultaneous? Or did he originally want to express other meanings, but in the end, due to the situation, he had no choice but to make the final revisions to the thesis written more than ten years ago according to other people's question and understanding?

How is the emperor's new clothes cut out?

To prove the theory of relativity, Eddington observed and measured the angles of light rays from stars deflected by the Sun during an eclipse. The quality of the observations was poor. Because his observations in Soral, Brazil, appeared to be closer to Newton's than to the relativistic model, the data were revised. And he used these revised data and his exalted position in academia to tailor the emperor's new mythic clothes for this century of relativity.

From the above, we can see that the research paper "On Electrodynamics of Moving Bodies" originally proposed for the theory of relativity presents a crude, non-rigorous and self-contradictory thesis.

But why is it that, starting from such a thesis, based on the notion of space time in this thesis, Einstein's thinking has structured how we view the universe for more than a hundred years, and more so now.

This is a question that every scientist needs to ponder.

Einstein's special theory of relativity is essentially mythic in nature. Nor is Eddington absolved by this recognition. Didn't Eddington modify his observation data in order to best conform to the theory? And when it was finally accepted as fact, no other scientist in the world was or is willing to criticize it by pointing out its opacity or the confusion it can create when seeking to understand it thoroughly. Even when experts identify mistakes in the theory, they believe it stems more from their own weakness in comprehension than not.

Just as we are writing about these criticisms, our hearts are shyly thinking: is it our own understanding that is wrong? If our understanding is wrong, please experts in the world who understand the theory of relativity provide the kind of criticism we can learn from.

.

Chapter Summary

We have studied relativity for more than ten years. **Finally, we found the truth. We should be happy, but we feel very sad: how can human scientific and technological thought be structured by a theory that is so obviously wrong over a century's length of time?**

What is the actual cause?

It took Einstein 15 years to prove and confirm that no system maintains absolute simultaneity in his relativistic world, which led to the denial of the existence of a system that maintains absolute simultaneity in the entire physical theory.

But it is easy to prove that there are a large number of systems in the world that maintain absolute simultaneity. In addition, we can also clarify in greater detail how to distinguish a system that maintains relative simultaneity from a system that maintains absolute simultaneity in a relative system defined by Einstein; that is, whether or not the two reference bodies in the relative system are independent. A relative system consisting of two reference bodies that depend on each other or are independent of each other, depicts a system that maintains absolute simultaneity or relative simultaneity.

Although Einstein repeatedly emphasized the non-existence of such a system, which has allowed modern physical kinematics to follow suit (denying it as well), we have proved the opposite with numerous examples. The relative system that maintains independence between two reference bodies is a relative system that conforms to Einstein's theory of relativity and also one that maintains relative simultaneity between reference bodies. It is a system that Einstein only acknowledges the existence of. However, the two reference bodies that remain independent of each other, because they are independent of each other, cannot affect each other. In this way, time dilation, the shortened moving ruler, and so on are not valid. The characteristics of the relative system obtained from the mutual influence of each other's motion are completely non-existent properties and are derived from the basis of what is, finally, mythic fantasy.

Readers are advised to go back and read Einstein's "On the Electrodynamics of Moving Bodies" at this time, and try to grasp and understand the theoretical basis of special relativity. When you do so, you will find that the §1 and §2 of the thesis use a ray and a rod plus various dazzling definitions and simple mathematics to define this base of great special theory of relativity that dominates our thinking to this day, and thus confirming the miracle of "moving clock slower" and "moving ruler shorter," Starting from §3 in the paper, there are a large number of mathematical derivations that are based on this illusion: of a ray and a rod with relatively independent motion. As a result, the entire special theory of relativity finds no root base and floats in the illusion provided.

Do human beings have to review why they have been ruled by such mathematical magic for a hundred years, even though they are struggling and unable to extricate themselves? So many scholars who opposed the theory of relativity have not seen through such simple mathematical magic?

For readers seeking further proof, please refer to the first chapter, "Is Mathematics Old?" in my book "誰有權談論宇宙 *Who Should Talk about Cosmos,*" published in 2005, and the section "Math is Old" in *Debate of Light and Dark* published in 2016.

In 2005, I began to question how mathematics played with physics, just as Hawking did. I was disgusted by the academic community (including myself) for maximizing the use of mathematics. But after ten years I came to a conclusion, however preliminary it was that "math is old." Even so, it was only a few years later that I finally saw the dazzling nature of the mathematical magic played by a ray and a rod.

When you think about it, it is no less than the myth of a universe, our universe, bursting out from an atom that explodes!

Chapter 8

What Is Space? Discussing General Relativity from the Viewpoint of Space

Ah, the world, is the hotel where all things are on their journey; and the time is the coming and going of the guests of hundred generations.

——Bai Li, Spring Night Banquet with Brother in Peach Flower Garden

The universe, like duckweed, wanders in the long river of time. Flow pushes galaxies form. Stars shine like shells, planets evolve... The brief existence of mankind is but an instant in this recurring flow. The one-hundred-year existence of a man but an instant of the universe's passing by; the billions of life years of the stars, are also only a finger flip instant of the universe of sea and earth exchange.

Time and space are the eternal, if mysterious, topic, that ever eludes our understanding. Two-thousand five hundred years ago, in the *Tao De Jing,* Lao Zi, its author, recognized this as he began writing. His response is still pertinent for us: "The way that can be told of is not an unvarying way; The names that can be named are not unvarying names. It was from the Nameless that Heaven and Earth sprang; The named is but the mother that rears the ten thousand creatures, each after its kind."

There are various ways to define and explain space, time and its properties. However, this explanation needs to be justified and persuasive, it must not contradict the empirical experience of human beings since ancient times, and it must not be mixed with nothingness or even modern mythology born from the imagination.

Here we must use the word "science" with caution, because in modern history we see that the Masters of Science—the masters of physics, mathematics, astronomy and cosmology—used "science" to mix up the concepts that have been clearly understood into a paste, and then used this paste to silence the mouths of other philosophers and thinkers in the name of science!

But what kind of concepts of space have our modern masters given us? Curved space, space mixed with time, time mixed with space, parallel space, multi-dimensional space and so on ...

There is a famous saying in physics:" Matter tells space how to curve, and curved space tells matter how to move." Place a mass down anywhere in the universe, and the space around it will curve in response.

We will fully prove that space is inflexible. We will continue to prove that time and space cannot be converted one into the other. But the mythic tales abound. For instance, in parallel space, another "I" exists. In multi-dimensional space many "Is" exist. These and similar conjectures, as vulgar as they are, and however woven from abstruse mathematics they were, are little more than scientific myths.

Thought is a journey on a post road. Go and walk, pick the scenery. Sometimes the willows darken the bright flowers, sometimes the mountains are at the end of the river, and there are also ecstatic, and there is also spring-like renewal

This chapter is a supplement to the discussion of space and time that I have in my books *Debate of Light and Dark - A 100-Year Bet with NASA* and *Clarifying Big Bang*. It is like checking the suitcase: take out the old and add the new.

Introduction to space

Frankly speaking, first, the space is empty, meaning there is nothing in the space!

Please think carefully about it first. Isn't this simple statement correct?

Space is empty. However, everything exists in space. Without space, there is nothing, and when there is nothing in space, space is still there.

If you agree that space is empty and nothing is in it, then **a very simple question ruins Einstein's general theory of relativity**: how can a space with nothing be bent? What is there to bend? Can you bend "nothing"?

If a space bends only when there is matter in it, then **it is not the space is bent, but the matter in it that is bent. After this matter leaves the space, the space still remains the same: empty.** Please consider all of this carefully.

The Sun rises from the east, and the sky where the Sun is located in the east has high temperature, strong light, gravity, geodesics, etc.; when the Sun sets in the west, all these high temperatures and so on are transferred to the space in the west with the Sun, and in the space to the east, the things brought by the Sun are gone. The important thing is that this space is still the same as when the Sun didn't pass by.

Space is like a bookshelf, and the bookshelf itself is a kind of boundary of the space, and the space within the boundary can hold books; if a magnet is placed in the space, there will be a gravitational field of the magnet; if a lamp is placed, there will be light; if without stuff, that shelf space is empty.

So, we see how important it is to figure out the right concept of space!

What is space?

We believe that space consists of the following three parts:

- **Ultimate Space**,
- **Subspace** or **Space,**
- **Subspace-Isolation** or **Isolation**.

We can imagine this: Take away all the celestial bodies, air, gas, energy, matter, information, thoughts, imagination..., everything in the universe, and what is left is space!

We have thus named this **Ultimate Space**.

A smaller space isolated from ultimate space according to a certain division we have named **Subspace**. Generally, it refers to the place where something occupies a part of the space. In ultimate space, a smaller space occupied by something is called the "**subspace**" that contains that thing. Generally, we call subspace also a space.

The boundary dividing the subspace from ultimate space we call "**subspace isolation**," or "**isolation**" for short.

Merging all subspaces together is the **ultimate space**.

If the ultimate space is likened to a huge cabinet for storing treasures, the subspace is any of the various grids, and the edge of the grid is the subspace isolation.

Depending on the purpose of our study, the division of subspace can also be changed.

For example, if there is a magnet in the space, then there can be multiple divisions for the subspace related to this magnet: one is to divide the subspace that accommodates its volume according to the volume of the magnet, and the other is to divide the subspace according to the magnetic field of the magnet. The range delineates its magnetic field space. The two subspaces are of course different. If the magnet is moved to another location, the subspace where it was in is still at the same location, but it is no longer the subspace that accommodated the magnet, and there is no magnet field anymore; and the new subspace that is accommodating the magnet and also magnet field, are already at the new location where the magnet is now inside.

The same goes for replacing this magnet with a celestial body. If there is something like a geodesic line due to the existence of this celestial body, then after the celestial body moves to another place, this geodesic line will also go with it. The original space is now empty, and nothing has changed there, certainly no bent of the space, and no geodesic line so that no light will transfer along this no existing geodesic line.

Generally speaking, we are talking about subspaces rather than ultimate spaces. For example, the space between the Earth and the Moon. The ultimate space is the whole of the space, and the subspace is a part of the ultimate space.

Ultimate space contains everything we know and what we don't know, but it's not crammed full. It can accommodate anything. But having this kind of capacity does not mean that you can put anything into the space at any time — it is conditional. For example, in the space occupied by the Earth, another planet cannot occupy the same space.

Can we imagine what the ultimate space looks like?

How big is the ultimate space?

Because the ultimate space is the space after removing all entities, imagination, and thoughts, **its scale is unknowable for humans**.

If there were a Big Bang in the universe, of course this Big Bang also happened in a certain corner of the ultimate space. Just like a nuclear test exploding in the desert, although it is powerful, it is only confined to a certain corner of the Earth. The Big Bang that occurred in the ultimate space, even if there is such an explosion, the space created by this Big Bang is the Big Bang universe that people can see, it just happened in a small corner in the ultimate space. Explosions in space, however large they are, are no different from ordinary explosions that happen on Earth, except the explosion range.

According to the Big Bang theory, it is believed that the boundary of the Big Bang is 13.7 billion light-years, which is the farthest distance the NASA telescope can see. In farther places, humans currently have no ability to see and perceive. But humans should not limit their thinking with such limited abilities.

Since scientists identified the space of the Big Bang several decades ago, the aforesaid 13.7 billion light-years big, with the help of dark energy and dark matter, celestial bodies, such as galaxies and nebulae, are also said to be expanding at a rate that, beyond the boundary of the visible universe, exceeds the speed of light. Thus, we can't see these bodies. However, these invisible celestial bodies that have escaped our observations have not vanished. They exist beyond our capacity to see them; in brief, they exist in the ultimate space.

Therefore, even according to the pseudo-god Big Bang theory favored by modern cosmology, in which the universe was produced by an atom exploding, the ultimate space is boundless. Are the celestial bodies that are more distant than our capacity to see them, having escaped the space created by the Big Bang, still moving at a speed faster than light? Does dark energy still permeate the space there? (Thinking of this, would it make people more suspicious of the Big Bang theory? And is it more suspicious of the so-called cosmic matter calculated by the great scientists?) The question multiplied.

Of course, as we don't know the starting point and ending point of the ultimate space, we cannot determine its starting time and ending time.

We humans, as tiny as we were when compared to celestial objects or cosmic space itself, should never refuse to imagine what we cannot see. Yet, by its given theories, including the Big Bang, the mainstream scientific community tells us that our capacity to understand the ultimate space is limited by the observable radius defined by the telescope, that is, within the scope of the Big Bang. And compared to the real space of the universe, the ultimate space, this observable radius is extremely small.

It is not terrible to be bound by a telescope to a certain range of perception. The terrible thing is to exaggerate this constraint then use it to constrain our thinking about, and understanding of, the universe, whether consciously or unconsciously. If what we know about the universe is constrained to the smoke and dust left by the pseudo-god Big Bang, then what we know and can know is tied to that constraint. The struggling creatures in this small space did not dare and did not want to even think about the wider world outside of the smoke and dust. What the universe may be, the ultimate space we have referred to, remains a mystery.

Wake up, the scientific world poisoned by the smoke and dust of an atom exploding Big Bang! The hypnotized and be hypnotized people!

The boundary dividing the subspace from the ultimate space is called "**subspace-isolation**," or "**isolation**" for short.

"Isolation" can be a certain substance, some physical thing, or an intangible product of thought or imagination, or a combination of the above.

"Isolation" has its own form, too, defined by the thing within it.

When we talk about space, however, be careful not to attach the "isolated" property of the space isolation to the space itself.

Houses are common subspaces separated by the substance of building materials. "Thousands of miles together share the light of moon lady" is an imaginary subspace isolated by moonlight. "Parallel space" uses imagination as "isolation." It has not been confirmed whether it really exists.

Space can hold anything. But if something exists in the subspace, many possible situations can arise. For example, if another thing enters that subspace and they do not conflict with each other in terms of position, then they can coexist. And this holds for any number of things.

If, however, the two things mentioned above cannot coexist in the same subspace, then what can exist there depends on which is more suitable for this subspace. For example, if a fluid encounters a solid entering the subspace occupied by the fluid, the solid will partially displace the fluid, which will overflow the subspace.

On the other hand, if the fluid meets a solid, the fluid will move around the solid according to the movement of the fluid.

For example, when starlight from one star encounters another star, where the starlight intersects with the star, the part of the starlight that intersects the star will be blocked and absorbed by the star, and the other part will flow around the star in a curved manner. Behind the reefs and beacons in the river, the flowing water can be seen moving around stones or buoys. This is the instinct of fluid motion and does not require any gravitational field or bending space to guide the change.

Subspaces can overlap. This can be a physical overlap, such as a small chamber in a big house or a Russian doll set within another set; it can be an imagined as well, such as the temple in an overarching palace; and it can also be physical and imagined, such as "white hair is three thousand feet."

All **subspaces are connected**. There is no material that can separate one subspace from another. The **entirety of subspaces constitutes ultimate space.**

So, when we talk about space in abstract terms, it is ultimate space, or general space. When we don't specifically state otherwise, we use "ultimate space" and "space" interchangeably.

1. The subspaces are part of ultimate space; the entirety of subspaces constitutes ultimate space.
2. Subspaces can be subspaces of other subspaces, and there can be multiple layers of subspaces.
3. Part of the space that does not contain anything can also be defined as subspace, according to our needs. There is nothing in this space, which

is divided by human thought with the help of things that exist around this empty space. For example, there is an empty subspace ES inside the solar system. There is no celestial body in this space ES, and it still belongs to the solar system.

4. The subspace is the same as ultimate space and has no properties. The properties exhibited by the subspace are those possessed by the objects in which it exists.
5. Subspace is a space inside the ultimate space that is divided into various shapes and sizes by human beings according to their own needs and conditions. To meet human needs, either naturally or artificially, subspaces have various forms, just like we separate the space inside a building into various subspaces as large as an auditorium, as small as a box, and even as small as the interior of an atom.

Generally speaking, a subspace is divided according to what it contains. The content and geometry of the thing determine the size and shape of the subspace.

The **Big Bang subspace** was generated based on the Big Bang theory. Whether it exists or not needs further study. In the description of space, temporarily think it is there. If it does not exist, it does not affect our spatial model.

The **Universe subspace** is the largest subspace that we can perceive, and it accommodates everything within the greatest possible perceptual range we possess: known and unknown celestial bodies, matter, objects, and their movements, etc. Cosmic subspace encompasses all other subfamilies within its space range. All celestial bodies that we can perceive, every development of everything, are all in universe space.

According to the current meaning, cosmic subspace is the subspace of the Big Bang space, and the Big Bang subspace is the subspace of ultimate space. Because it will not cause misunderstanding, we will refer to cosmic subspace as cosmic space or universe space and the Big Bang subspace as Big Bang space.

If we follow the Big Bang theory, we can only conclude that Big Bang space is larger than cosmic space. This is because the space created by the Big Bang has been expanding beyond the observable limits particular to humans. Many years ago, we defined where the edge of the universe is, and noted its expansion rate as equal to the speed of light. Beyond the edge of the universe, where our ability to observe vanishes, the scope enlarges; it is beyond universe space, but we cannot observe it anymore. However, it still belongs to the Big Bang space. It's like a nuclear explosion in the desert. Assuming that the maximum distance that we can observe the dust of a nuclear explosion is ten kilometers, when the range expands beyond this, such as at 12 kilometers, the explosion space between 10 kilometers and 12 kilometers is outside observable space but still within explosion space.

The **cosmic subspace** is determined by the celestial bodies that humans can observe (or perceive). If the universe is defined according to other definitions, the 13.7-billion-light-year space range will change. For example, if we define the cosmic universe as being within the maximum range that known celestial light can possibly reach, then according to the new definition, the range of the new universe will be greatly expanded.

Figure 8.1 presents a pertinent example.

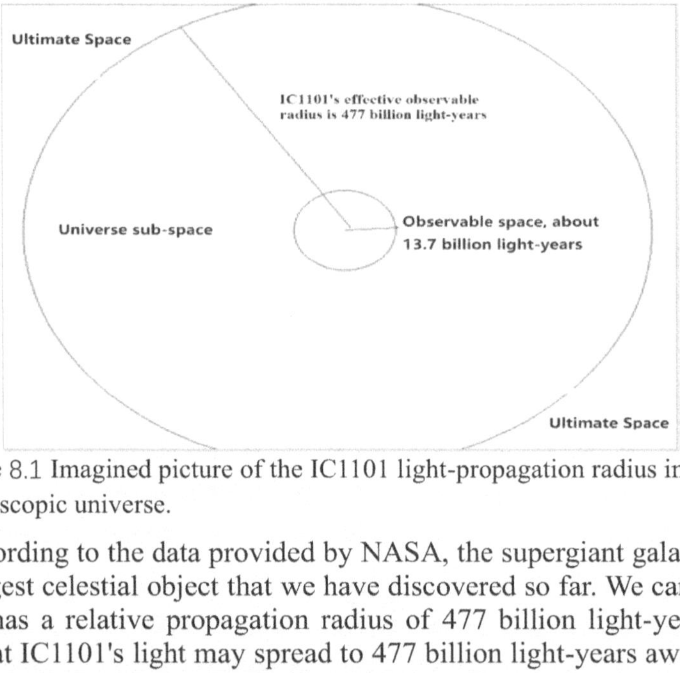

Figure 8.1 Imagined picture of the IC1101 light-propagation radius in the macroscopic universe.

According to the data provided by NASA, the supergiant galaxy IC1101 is the largest celestial object that we have discovered so far. We can calculate its light has a relative propagation radius of 477 billion light-years, which means that IC1101's light may spread to 477 billion light-years away. We use this as the maximum distance that can be inferred. Although we can't observe such distance, we have reason to infer it. Based on that inference, we use 477 billion light-years as the scope of the universe. This range is much bigger than our current range of perception: 13.7 billion light-years. It is many times the scope of the universe NASA defined that human can perceive.

But according to the above definition of universe space, Big Bang space has become a subspace of the newly defined universe space.

Cosmic space is all connected subspaces. The movements of objects within a local subspace cannot be freely extended beyond it. This is not a limitation of this space, but the limitation of the object itself. All movements and the promotions of movement need to consume energy.

Any feature other than the "emptiness" exhibited by the space

belongs to the things it contains, and the space has no other feature except being "empty." People often mistake the events or material properties contained in the space as the space itself. Attributes such as the aforementioned length, width, and height are measurement attributes that describe the substances that exist in space, and they have nothing to do with space itself. Space has no shape, and space does not require measurement.

The **space has no shape and cannot be changed**. It's like a box with no cover, no bottom, no borders. Anything explodes inside, the thing that explodes has nothing to do with the box. What changes are the isolation boundary conditions of the subspace's morphology, volume, composition, etc., not the space itself.

Space is eternal; it does **not increase or decrease**. The volume of a subspace has changed not because the space itself has changed, but because the "isolation" that defines the subspace has expanded or contracted.

Of course, after the isolation of the subspace changes, the shape of the subspace changes accordingly.

Space cannot be curved, and no way to bend!

Can space be curved?

There is nothing in the no boundless ultimate space. If it can be bent, how can it be bent? How much can it bend? In what direction? Is it one bend or multiple bends? How then can celestial bodies enter a curved space?

If there is no way to answer these questions, how can it be determined that the ultimate space can be bent?

Therefore, the ultimate space cannot be bent.

Then, all subspaces are part of the ultimate space, and of course they also can't be bent.

Imagine a subspace containing only starlight. How would it bend? Starlight comes and travels from and to all directions, so does this subspace bend in all directions?

As shown in the left image of Figure 8.2, whether the Earth is in the space shown at point A or B, or at any other location, the space of the Earth's orbit around the Sun is always there; points A and B and all others are always there. The space will not move, it will not disappear, it will not def.

As shown in Figure 8.2, suppose we define the orbit of the Earth as the "orbit space" of the Earth's movement. This orbital space is roughly an elliptical cylinder surrounding the sun. This is actually only drawn by our imagination based on the trajectory of the Earth after it passes. In reality, the rest of the trajectory is indistinguishable from the surrounding space except the space where the Earth is. Now let's take two locations in the orbital space that are far apart from each other, indicated by points A and B in Figure 8.2.

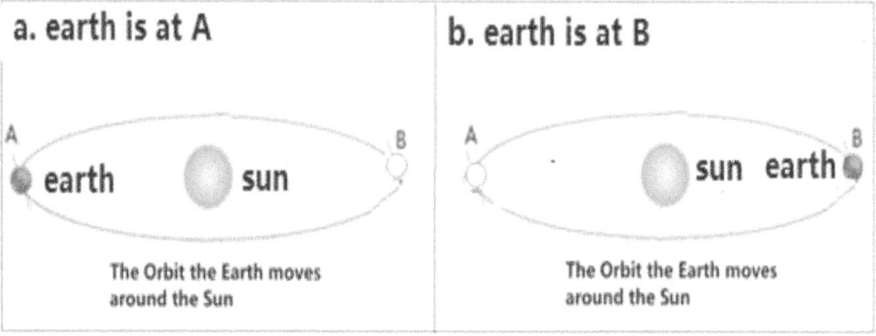

Figure 8.2. Earth's orbital space around the sun. By definition, the Earth passes through points in orbital space in turn. Left: When the Earth is located at Point A of the Earth's orbital space. Right: When the Earth is located at Point B of the Earth's orbital space.

What we want to examine here is the different states of the spaces when the Earth is located at two points as far apart as possible in this orbital space and then the changed state of spaces indicated by the two points when the Earth moves from one point to the other.

As shown in the left picture of Figure 8.2, whether the Earth is in the space shown at point A or B, or at any other location, the space of the Earth's orbit around the Sun is always there; points A and B and all others are always there. The space will not move, it will not disappear or deform.

When the Earth is at position A, the space shown at position B is an empty space, and the Earth has not yet reached position B. At this time, the space shown at position A of the Earth contains the Earth, the gravitational force of the Earth, the gravitational field generated by the Earth, which holds people on the Earth, and so on. And the space shown at position B, because the Earth is not in it, is empty, so there is no Earth gravity, and people here will not be affected by gravity and are in a state of weightlessness

Now the Earth moves to the space position B, as shown in the right image of the Figure. When the Earth is at position B, the state of space represented by points A and B is completely reversed. Position B has the original state of position A, where people will be limited by gravity on the Earth, and position A has nothing, and people will be in a weightless state there. This fully shows that the gravitational force exhibited by space is an attribute of the Earth that the space contains, not an attribute of the space in which the Earth is located.

For all celestial bodies, at any position where the celestial body is located, such as the M point, the space near the point will generate the properties of the celestial body due to its presence, but after the celestial body leaves the M point and moves to other positions, this M position has no celestial body in it.

This results in the loss of the gravitational field of the celestial body. **Gravity and other properties follow the celestial body, not space** it stays inside. The gravitation of bending still belongs to the celestial body, not to the space where it stays.

When the Sun moves to a certain space, the matter in that space burns and melts, and no other place in its orbital space emits heat waves. One cannot say that the space where the Sun is housed generates heat waves, because the heat waves move with the sun, but space does not move with the sun, and solar heat waves do not follow space. When the Sun moves to aother space, it emits heat waves in that space, but it no longer emits heat waves in its old space. This fully shows that heat waves are emitted by the Sun, not by the space where the Sun is. Heat waves have nothing to do with space. Here, the same conclusion can be reached by replacing the heat wave with the gravitational wave generated by the sun!

This shows that the space is empty, and any measurable property, such as gravity, is brought by the objects contained in the space. After an object enters a subspace, the subspace exhibits the properties of the object; when the object leaves this subspace, the subspace no longer exhibits such properties.

It cannot be said that lightning emanating from dark clouds was sent by the space hosting the cloud; it cannot be said that singing in the air was caused by the space. If an object in space is affected by gravity, it means that the thing existing in the space is affected by gravity, which has nothing to do with the space in which the object exists.

Examine a certain space above the track of a high-speed railway. When a train passes through this space, it will bring up a strong wind and form some kind of air flow. One cannot say that this space is flowing because the space remains there after the train has passed.

When a car hits a person, it can't be said that the space that houses the car hits a person. After the car leaves, that space will lose the ability to impact the person.

The reasoning can be reversed: if the gravitational force that exists in space can be regarded as the attribute of space, then the light that exists in space can be regarded in the same manner as can the color of space and magical dark energy and dark matter … one can draw the attributes into space in the same way; according to this logic, the attributes of everything in the space can become the attributes of space.

If so, the attributes of this space become meaningless: having all attributes is equivalent to having no attributes. If you can change one attribute freely, you can change all attributes; if you can add one attribute at will, you can add countless attributes. Therefore, the space will have a variety of countless attributes belonging to different subspaces, and the attributes of each subspace will be different. There is no way to describe different substances in space and their interactions. Such a definition of so-called spatial attributes is meaningless.

Therefore, it is wrong to give to space any property possessed by the things it contains. This is basic common sense.

Some people will argue that the space in which the Sun is located curves; when the Sun leaves that space, that space is restored. If so, what does the curved space rely on to recover after the Sun leaves? Where does the energy it consumes come from? Can a small piece of space separate from the Ultimate Space? What theory supports this argument?

By the way, here are some other questions: What does distorted space look like? What does the space look like without distortion?

The imposition of attributes that belong to celestial bodies in space, which become the attributes of space, highlights the inadequacies of Einstein's curved space; that is, the basic reason why general relativity cannot be established.

Some experts, however, may pose arguments against our position by using Einstein's field equation. May we ask, has anyone proved that Einstein's equation correctly describes the universe? Can the universe be described by only one equation? We will discuss this issue in more detail later in this chapter.

Why did the soft fluttering balloon bend the sunlight?

It is true that the starlight bends as it passes near the Sun. But it is wrong to say that the bending was caused by the mass or gravity of the Sun.

Let's explore the real reason why starlight bends as it passes near the Sun by investigating sunlight blocked by a balloon in Figure 8.3.

This problem can be rephrased: if the Sun is replaced with a cardboard ball of the same size cardboard-Sun, will the starlight be blocked by this opaque and almost weightless cardboard-Sun when the starlight passes near the cardboard-Sun? Or will the starlight be bent?

On a sunny day, there is a big balloon overhead that covers the Sun. After the Sun shines on the balloon, the bending of the sunlight is, of course, not caused by the gravity of the balloon, because the gravity or mass of the balloon can be ignored in this case. Actually, this balloon blocks the sunlight, so that the curved sunlight leaves a shadow on the ground, and it can only be said that this has nothing to do with the gravity or mass of the balloon!

If we gradually increase the balloon size in Figure 8.3, it will eventually be larger than the moon. In this process, the balloon will always block and bend the sunlight and form a conical dark space behind itself.

Figure 8.3 A balloon with negligible gravity or mass. How does the balloon block and deflect sunlight passing nearby it, leaving irregular shadows?

If we slowly fill the expanded balloon with material, its weight will slowly increase and eventually exceed the weight of the moon. At this point, the balloon, which has both gravity and mass, will still block and deflect sunlight just as it would without mass and weight.

Simple Mathematical Proof that Space Cannot be Curved

Simply put, mathematically, if we define the empty set SU = { } as the Ultimate Space set, SU can accommodate all things and subspaces, all beings, and everything that does not conflict with something else.

An arbitrary subspace S of SU is a bounded set S = $\{x_i \mid x_k$ and x_j do not conflict with each other$\}$. The position of S in space is fixed. It contains a certain or limited number of things, concepts, subspaces, etc. that can coexist without conflict.

The boundary of S is defined by the members it contains.

S itself has no attributes. The attributes represented by S are attributes of the members contained in S. When a member contained in S is transferred or detached from S, the attributes brought into the subspace by the member are transferred or disappear from the subspace because, if any attribute of member x in S is said to be an attribute of S, then, since S can accommodate everything in the universe, S will have all attributes. This does not correspond to any subspace we are familiar with.

For example, if m1 = TV, m2 = plough, m3 = cow…, they all belong to the space set S1 = farm-household, that is, S1 = {m1, m2, m3, ...}. When all these members are working, we can't say that the S1 collection itself will show TV shows, will plow the fields, and will produce milk… Otherwise, if space S1 = delivery room, can we say that S1 will produce children?

Of course, people have the freedom to treat the attributes of the members contained in the space as attributes of the space and use them to define the space and its attributes. But what is the practical significance of such a definition besides confusing people's understanding of space? What we need is a theorem that can reasonably portray natural facts, not just mathematically beautiful things.

The imposition of attributes that belong to celestial bodies in space, which become the attributes of space, highlights the inadequacies of Mr. Einstein's curved space, that is, the basic reason why general relativity cannot be established.

Some people will point to Mr. Einstein's equation of the universe, the field equation. May I ask, has anyone proved that does Mr. Einstein's equation of the universe correctly describe the universe? Can the universe be described by only one equation? We will discuss this issue in more detail later in this chapter.

Explain from Multiple Angles Why Light Does Bend Without Gravity as It Passes Near a Star

Sometimes we really don't understand the profound ideas of great scientists.

For example, there is the gravitational cone.

What kind of opaque object does not produce a cone-shaped shadow when it blocks the sunlight? But between celestial bodies, why does a normal shadow have to rely on the gravitation of the celestial body to form?

There are several methods that can be used to explain why the light from a celestial body bends after passing near a massive other celestial body. And each explanation is more reasonable and scientific than the Space-Time curvature that Einstein worked hard to figure out. Here are two of them.

First, let's compare it with the pictures of light diffraction in any physics book (Figure 8.4).

The light rays irradiated in parallel will gradually diffuse into the shadow space of the backlight of the obstacle after encountering the obstacle.

Here, the reason why light waves bend after encountering obstacles has nothing to do with gravity. It is a basic common sense in classical physics theory.

When we wrote this, we wondered: This diffraction theorem has appeared in junior high school level textbooks. It is used to explain how opaque objects block light waves emitted by the Sun or other light sources. As the light waves cannot pass through the opaque object, behind the object a dark region without light waves will form. The light waves passing around the edge of the obstacle, however, will gradually spread into this dark area. Therefore, behind any obstacle, a **Conic Dark Space** will be formed because the obstacle itself blocks the light wave. However, due to the diffractive nature of light, this cone-shaped dark space will gradually be eroded by the surrounding light waves that are not blocked by the obstacle, making this conical dark space gradually diminish and eventually disappear.

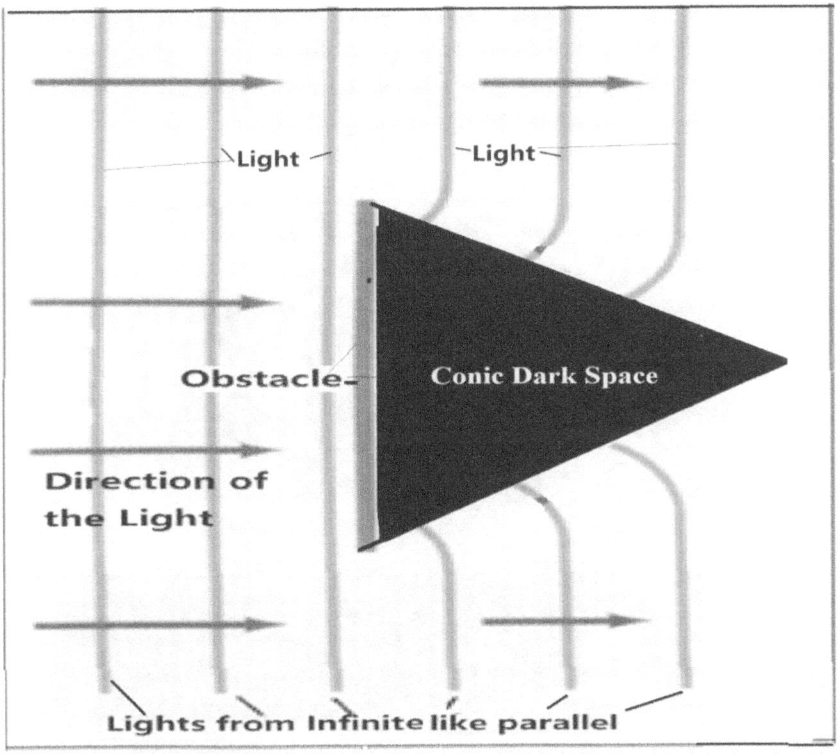

Figure 8.4 **Diffraction** diagram of light. Copied from physics textbook. The upper right arrow shows the course of light deflection under the effect of diffraction, form a **Conic Dark Space**. It has nothing to do with the gravitational field.

This is basic common sense. Why are both Einstein and contemporary physicists turning a blind eye to it? In the space between the light source and the observer, whether it is a large-mass celestial body, or a cardboard ball without weight, or any opaque object, if they are big enough, there will be dark shadow space where light waves gradually shrink inward. This dark cone-shaped space is obviously not caused by gravity.

Yes, Einstein turned a blind eye to this common phenomenon. When discussing this shaded space, he first said that it was bending light solely because of gravity, and later attributed it to the curvature of space-time near a large-mass star. The obvious fact is not considered at all; that is, the diffractive deflection of light between a celestial body and an observer, without considering its mass and gravity, but as a non-gravitational effect produced by a pure obstacle.

If the obstacle was bigger and heavier, then would it not block the light and change its course?

The light wave of any star is too large for the Sun, so large that the ratio of the stellar light wave to the Sun is smaller than the ratio of a man to a haze-particle. Can a particle smaller than the haze affect the light waves?

Taking the light wave emitted by the nearest star Centaur Proximal to the Sun as an example, let us examine it from far to near, from whole to partial, from the whole picture to the details. (Figure 8.5)

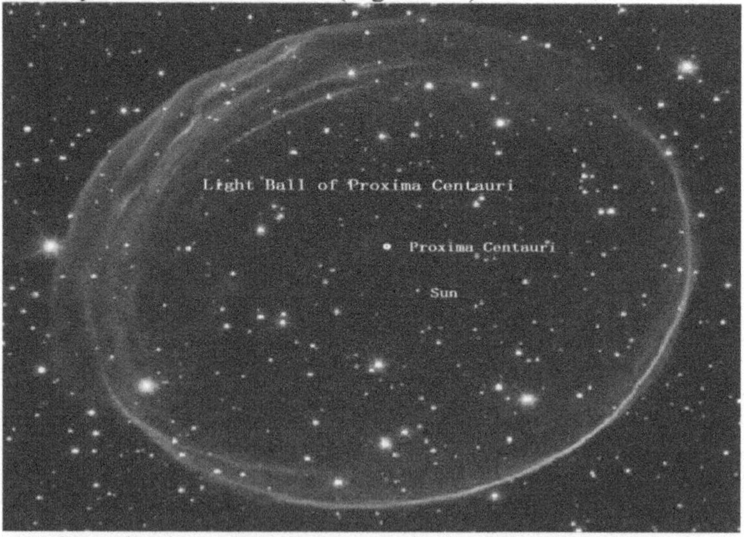

Figure 8.5 A schematic view of the light wave ball emitted from the Centaurus star. Relatively, the Sun is only a tiny dust in this ball. The figure is a completely out-of-proportion drawing. In fact, it cannot be described because relatively the Sun is so small that to be completely invisible.

Figure 8.5 is a macro view of the light wave sphere sent by the neighboring star of Centaurus towards the Sun.The Sun is 4.2 light years from Centaurus. The diameter of the Sun is about 1.39 million kilometers. Therefore, for the Sun, the radius of the sphere of light emitted by Centaurus to the Sun is 4.2 light-years.

As shown by the small circle in Figure 8.5, Let's zoom in on this small circle, as shown in Figure 8.6 below.

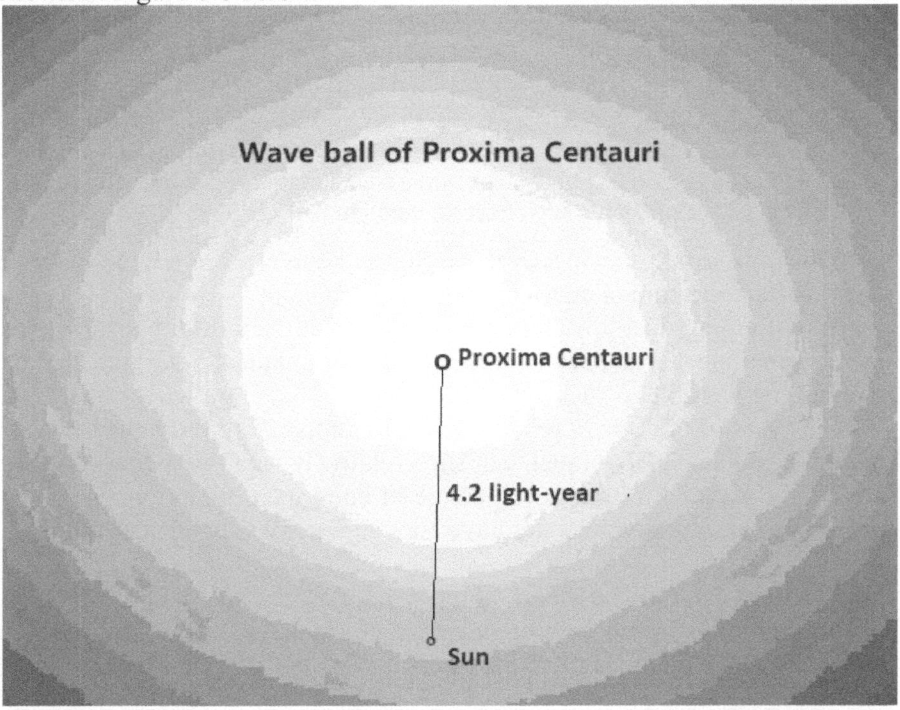

Figure 8.6 The figure, enlarged from the small circle, relative to the Sun, the light wave of the Centaurus is full of space.

The diameter of the Sun, relative to the radius of the light ball from the Centaur emitted to the Sun, is 1: 6.8 million.

How big is this ratio? Assuming the height of a person is 1.7 meters, haze particles are 2.5 microns = 0.0000025 meters, so haze particles and height ratio is 1: 0.68 million. Please carefully compare the above two figures. That is, if we increase the diameter of the Sun by a factor of 10, the ratio of the diameter of the Sun to the radius of the light ball from Centaur, corresponds to the ratio of an adult to a haze particle!

Then enlarge the Sun and the surrounding light from Centaur is like shown in Figure 8.7. We can imagine the Sun in the picture as a dust, what ability does it have to bend space? Although the figure is actually far from depicting the Sun's tiny real scale.

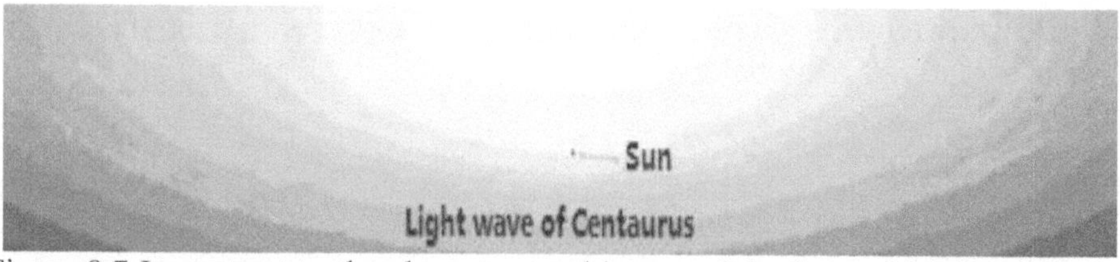

Figure 8.7 In contrast to the above zoomed image of the Sun, which is relatively smaller than a haze. The light waves of the Centaur almost become a plane over the Sun. The Sun cannot have any effect on the light ball.

So, from a ratio perspective, the Sun is equivalent to a small haze, how does it affect the equivalent of ten adults of the giant?

One floating dust ups and downs, can affect the world that surrounds it? We can imagine that an ant trying to bend the mountain winds of the Himalayas, would it be possible to succeed?

The size of the Sun is relatively small compared to the light waves from the nearest galaxy. How much can such relatively small particles bend such a huge light wave? From the perspective of energy consumption, such a small proportion of particles does not have enough energy to bend such a huge light wave.

Confetti can only be messy in the wind, and cannot change the trajectory of the wind; people can only be swept away in the tide of the Qiantang River, and cannot change the pace of the tide.

Figure 8.8 Thin legs have no effect on giant waves

When star rays were transmitted to be received by the telescope, how many celestial bodies' gravitational fields were passed by them?

Einstein's Cosmic Field Equation that Needs to Be Re-evaluated and Justified

It is sad to rely solely on mathematical models to describe space without considering the reality of the universe. But this is the foundation and core of the modern universe "science" view.

When Mr. Einstein wrote the field equations of the universe that he could not solve using Riemann geometry he just learned, the very complicated and scary thing named the "cosmic" equation ruled the theory of the universe because of his reputation. Because of Einstein's fame, no one has heard anyone question this cosmic equation. No one asked to explain why this equation could be an equation describing the universe!

Mr. Einstein specialized in Riemann geometry, and then wrote the following field equations and called them "universe equations" (if you don't like mathematics, you can skip the formula portion, and can continue reading without affecting understanding. To be honest, I don't understand it very well, I don't even want to look at it carefully. But even Mr. Einstein himself didn't understand it very well and didn't find a solution to this partial differential equation. So, it doesn't matter that we don't understand it well, right?)

Mr. Einstein's field equation is the most important equation in Mr. Einstein's general theory of relativity. It is also the most important foundation of Mr. Einstein's theoretical system. Mr. Einstein's gravitational field equation is a slightly modified very complex second-order partial differential equation modified from Riemann geometry. A specific form of it is as follows:

$$G_{\mu\nu}+\Lambda g_{\mu\nu}=\frac{8\pi G}{c^4}T_{\mu\nu}$$

This second-order partial differential equation is very complicated, and Mr. Einstein himself did not know how to solve it. Since Mr. Einstein proposed this weird and complicated field equation in 1915, the development of general relativity has concentrated on solving this equation. The physical interpretation of the solution and the search for possible experiments and observations also account a big part for it. But because the field equation is a nonlinear partial differential equation, it is difficult to get a solution, so before the computer began to be applied to science, only a few results were solved. However, the earliest of these solutions is one year after the equation was invented. The most famous of these are three solutions: Schwarzschild's solution, Ressler-Nordstrum solution and Kerr solution. But no Einstein solution.

In Mr. Einstein's equations, when the gravitational constant G is zero, the field equation basically fails. However, after the light passing by this zero

G flat and no space-time bending cardboard ball or obstacle, the shadow cone behind the obstacle will still appear, although the deflect angle θ was calculated as zero. But such shadows and similar things were once attributed to be caused by the gravitational field.

So when the light of the star in Figure 8.9 passes near the sun, it is not bent by the sun 's gravity, but by the mass of the sun that distorts the space and time near it, and causes the star light to travel along the curved space.

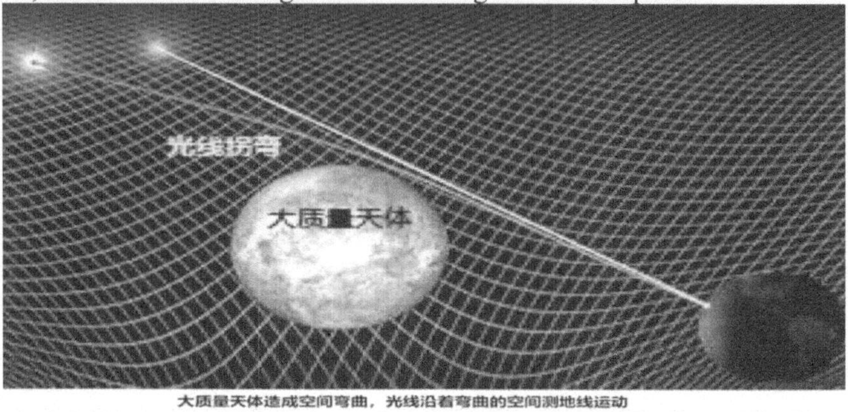

Figure 8. 9 Mr. Einstein's point of view: The space near a massive celestial body is curved, and light will move along the curved space geodesic when it passes nearby the curved space

After Mr. Einstein's interpretation of the curved space came out, various similar explanations and images came out. For example, the following image below in Figure 8.10 is just one of thousands of similar pictures:

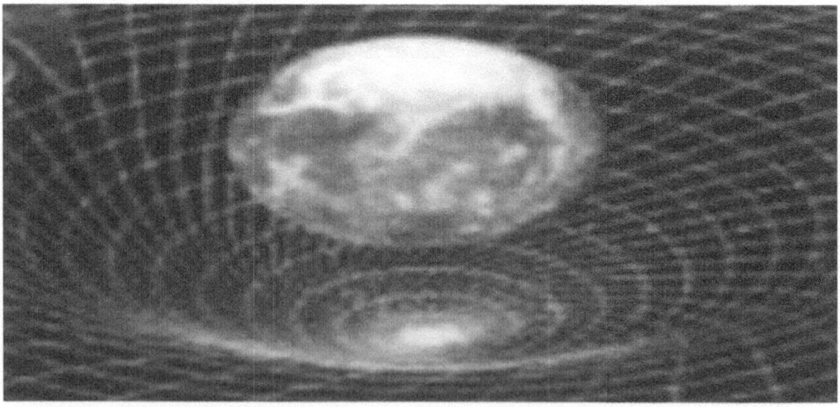

Figure 8.10 Typical image of Mr. Einstein's curved space

The biggest problem caused by such numerous and similar images is that people completely forget the original problem, and their thoughts completely enter into the category of Mr. Einstein's thought, considering the macro images of the universe, celestial bodies and space. the relationship between light and space, how light moves in such a mass and curved space, etc....

But why did Mr. Einstein propose this curved spacetime? What problem was he trying to solve?

Please cover up your reading and review the question.

Mr. Einstein's Twisted Curved Spacetime

The above briefly reviewed the development process of Mr. Einstein's space-time field. In simple terms:

- Mr. Einstein described the phenomenon of celestial bodies blocking stars (like the moon blocking sunlight) as "only gravitation of a celestial body bends the starlight." So, delve into how the gravitational field or gravity of a celestial body makes starlight move along a curve.
- Mr. Einstein first predicted that when the starlight passes by a massive celestial body, it will be bent by the celestial body;
- Through his observations of the solar eclipse on Earth, Eddington found that at that time, many stars, their positions were behind the sun and blocked by the sun, but in the eclipse, these stars that could not be seen on the earth now could be seen, so he announced to the world: because light has weight, so the light of the stars passed near the Sun was bent, and was attracted by the gravitational pull of the Sun. The starlight, which was originally blocked by the Sun and was not visible, was observed by him. He believed that this showed that Mr. Einstein's prediction was correct.
- But Mr. Einstein later discovered that his theory that the light was bent by the celestial gravitational field conflicted with his own special relativity theory. He needs to find a way to resolve this conflict theoretically. So thought out another round way: the theory of curved time and space was born!

Think of a big basketball that gradually increases weight and volume! Hold it up to face a light source and see what happens.

Link to Figure 8.11 below, we can see some more important things.

Looking at the equation used to calculate the angle θ, we can find that θ is fully dependent on the gravitational G and mass M of the Sun. But this is doubtful. Imaging we gradually take out the mass from inside the Sun, until it is like a balloon ball sun, without or with very little of mass and gravity. The light will still be bent by this balloon ball sun, like that in Figure 8.3.

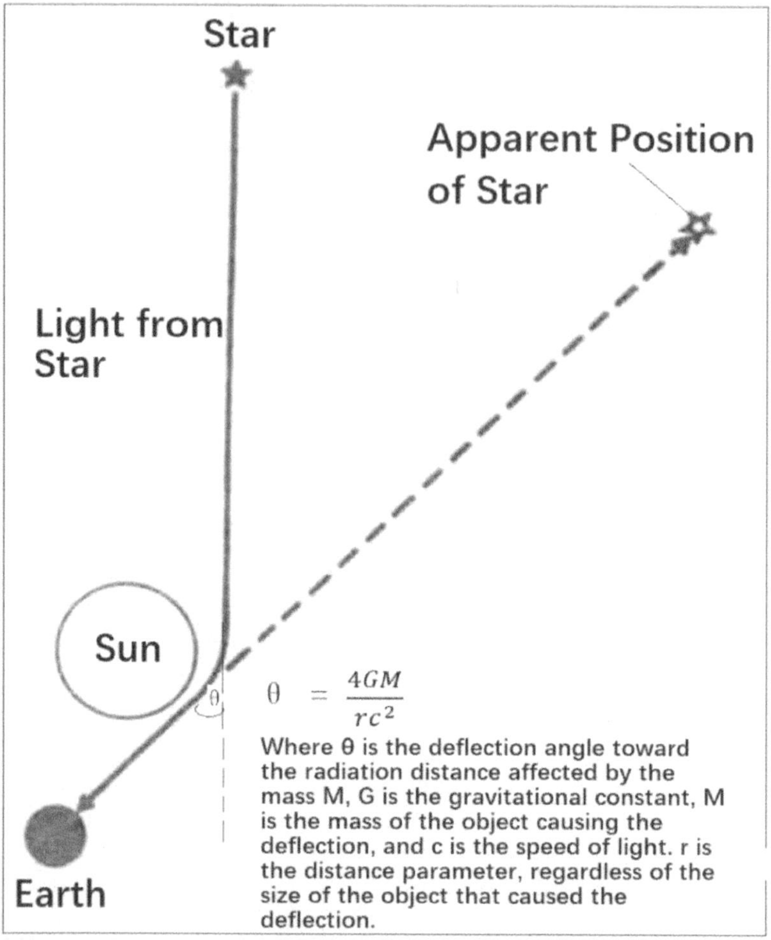

Figure 8.11 Calculation of the angle θ shifted by light ray in gravitational field

It can also be considered that the farther away from the surface of the celestial body, the smaller the gravitational force, and the smaller the gravitational force can only attract less light, which is exactly the opposite of the shape of the gravitational cone defined by Mr. Einstein. As shown in Figure 8.12, if gravity moves the surrounding light to a dark area, the closer it is to the celestial body, the greater its gravitational force. The gravitational force on the surface of the celestial body is the largest, and the amount of light attracted by gravity should be the most. If the gravitational cone is caused by gravity, then the earth's surface should have the strongest light, as shown in the picture below on the left. The actual Moon's shadow appears to an observer on Earth as shown below on the right.

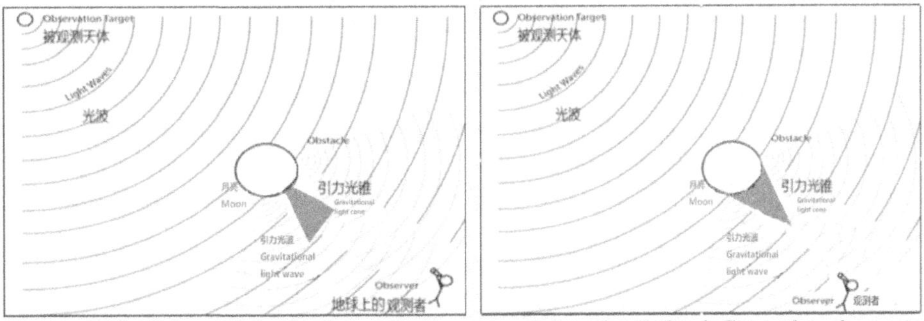

Figure 8.12 If the gravity is the main reason causing the light deflect, then because gravity gradually decays with increasing distance from the moon's surface, the closer it is to the moon's surface, the greater the gravity and the lighter it should be able to capture. As the distance from the moon's surface increases, the gravitational force decays and the captured light decreases. Of course, the inverted shadow cone in this figure is ridiculous, but if you follow the principle of attracting more light with greater gravity, it should look like this strange look. It can be seen that gravity makes space and time bend or light bends are weird words like this picture.

We can also use light pressure to explain the phenomenon of light deflection when encountering obstacles.

When a shadow appears on the back side of the moon the light pressure in the shadow space will be lighter than outside the shadow. As a result, the light gradually erodes from the place where the light has higher pressure and enters the space where the light-pressure is lower, molding the shadow on the back into a cone-shaped space.

Similar physical phenomena are often seen in the currents of daily life. Behind the midstream stone in the rapids, a small dry space often forms. Sometimes this phenomenon can be seen even behind the buoys in the rapids on the river.

It can also be explained from other angles, which I explore in my other books. But we find it boring to continue talking about it here. It is an obvious thing. Nonetheless, Einstein came up with all kinds of tricks to solve the contradictions inherent in his theory of relativity, which scientists the world over have studied to repletion.

Math is Not Reality

The connotations derived from mathematics and real physics are not the same. Roughly speaking, we can regard the dimension of mathematical space as containing the number of variables in mathematical formulas. The strange thing is that people are hardly willing to face up, and rarely discuss the differences involved. Moreover, scientists these days – and of course mathematicians — will commonly take mathematics as more real than reality. The most obvious example is the mythical mathematical game in which Einstein used a ray, a rigid rod, and a lot of dazzling definitions, concepts and formulas to build up his special relativity!

This is the evil result of the aging of mathematics made by human beings for hundreds of years!

In 1908, the German mathematician Minkowski published the paper "Space and Time (Figure 8.13)," which combined the space described by three-dimensional (X, Y, Z) vectors with the time described by one-dimensional (T) to form a four-dimensional (X, Y, Z, T) mathematical description of time and space. Later, Einstein introduced four-dimensional space-time into his theory of relativity in his works, such as "Cosmological Considerations about Relativity," published in 1917.

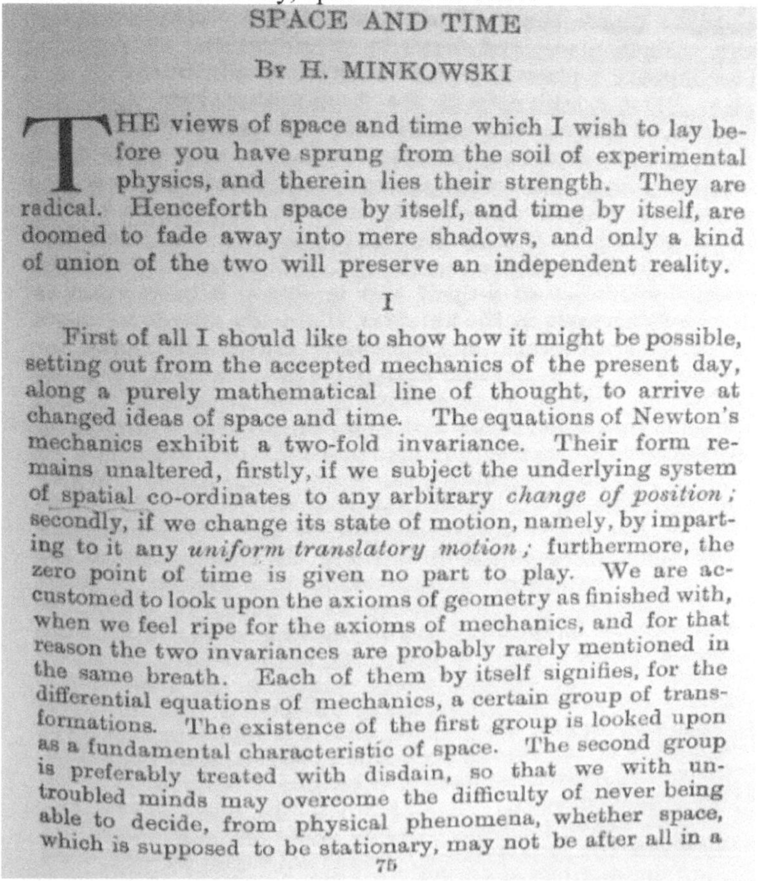

SPACE AND TIME

By H. MINKOWSKI

THE views of space and time which I wish to lay before you have sprung from the soil of experimental physics, and therein lies their strength. They are radical. Henceforth space by itself, and time by itself, are doomed to fade away into mere shadows, and only a kind of union of the two will preserve an independent reality.

I

First of all I should like to show how it might be possible, setting out from the accepted mechanics of the present day, along a purely mathematical line of thought, to arrive at changed ideas of space and time. The equations of Newton's mechanics exhibit a two-fold invariance. Their form remains unaltered, firstly, if we subject the underlying system of spatial co-ordinates to any arbitrary *change of position*; secondly, if we change its state of motion, namely, by imparting to it any *uniform translatory motion*; furthermore, the zero point of time is given no part to play. We are accustomed to look upon the axioms of geometry as finished with, when we feel ripe for the axioms of mechanics, and for that reason the two invariances are probably rarely mentioned in the same breath. Each of them by itself signifies, for the differential equations of mechanics, a certain group of transformations. The existence of the first group is looked upon as a fundamental characteristic of space. The second group is preferably treated with disdain, so that we with untroubled minds may overcome the difficulty of never being able to decide, from physical phenomena, whether space, which is supposed to be stationary, may not be after all in a

75

Figure 8.13. Minkowski is "along a purely accepted mathematical line."

However, from the beginning, they did not distinguish the concept of the "space, time" of mathematics from that of the physical world. To this day, there has not been any in-depth discussion on the issues of "space and time of matter" and "space and time of mathematics." Searching all their related works, we found that before they discussed the space and time of mathematics, they did not have a definite or comprehensive and correct understanding of real-world physical or living space and time. In other words, the space and time of Einstein and Makowski are pure mathematical models or universes constructed mathematically, and completely different from physical or living space and time. Because this situation continues, we will discuss the difference between the living space and time of matter and the space and time of mathematics to gain a deeper understanding of general relativity.

From the first two sentences, we can see that MINKOWSKI did not consider material space and time at all, but directly used mathematics to construct his four-dimensional space-time model.

These two sentences follow:

"Space and Time

The views of space and time which I wish to lay before you have sprung from the soil of experimental physics, and therein lies their strength. They are radical. Henceforth space by itself, and time by itself. are doomed to fade away into mere shadows, and only a kind of union of the two will preserve an independent reality."

In fact, because the space and time of mathematics are not equal to those in physical reality, it is not appropriate to use pure mathematics to explain and reflect the living space and time in physical reality. The concept of mathematical space and time is clearly defined, while the concept of living space and time in the philosophy of reality includes a variety of simple, obscure and inaccurate understandings. On this basis, directly and crudely applying the purely mathematically defined space-time concepts to the vague and narrowly defined living space-time concepts of matter, of course, various conflicts and contradictions will arise.

In order to avoid using "physical space" – in some situations an unclear and easily misunderstood phrase — "living space" is used here. This is easier to understand. And the only space that can live is three-dimensional space.

Mathematical space concept

The concept of space in mathematics is very complex and all-encompassing.

Complicated mathematical space can be abstracted to the extreme. For example, various math functions can form the function space; algebra, geometry, trigonometry, etc. can form the mathematical discipline space;

industry, agriculture, economics, etc. can form the different categories……

A simple mathematical space may be regarded as a linear space. In this space, a variable is one dimension, and if there are n variables, the space is called a n-dimensional space.

In general, mathematical space consists of selected mathematical objects that are regarded as points and selected relationships among these points. A point can be any simple or abstract object, ranging from a "point" without shape in geometry to as large as galaxies, countries, disciplines, math functions, and so on. The relationship between points is the main object of mathematical space research.

Living space concept in physical reality

Human living space, which is consistent with physical reality, is limited by the material world and the ability of human beings to understand and transform the world.

Usually, people can only intuitively feel the three-dimensional living space where they are. When people want to visually study images in four or more-dimensional space, until now at least, there is no way to do it.

Because of the ubiquitous influence of mathematics on modern people, people often confuse mathematical space with the living space.

For example, people often use the concepts in Figure 8.14 below:

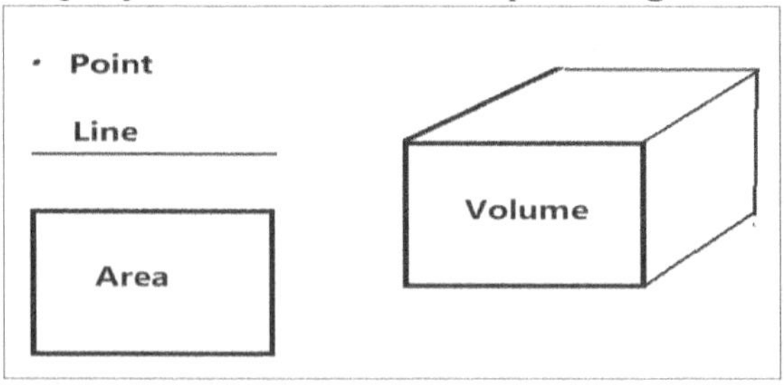

Figure 8.14 Basic elements of Euclid geometry. Humans has no way to draw 4 or higher dimensional figures.

This is the most used concept learned in elementary school mathematics. It is drawn according to the Cartesian coordinate system.

But even with such a simple thing, when thinking about it carefully confusions can arise.

Of course, the volume represents a three-dimensional space, which can be represented by three coordinates (x, y, z). The surface is a two-dimensional space and the line is one-dimensional. Then the confusion comes:

What is the dimension of a point without dimension?

Countless lines in one dimension are stacked to form a two-dimensional surface, and countless surfaces in two dimensions are stacked to form a volume. Then, if countless points of zero dimension are piled up into a line, will the line be born out of zero or nothing? In addition, a point in the body often needs three numbers together in the coordinate system to represent it. At this time, how many dimensions is this point?

People often use rectangular coordinate systems to represent geometric images in mathematics, so they always think that the two-dimensional plane is like the one in Figure 8.14 but forget that it is only a special case of mathematical representation. For example, the surface of the human body is also a surface. It is more troublesome to use mathematics to describe this aspect.

How many dimensions is the curve that is composed of points? How many dimensions is the surface? ...

It seems that things are getting more and more complicated.

Therefore, it is better not to involve mathematical space when we are talking about physical living space.

Living space is relatively simple, but it also annoys mathematicians when studying complicated situations.

Simply put, the living space should be a space that people can recognize, can intuitively see and feel. It has only three dimensions at most. Complexity means that even objects in three-dimensional space, if we want to describe them in mathematics, are sometimes so complicated that it is very difficult to describe them with a combination of formulas. For example, the mathematical description of the human body surface mentioned above.

We often notice people talking about four-dimensional space, or six-dimensional images. Doing so, they are confusing a pure mathematical concept with the real-world living space in which they are.

The spatial images that people can visually see and visually express are at most only three-dimensional.

The concept of time is treated in the same way as the concept of space in the theory of relativity, as one dimension of the space-time model. Since people cannot depict real four-dimensional images, they can neither see four-dimensional geometric space nor draw four-dimensional images.

In view of this, celebrated scientists such as Minkowski, Einstein, Hawking, and others had to replace one of the three-dimensional spaces with time-dimension. Thus, they also turned the three-dimensional space into a space-time visual space composed of two-dimensional space and one dimension of time. They did this to enable people to intuitively observe the relationship between space and time. But this method has serious flaws and errors, and causes, when applied, numerous conceptual and computational errors.

For example, let's take the depiction of the world line to identify where

Minkowski and Einstein's four-dimensional space-time went wrong. Then we'll look at our correct four-dimensional space-time representation.

Spacetime Concepts

We cannot accurately observe and directly describe four - dimensional space-time as we do with a three-dimensional coordinate system, which causes big problems, particularly in how we describe in four -dimensional space-time the law of motion for objects; limiting our understanding of, and research on, the objective world.

Minkowski's space time diagram cannot express the movement of things in four-dimensional space

In 1908, Minkowski used two-dimensional graphics to express light cones, world lines, events, inertial systems, etc., and gave intuitive illustrations of the spatial and temporal properties of relativity. His purpose was to illustrate the phenomena he wrote about for people who do not understand mathematical concepts or their formulas, such as time dilation and length contracting. This kind of graph, the Minkowski graph, is also called the space-time graph. It is still in use today.

However, studying Minkowski's space-time diagram and light cone diagram, as shown in Figure 8.15.

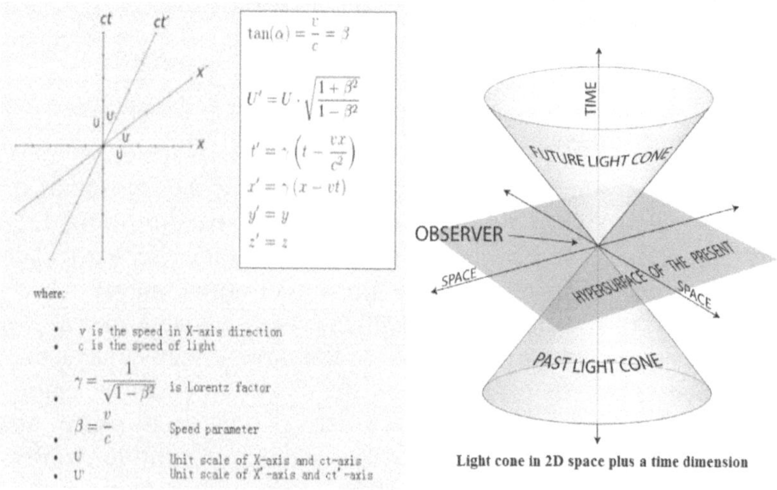

Figure 8.15 Minkowski's space-time diagram

We can find that these expressions have great limitations. Many things that should be expressed cannot be intuitively expressed, and some expressions are misunderstood because of being too obscure, and some were totally wrong like the spacetime light cone imaged in the Figure.

Even experts often make various mistakes in its application. We will discuss this again in the next chapter.

At this point, let us review the various mistakes made by Stephen Hawking in his book, particularly in regard to Minkowski's space-time light cone.; Our discussion thus far clearly shows that Minkowski's diagram will not help us understand the concept of time and space clearly and intuitively. Even celebrated scientists like Hawking can't understand space-time light cones correctly.

Minkowski defined the four-dimensional space-time model (X, Y, Z, T) used to represent world-point and world in his paper, "Space and Time."

However, since we are only be able to draw three-dimensional images, the four-dimensional space-time was compressed into three-dimension, with the Z axis of the space in the three-dimensional space (X, Y, Z) being replaced by the time axis T. So, there is a three-dimensional visual space-time model (X, Y, T).

The space-time light cone is an intuitive image drawn according to the world-point and world as Minkowski defined it in his paper.

Again, Minkowski came up with his light cone diagram to enable us to more intuitively grasp and express the concepts related to the space-time light cone. But the actual situation is just the opposite: the space-time light cone diagram does not make it easier for us to intuitively grasp or understand the related concepts.

If we consider real-world applications here, the space-time light cone will not work as Minkowski hoped it would. The only exception is the "now" point! This will be discussed later in the next question. Even in a perfect mathematical world, the defects involved in using the light cone are clear, and noted in the following points:

1) The space-time light cone has limited expressive ability

One light cone can only express one event. If we need to study two or more interacting events or world lines, the light cone cannot express them.

2) The relationship between two events cannot be expressed

If we need to study two events or world-lines that affect each other, the light cone has no way to express them.

3) Most images of space-time light cones are wrong and useless

Of course, a future light cone possibly contained in the given space-time light cone has not happened yet, and does not yet exist. A future light cone is also just an estimate of the maximum range that light can propagate in this instance. It is the same for any future light cone formed by other events, and has little significance on its own. And, if an event concludes or stops at the "present" moment, a future light cone will not appear or exist in the future.

The past light cone in the space-time light cone, as Minkowski said, is "*the front cone of O, [which] consists, let us say, of all the world-points which send the light to O.*" "*The light emitted by any world-point will only diverge with the elapse of time, and will not converge to point O. The light emitted from a big portion of world-points cannot reach the O point.*"

4) The image of the space-time light cone is obscure and difficult to understand, let alone apply.

What we see in Minkowski's diagram is an image that cannot express the relationship between events for different objects. It is the speed v that obscures the meaning of space and time, so that these two completely different physical quantities can be converted into each other. We've covered this in detail earlier when discussing Einstein's original paper

The reason for the above defects is that Minkowski's Space-Time image lacks an intuitive means of expressing four-dimensional space with image, has to be based on three-dimensional images that consists of two-dimensional space and one-dimensional time. The time axis of this three-dimensional image is upward, and the space is represented by a horizontal two-dimensional space axis.

Since the expression space is only two-dimensions, the expression of events is greatly restricted. Because of such a method, there is no way to truly express the three-dimensional spatial location of events, let alone the spatial law of the occurrence and development of things. These deficiencies also greatly weaken the role of Minkowski's Space-Time diagram as a tool for correct understanding of related concepts, and even cause unnecessary misunderstandings.

To solve the space problem, a three-dimensional light cone is defined. However, although this light cone, which people used extensively later, looks three-dimensional, it is actually two-dimensional, and it is more likely to cause people to misunderstand related concepts.

What's more serious is that the description of this three-dimensional light cone is completely wrong. It can't express the different world lines correctly. Below we illustrate this problem through a simple model to see how Minkowski's space-time model can't express different world lines correctly.

Example: below there are N curves that are **different** in space (X, Y, Z)

$X = x,$
$Y = y,$
$Z = K_i (x + y), \quad Ki = 1, 2, 3, \ldots, N$

When the Z-axis is replaced by the time axis T, these curves become as following:

$X = x$
$Y = y$
$T_i = t, \quad i = 1, 2, 3, \ldots, N$

All these N curves **become exactly the same** in (X, Y, T) spacetime.

We can clearly see that the curve compressed in (X, Y, T) space-time cannot correctly describe the original world line in (X, Y, Z) space.

For any curve in three-dimensional (X, Y, Z) space, if there are different changes in the replaced Z space, it is impossible to correctly display that space in a three-dimensional (X, Y, T) space-time model.

For clarity, we further illustrate this problem with the following simple example.

The following two curves are completely different in (X, Y, Z) space:

Curve 1: $X = x,$
$Y = y,$
$Z_1 = (x + y)$

Curve 2: $X = x,$
$Y = y,$
$Z_2 = 0.5 \ (x + y)$

Their images are two different lines in the (X, Y, Z) space as shown in Figure 8.16 left frame below, Curve1(x, y, (x + y)) and Cure2(x, y, 0.5 (x + y)) are two different curves. After removing the Z axis shown in the left frame of Figure 8.16, and replaced Z axis with T axis as shown in Figure 8.16 right frame, the two curves Curve1 (x, y, t) and Cure2 (x, y, t) are the same now, there is no different between them.

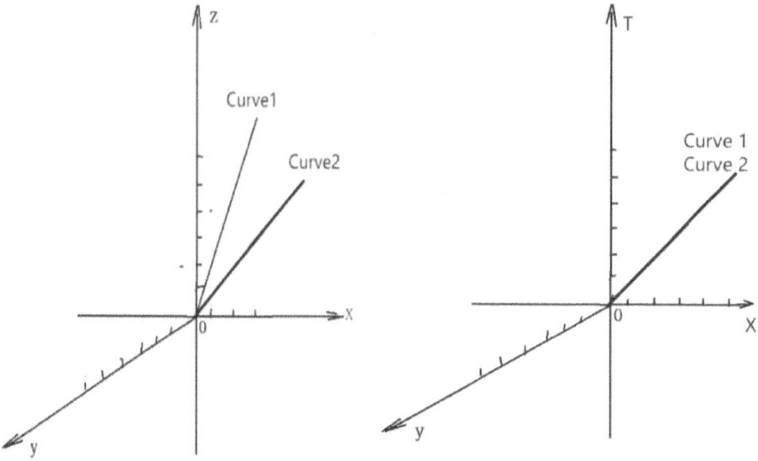

Figure 8.16 Left frame: In (X, Y, Z) space, Curve1 and Curve2 are different curves. Right frame: Replace Z to T, then in (X, Y, T) space, Curve1 and Curve2 are same.

In order to correctly express (X, Y, Z, T) four-dimensional space time, we defined a four-dimensional space-time image model that depicts the world line of object events. Comparing Minkowski's three-dimensional space-time diagram in the application example, we can see that the four-dimensional space-time diagram is clear, simple, and expressive in expressing the occurrence and development process of the object event. It is easy to understand the part that people care about and the process of multi-events. The event occurrence and development process of each object event, and the relationship between the dynamic development of events between object events.

4-dimensional space-time diagram that expresses the world line of the object event

Minkowski said in "Space and Time":

"Let x, y, z be rectangular co-ordinates for space, and let t denote time.

*But I still respect the dogma that **space time have independent** significance.*

It is only in four dimensions that the relations here taken under consideration reveal their inner being in full simplicity, and that on a three-dimensional space forced upon us a priori they cast only a very complicated projection."

These statements have become our basic principles for designing world-line image representations in four-dimensional space time. Let's determine how we can correct them.

1) A Conventional 3D space image

Figure 8.17 presents a typical conventional trajectory diagram of a celestial body, StarA, in a three-dimensional (X, Y, Z) space.

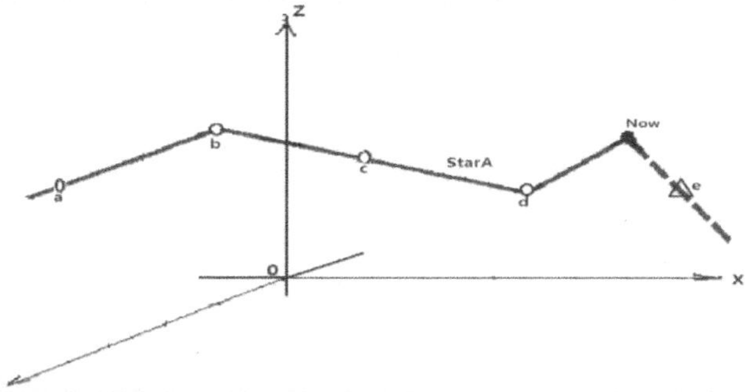

Figure 8.17 is a typical conventional trajectory diagram of a celestial body StarA in a 3-dimensional (X, Y, Z) space.

The **empty dots** on the line are **focus-points** of the events (or objects) of StarA. The **solid dot** is the present "**Now**" moment. The dotted line beyond "Now" is the estimated future path of StarA. The focus-points are numbered sequentially from the beginning of the world line to the "Now" moment of the world line. Beyond "Now" is the future, which does not yet exist. So, this dotted line is only an estimate of where the StarA perhaps would go

Since **there is no corresponding time, the information expressed by this graph is not clear.**

But we have not been able to add the timeline to Figure 8.17 so far. Because it becomes a four-dimensional image, people don't know how to draw the four-dimensional image intuitively.

For his part, Minkowski removed the Z axis and replaced it with a time axis T. This can allow us to draw a three-dimensional image consisting of two-dimensional space and one-dimensional time, But we have fully discussed the various errors and disadvantages that this method brings.

Then, how do we now add the time axis to Figure 8.17, so that it cannot only avoid the various problems discussed above, but also accurately and clearly express the time on the trajectory of StarA?

2) Our 4-Dimensional Space—Time Worldline

Our solution is to add a time axis corresponding to the StarA trajectory below the three-dimensional StarA space trajectory graph as

shown in Figure 8.18.

Figure 8.18 now represents the four-dimensional space-time of the object event world line, called the four-dimensional space-time diagram of the object event world line, and further abbreviated as the object space-time diagram.

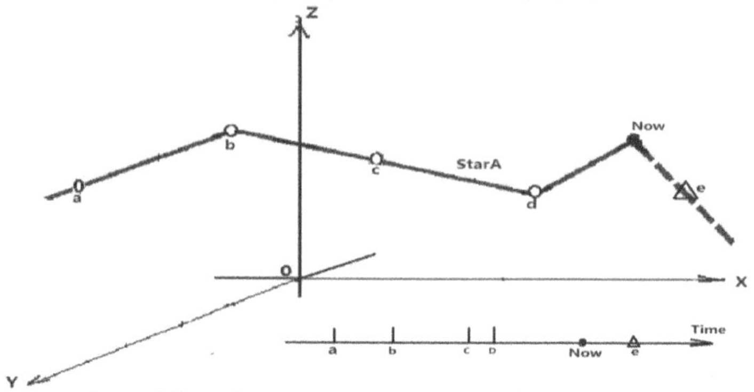

Figure 8.18 After adding the time axis to the 3—Dimensional (X, Y, Z) graph, the trajectory of StarA becomes a Space—Time graph in a virtual 4-dimensional (X, Y, Z, T) space. Each point on the trajectory of StarA corresponds to the same point on the time axis, a to a, b to b, etc.

A Space — Time graph means that the spatial scale is above and the time scale is below. The time scale corresponds to the spatial process of event development.

The space in Figure 8.18 is represented by traditional (X, Y, Z) coordinates.

Then draw a time axis parallel to the X axis from the past to the future below the traditional space coordinates. This time axis takes the "now" coordinate as the origin and is named as T_{NOW}, and the direction of the arrow corresponds to the order of the events of the object events in the order of occurrence. Time units and scales can be specified above.

The name of the object event's worldline is the name of the object event.

The **origin** of the event axis is the "**now**" moment of the world line trajectory of the event. The time of the occurrence of each event is clearly marked on the relative position of the time axis with the same point identifier.

In this way, the space trajectory of StarA has a one-to-one correspondence with the corresponding time. From Figure 8.18, we can clearly see all the information we need.

This method seems very simple, but it is not. It combines the time each event occurs with its spatial motion accurately; the activity of each world point in three-dimensional space combined with the spatial motion of each world point. It enables us to use the diagram intuitively to draw the entire world line clearly and accurately.

The hollow dots on the world line represent the spatial position of the object event at a certain time in the past, and it is also the position where the object event once existed, which is called the historical event of the object event, or the event for short; that is, points a, b, c, d of the world line StarA in Figure 8.18.

The event object itself is represented by solid circles at the top of the solid history line. Time position "now" represents the spatial position of event objects at time "now," as shown in Figure 8.18, the "now" point at the intersection of the solid history line and the dashed future line.

The potential future path is represented by dotted line segments extending from the object event solid circle. Typically, this imaginary dotted line is very short because it is denoting a potential or expected development only, not a fact. The potential future event points noted by the dotted line and our potential or expected future positions are represented by two hollow triangles with an "e" identifier; noting this event will potentially take place, or is expected to take place, there and then.

The time axis below is no different from the usual one-dimensional axis. Note that "now" must be set at the origin. We will discuss this in detail later.

With this method of combining three-dimensional space and one-dimensional time, we can correctly express the intuitive image of four-dimensional space and time.

Although this method seems simple, it can function is in significant ways. It can complete and express a great deal of information that the original three-dimensional space-time light cone could not express at all. And it can inspire us to a new understanding of the laws of motion, including the "law of existence of things" to be discussed later.

3) Simultaneously express and compare the world lines of multiple objects in the image

Things in the world are complex. It is often necessary to express and compare the motion states of multiple objects at the same time. This makes the space-time light cone completely useless while the new expression of the world line that we designed shows its powerful functionality.

Combining all we have learned from our discussion of the above two figures, let's now consider a situation — where the world lines that intuitively express multiple objects intersect each other — from the perspective of practical application. For example, two air objects collide in motion, or two objects arrive at the same place in space one after another. Of course, depending on the specific situation, various applications can also be considered.

Figure 8.19 shows how the world lines of three objects can be represented and compared in one diagram at the same time.

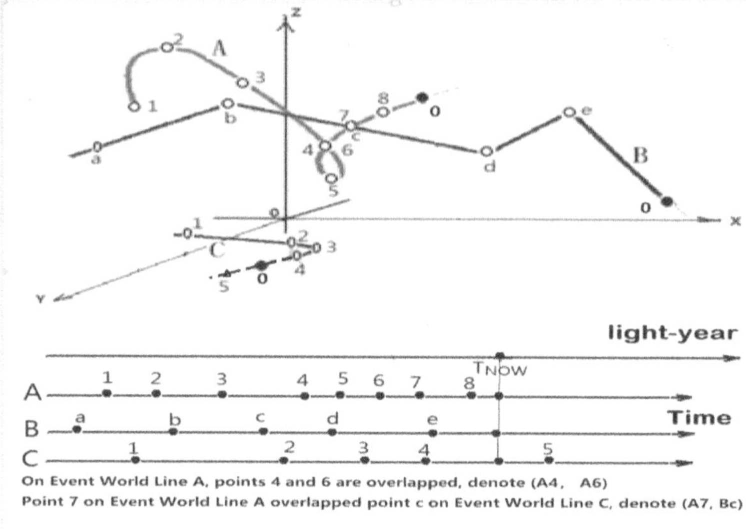

Figure 8.19 A four-dimensional space-time graph of object events

Figure 8. 19 uses a four-dimensional space-time graph of object events to express the relationship between object events. We also noticed that the "Now" moment of all worldlines happens at the same current moment 0.

In Figure 8. 19, the upper part of the figure shows three world lines A, B, and C, which represent the spatial motion trajectories of objects A, B, and C. Below are timelines A, B, and C, corresponding to each world line.

There are two interesting points on the world line of the A object events marked 4 and 6. Their spatial coordinates coincide successively at the same point in space, but at different event occurrence times. This can be seen from the corresponding timeline. The time corresponding to point 4 is TA4, and the time corresponding to point 6 is TA6. Such spatially overlapping but not temporally overlapping points require two-point names to represent their overlap, so we mark them as such with one addition (the object name): A (A4, A6) — the A outside the parentheses indicates the object name; the A inside the parentheses indicates the timeline name. Here, the time axis has the same name as the event, which can be simplified as A (4, 6).

The A represents the world line curve and the time axis A has the same name. The 4 and 6 respectively represent the event names; and the times with the same name on the corresponding timeline A below. That is, time 4 and 6 respectively represent the occurrence times of events 4 and 6.

It can also be seen from the figure that the event point 7 of object A overlaps with the event point c of object B, which can be recorded as: AB (A7, Bc). Where AB represents objects A and B, times are in brackets, A7 is point

7 on A's time axis, and Bc is point 'c' on B's time axis.

In this way, we have basically completed the design and drawing of the 4-dimensional object event space-time world line representation diagram. However, the story is not over yet.

Application of the Four-dimensional Space-time Object Event World Line Diagram

In Hawking's *Brief History of Time*, there is such an example as shown in Figure 8.20 below.

We do not know how to read this Figure 8. 20. we do not fully understand it either. How did the Earth enter this dark event? Why is there no point and event identified before the time sequence involved begins?

> For example, if the sun were to cease to shine at this very moment, it would not affect things on Earth at the present time because they would be in the elsewhere of the event when the sun went out (Fig. 2.6). We would know about it only after eight minutes, the time it takes light to reach us from the sun. Only then would events on Earth lie in the future light cone of the event at which the sun went out.

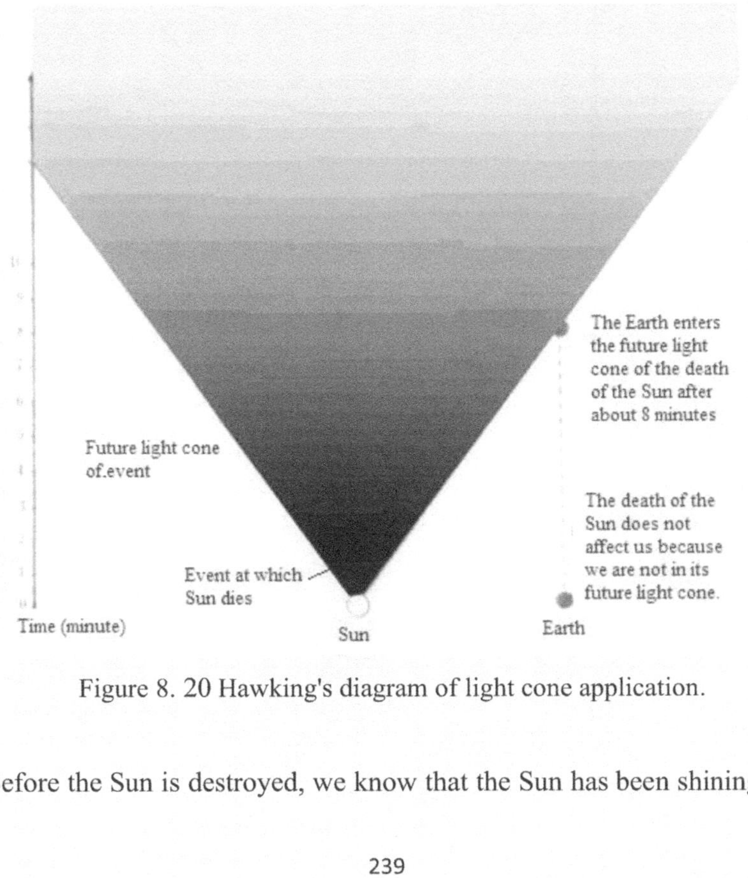

Figure 8. 20 Hawking's diagram of light cone application.

Before the Sun is destroyed, we know that the Sun has been shining on the

Earth. After the Sun dies, its light will continue to shine on the Earth for eight minutes. Only when the Earth receives no sunlight can we know that the Sun has died. This kind of causality is not shown at all in the light cone diagram in Figure 8. 20, as devised by Hawking.

The following proposition is expressed by the four-dimensional space-time world line image, which is called the **dark proposition**: *If the Sun goes out now, people on earth will know it at 8th minute.*

Figure 8.21 below represents three-dimensional space coordinates.

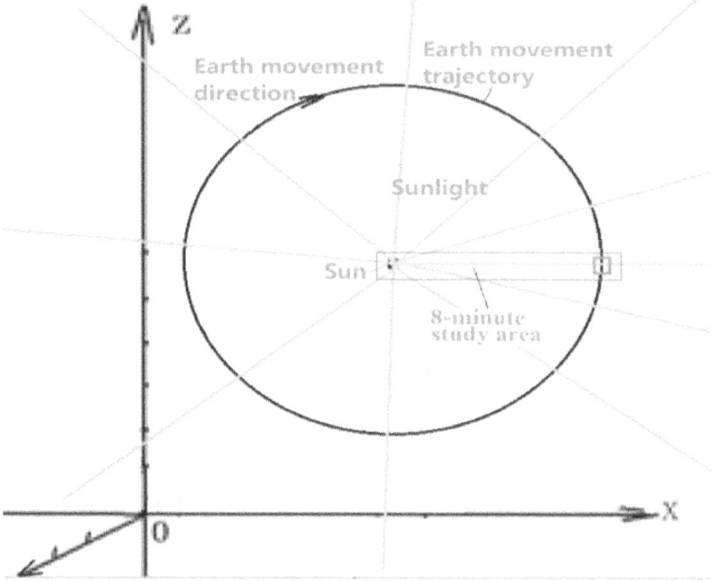

Figure 8.21 Solve the problem in above Fig.8.20

In Figure 8.21, the Sun sends sunlight to the surrounding space; the Earth moves in a clockwise direction. We mark the 8-minute interesting trajectories of the sunlight and the Earth with a rectangle frame, that is, noted as the "8-minute study area."

The object we want to study is the death of the Sun as sunlight continues to fall on the Earth. If the event occurs now — the Sun goes out — then the black event will reach the earth after it propagates for eight light minutes; that is, the world line of "the Sun goes out event" will overlap the Earth's orbit after eight minutes.

Now we enlarge and copy this "eight-minute study area" from Figure 8.21 to Figure 8.22 as frame 1 for research purposes. Here, the orbit of the Earth occupies just a small section of the plane.

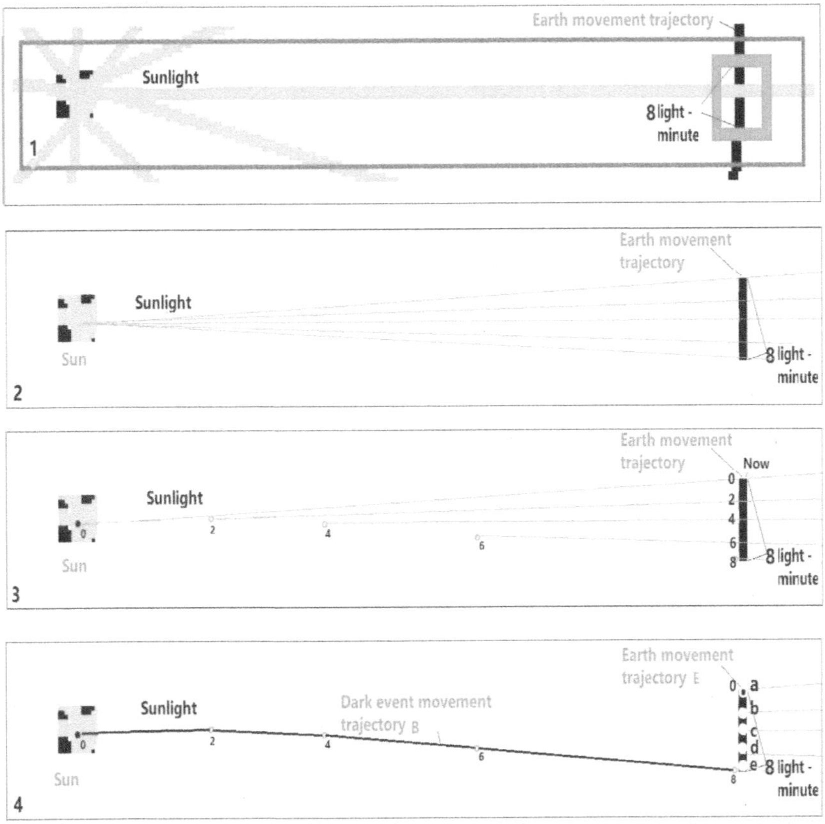

Figure 8.22 Detailed step-by-step study of the "8-minute study area" by frames 1, 2, 3, 4 that are copied and enlarged from Figure 8.21.

Frame 2 is copied from frame 1, but we remove the irrelevant things in frame 1, and divide the eight-minute orbit of the earth's trajectory into four equal parts with each part illuminated by the Sun.

In frame 3,

- The "now" moment when the Sun goes out is marked as dark event start time 0, when sunlight is still intact from the Sun to the Earth, and the dark event just begins to spread after the Sun goes out;
- At 2 minutes, the Earth moves clockwise for 2 minutes, and the dark event of the Sun moves to the position shown by 2 in the figure;
- At 4 minutes, the dark event moves to the position shown by 4 in the figure;
- At 8 minutes, the dark event moves to the position at 8 in the figure.

In frame 4, we removed the redundant information in frame 3, and then named the earth's orbit at 0, 2, 4, 6, and 8 minutes as a, b, c, d, and e, respectively. The Earth's world line track is named E. The events of the dark event world line track are named 0, 2, 4, 6, and 8 respectively according to frame 3, and the dark event world line itself is marked as B.

We copy frame 4 of Figure 8.22 back to Figure 8.21, get Figure 8.23.

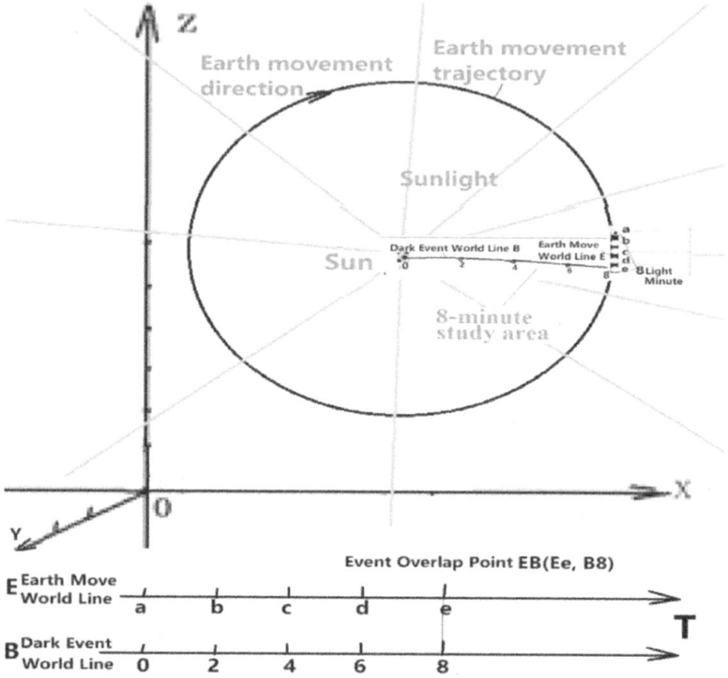

Figure 8.23 Schematic diagram of the four-dimensional world line for solving the "dark proposition". The point of event overlap is the time that Earthlings begin to know the Sun goes out.

In Figure 8.23, we see the Earth trajectory world line E and the dark event world line B in the "8-minute study area". The overlapping points of these two world lines is the “8” event point on B and the “e” event point on E, denoted as

EB (Ee, B8)

Figure 8.23 completes the dark proposition with the addition of the relevant timeline.

Of course, our four-dimensional worldlines can solve far more complex problems than this proposition.

We now summarize the definition of the four-dimensional image of the object event space-time world line in Figure 8.24.

Four-dimensional Space-Time Diagram of Object Space-Time World Line

Space: Usually using orthogonal X, Y, Z coordinate system of the conventional Euclid space

Time Plane: parallel to the XY plane and space, perpendicular to the Z coordinate space, and space systems disjoint Independent flat

Basic Timeline: the traditional one-dimensional time ruler with arrow, and even the definition of the unit of time scale and the "Now" moment as origin point.

Object: the presence of a substance in the universe, the celestial, people, matter, particle, etc. we are interested in our study.

Object World Line, a line of an object in 3-dimensional space with each point corresponding to an event of the object. This point also related to one point at Time-Axis. The points are arranged strictly sequentially to the time from the past to the future, with the original point on the Time-Axis related to a event of the "Current" moment of the line, past events are happened before the "Current" moment and future events may perhaps happen or not happen after the "Current" moment. Past point is denoted by solid circle and future points are denoted by dotted circle.

Time Line of the world line, one-dimensional way axis, is the projection of world line on the time plane in time sequential. A world line corresponds to a time line. If the image is not confused, the basic time axis can also be use as time line of the world line.

Events on the world line:: the point we are interested in the world online, also known as event points, or event.

World line event collection: a collection of all points of events on the world line studied.

World line collection: all of researching world lines collection.

Time Line set of world line: all of researching time lines of world lines collection

Time line plane (five-dimensional space-time diagram): define a time plan parallel to XY plane but without any intersection with the world line space. Extend the time line to the time plane, project events on different object world lines on the time plane, with different angles in different time axis to represent the concepts to be expressed.

Six to nine-dimensional space: on a plane vertical X axis and parallel to the YZ plane, draw an E axis parallel to Y axis, thus construct an XYZ-TE-E 6-dimensional space-time diagram. Following the same method, the E-axis can be extended to EE plane, thus gets XYZ-TT-EE 7-dimensional space-time diagram.

Figure 8.24. The definition of the four-dimensional image of the object event space-time world line

The Law of Existence of Objects deduced from the four-dimensional space-time diagram

Looking back at Figure 8.19, we may notice that the origin of any world line is at the same position on the time axis, that is, the "now" moment.

If we admit that the world in front of us is composed of countless world lines, or that any huge or tiny object in this world can be represented by a *world line, and even the world itself can be represented by a world line*, then we can deduce an astonishing fact from the "now" moment of each world line:

Figure 8.25 Time is like passing water running to the east. Though we call and beg, it will not come back. Confucius said by the river:" Time is flowing away like the running water.

The **law of existence of objects** says that: **things exist only at the current moment.**

When we stand by the river, we can see the water flowing, and we might feel that the law of existence of objects is ridiculous.

When we look up to the Sun and feel its touch, will might have the same feeling that the existence of objects theorem is ridiculous.

When we are standing on solid ground, with yesterday's smile still in mind, when today we are enjoying a better life than we did before, and tomorrow holds for us the same promise, why just it then we only exist in the current moment?

Isn't that a ridiculous idea as well?

Earth is spinning, and it will nurture us for hundreds of millions of years.

The Sun shines, and it will shine on us for a billion years.

For their part, the cosmologists have given us a dream of traveling at the speed of light. They tell us that if we can travel faster than light, we may return to the moment that we have not yet been born to kill our father or mother. Of course, we will not do this; it is too cruel, too incompatible with the values we prize. We do some happy things, such as going to chase Yang Yuhuan; or to find Zhao Feiyan; if Yang Yuhuan is a little bit fat. The girls can go to find Pan An or Song Yu. Or we can raft with Li Bai or take Xu Zhimo's poems to discuss with Shakespeare......

Yet we are one with heaven and Earth; we coexist with time.

How is it then that we exist only now can we only exist for only one moment?

But we still must whisper and stubbornly say: "We only exist in the 'now,' in this moment!"

Our past is only the evolution of objects that have left us some form of memory; we cannot return to the real past.

The reality of our future does not exist in the "present" moment.

Our subjective future is our expectation of the evolution of objects that will develop in the future.

We never have the opportunity to go back to change the past. Whether you can travel at the speed of light or faster, you just cannot change the reality in a movie. We have already discussed this cruel, no poetic fact. These ideas are entirely based on general science, and do not need to use the existence of object theorem.

Before we give proof of the **existence of object** theorem, we would like to ask you to think about a few questions:

If we examine every drop of the river that Confucius stands by, is it the same river now and then? Or does every moment the river only exist in a different fixed point in space? Where was the river in the past? Where is it in the future?

What is the relationship between the Sun and sunlight? Is the sunlight shining on our cheeks the same every moment? Are the photons of sunlight all the same all the time?

If we use a time axis scaled to a hundred years, near its beginning we enjoy youth, but at the next point where are we? If we use one day as the scale, today will not be much different from yesterday, but where will our ephemeral today be tomorrow?

When considering such issues, perhaps few people would deny the existence of the object theorem.

In fact, the idea of the existence theorem of objects is similar to the view of the ancient Greek philosopher Parmenides, who said, "We can speak and think only of what exists. And what exists is uncreated and imperishable for it is whole and unchanging and complete. It was not nor shall it be different since it is now, all at once, one and continuous."

The ancient Greek philosopher Heraclitus said that people cannot enter the same river twice.

The Existence of Objects Theorem

The definition of **The Existence of Objects Theorem：From its birth, through its past, to the present, and into the future, an object only exists at the "present" moment.**

The **past** of an object simply resides in our memory of the sequential incidents that the object's evolution has left us. As to other temporal values important to our discussion here, there are two: relative and absolute future of the object.

The **relative future** of an object consists of events with respect to the object that have already happened, but which we have not yet received notice of; that is, the information that identifies the event or events, and that depicts them as ordered and irreversible, has yet to arrive at our reception point.

The **absolute future** of things does not exist in the "now" moment, but is merely our prediction or expectation of events in the evolutionary trajectory of things as they develop to certain future moments.

The **completed future of a thing** is the trajectory of a part of the thing's information that has passed through the "now" for a period of time.

Watch something external to the system from any place to see the object itself exists only in the corresponding system itself at the "now" moment. All the rest occurred in the history of the event recorded before this "now" moment, and events after this "now" moment do not exist, cannot be seen, and possibly might happen in the future. Things can be broken down as well as combined.

This is the **Existence of Objects Theorem**, referred to as **Existence Theorem**.

We will now define two new concepts important to our discussion: the **relative future Event** and the **absolute future event**. For example, working in the city, Peter wrote a letter to his wife at home. When he put the letter into the mailbox, the event — "Peter's wife is getting a letter from Peter —" became a relative future event for his wife because he has written the letter but it has not yet arrived. If Peter has not written the letter, then the event —

"Peter's wife is getting a letter from Peter" — is an absolute future event because it has not happened yet. But we know that, based on the historical record, there is a great possibility that it will happen under normal circumstances.

The relative future depicts an event that has happened but which we have not received notice of; its information has yet to arrive. If there is no problem with the transfer process involved here — from absolute to relative future event — the event will finally occur, but there is a possibility that it may not occur. If Peter wrote a letter to his wife, the event of "Peter's wife is getting a letter from Peter" becomes a relative future event. But it is also possible that Peter's letter will be lost in the delivery process and the event of "Peter's wife is receiving the letter" will not occur.

The absolute future belongs to the event that did not happen. Peter has not written the letter yet. However much it is likely to occur, it also may not occur.

The concepts of relative and absolute future are important when discussing the issues pertinent to our understanding of the universe, and how it works. The light that the Sun emitted five minutes ago, to an observer on Earth, is a relative future event because it will be received and observed on the Earth three minutes later. The light that the Sun will emit ten minutes later, to an observer on Earth, is an absolute future event. We believe that the Sun will continue to exist after eight minutes and continue to send its light to Earth; there is a great possibility that an observer on Earth will see the sunlight emitted ten minutes later, but it has not yet happened. The Sun may have been destroyed at the ninth minute

Visual Evidence of the Law of Object Existence

The law of object existence only explicitly introduces time coordinates into the definition of existence, for example, that Parmenides conceived.

In order to understand this a bit more, consider the following. You can chart a person's life from birth to death, in a diagram with a minimum unit of one year, sequenced by a factor of ten. For every ten years this person has lived drawn a picture. It's okay to paint inaccurately here. Suppose this person has lived 100 years. Then, in accordance with the above provisions, you need to draw ten pictures to chart in sequence this person's life.

If a picture is drawn each year, 100 pictures are needed.

How about drawing one picture per hour? Per minute? Or per second? How about drawing a picture every 0.001 second? Or every 0.0000000001 second, and so on,

Figure 8.26 A man's whole life (70 yrs.) on display (Painted by Wu Yongjian).

At infinitely small intervals, of course, there will be unlimited images. This also means that a person's life can be an infinite number of snapshots. At every given moment, there is an entity that exists at that moment. And in that moment before or after the moment, the corresponding snapshot of the person has changed.

To make it clearer, let's magnify the snapshot interval of a snippet of a man's life. (Figure 8.27)

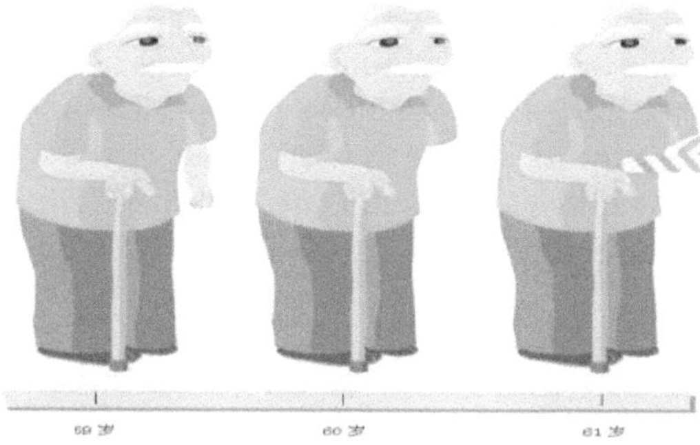

Figure 8.27. Schematic diagram of the time instances of "past, present, future" of a 60-year-old person who suffered a car accident at age 60. (Painted by Wu Yongjian).

Mark any three moments (or any moment in the life of the person), such as 59, 60, and 61 (if you have studied algebra, then we use x, x -1, x + 1 three arbitrary moments, not necessarily with x = 60, nor necessarily in units of years). To put it more clearly, let's suppose this person had a car accident and lost an arm at 60 years old. In that case, his 60-year-old status should be the middle figure with no arm, his status at 59 is normal, and his status at 61 is that he has a prosthetic.

What we should note is that according to the law of object existence, at the very moment when this person has lived to the snapshot of 60, the "he" indicated by the 59th snapshot has ceased to exist, and his good arms are just a memory; his history captured in some form of photograph. In the snapshot of him at 61, he has lost his arm. But this event at 61 has not happened at the snapshot moment at 60. That's just a possible future condition of this person relative to the moment of the snapshot marked 60. At the moment when this person has lived to the snapshot at 60, or at a later time, we may speculate based on a variety of information, such as current medical standards, that it is probable that the individual will be alive at the moment of the snapshot at 61. If this person lost his arm during the Han Emperor Wudi period, no new prosthetic limbs could be drawn on the snapshot; perhaps a stick would be attached.

This shows that a person's life consists of innumerable moments. But in every moment, people only exist at that moment!

The moment when the present moment exists, we call "the present."

In Haikou, we saw the statues of Past-Buddha, Present-Buddha, and Future-Buddha in the temple. We just couldn't help but feel a great deal that the ancient Buddhists saw the world more thoroughly than some modern cosmic experts. Ah!

The Object Existence Theorem Described by the World Line

A boat glides slowly in a calm lake. The boat is the point where the object now exists. The wave behind the ship is the path the ship took and indicates where the boat once was. Where the boat will be in the future, we can probably calculate, according to its current position and trajectory, but reaching that future place has not happened yet. Perhaps the boat will sink or change its direction.

The same reasoning and the same proof can be extended to everything we know in the universe.

Take, for example, the Sun. It is always creating and releasing energy, and it is not the same every moment.

Another example is a photon from the Sun. At one moment, it is on the surface of the Sun, and the next moment it is a few hundred thousand kilometers closer to us. And as it travels hundreds of thousands of kilometers, every place it passes through is its trajectory, denoting a moment of its existence. The sum of all the existing moments constitutes the trajectory of the photon; that is, its world line. But the photon is not the world line; the photon is only one point. It is the unique operation of the trajectory of the photon that constitutes its world line; the track of each point corresponding to a different moment, with this photon at the head of this world line

A river is made up of countless water droplets. At every moment, a large amount of river water flows into the sea, and water from the source is injected into the river. No river at one moment is exactly the same at another moment.

The world line concept we are using here has a clear definition: an independent object represented in a world line, with the beginning of the world line corresponding to the event's starting position and time. The moment "now," which corresponds to the world line's forward vertex, also denotes the location of the object. All event points on the world line are strict, not chaotically arranged, from the beginning to the endpoint between the vertices, from the past to the present moment in an orderly sequence. Each point, in turn, is the expression of the state of the object event at that time, and from the moment they occur to now corresponds to each point on the timeline.

From the viewpoint of different objects, a planet, for example, can express its entire trajectory when moving through a part of that trajectory. The whole time, though, it only exists at the forward end of the trajectory; that is, where it is at that moment in its trajectory. Its past is the memory of the locations that it has moved through on its trajectory, and its future, based on expectations made from its past and present, still does not exist in reality.

A planet can be broken down into mountains, rivers, humans, animals, birds, insects, vegetation, sediment, and so on. A river can be broken down into countless water droplets. For every one of these, we can depict the object based on knowledge of its world line. At the forefront of the world line is the "now" moment when the object exists, including the planet itself – composed by collecting all the individual object world lines that constitute a planet into that planet's world line "now."

We can gradually and finely decompose, and also gradually polymerize, from large to small, or small to large, but there is only one set of world lines, and their apexes represent the existence of the real objects!

When we call for spring in the winter of 2022, we expect the flowers of 2023. Spring is there year after year, but no one would say that flowers in 2023 spring are those that bloom in 2022.

Space, time, is forever a mysterious theme.

Chapter Summary

An ordinary ray cannot affect the length of a rigid rod and the flow rate of time on the rod.

Space cannot be bent, neither produced nor destroyed. It is the things arising and dying, contained in the space that are bent, not the space itself.

All of the above are intuitive common sense, but they are things that contemporary physics is inexplicably confused about.

The root cause of this confusion goes back to the great master Einstein.

Disregarding the rationality of the physical model, Einstein bluntly laid the foundation of the special theory of relativity with simple elementary school arithmetic but related to the wrong physical model: "a ray affects a rod." Then on this foundation he constructed the special theory of relativity.

Unfortunately, in contemporary astrophysics a great fuss can be made about an irrational, unreasonable, or inaccurate physical model without first carefully studying whether the physical model is reasonable or not. It is also the root cause of Einstein's continuous use of the wrong physical models in the theory of relativity.

Neither did Einstein understand the two concepts of simultaneity and synchronization, despite the fact that he used them to write his famous paper. Nor did he understand the respective properties of space and time, as he conceived them in general relativity.

Einstein did this in both his Special Theory of Relativity and General Relativity!

When doing scientific research, the methodology is clear: first, subdivide the research objects, study each of them carefully, and then conduct a reasonable comprehensive study of multiple objects according to your needs.

If someone combines several concepts together without careful study, then this approach is basically not scientific, and its results are often questionable.

This also speaks to combining the concepts of "space" and "time" – into "space-time" or "time-space" – or even to saying that time and space are interchangeable.

Question: Have you figured out the concept of "time"?

The truth is: time and space have not much to do with each other. Time is only about what is contained in space!

Let us explain this amazing fact below.

Chapter 9

The Concept of Time and the Myths Attached to It

What is the time?

Someone might dream of stepping into the old footprints he left on the road, to meet his first love hidden in some corner of life's memory. Somebody else might want to look into the future, to know the road the world will be traveling, or he can take advantage of what will happen.

Time is like the rings of an old big tree, circles retaining the lonely, silently flowing time to stay. If there were no time rings from small to large, it would be like time itself that is not clear in its meaning and what it is describing.

From ancient times to the present, so many pioneer philosophers, religious leaders, and scientists from different countries and cultures have commented on time that the differences in the perspectives they bring to time, whether large or small, can be confusing. Be that as it may, a commonality presides: none do more than define time based on concrete life, the passage of the Sun, the Moon, the stars, the seasons, and so on. The passage of time is elusive, and it is gone without any trace.

In ancient times, before mentioning their names emperors were hailed by "long live," but none lived for a hundred years! In modern times, it is much the same, especially for those who invented "scientific" time and tales of space travel that intrigue and entertain. Notwithstanding when they were born, we do hope they can travel through time, so that we can meet them again in the future. We also want to time travel, with countless numbers of "I" in parallel worlds, so that we can attain immortality. How much beauty could we experience, as much in the past as in the future? Look at thousands of time-travel storybooks to know what a deadly and wonderful dream elixir of going back to the past is!

Unfortunately, this is all the stuff of dreams and fantasy.

Those bones in the cemetery tell us what time really means.

Modern science, however, disagrees, replacing **empirical experiments** with imaginary experiments, violates the spirit of science, and branded the time travel with the label of science.

For example, in the prestigious journal *Physical Review*, an article entitled "*Time Travel Using Degenerate Metrics*" was published in June of 2018. In the article, the authors used all-powerful mathematics to discuss how to travel through time.

The abstract of this paper follows:

The possibility of time travel is explored within the framework of first order gravity theory in vacuum. We present explicit solutions (to the field equations), whose geodesics allow the proper clock to run backwards. These four-geometries contain degenerate as well as nondegenerate tetrad fields that are sewn together continuously over different regions of the spacetime. Being devoid of matter and also of torsion by construction, these solutions have no exotic field content.

The thesis is based on Einstein's field equation, and theoretically explores how to realize time travel, and travel back to the past. Why bother when time travel itself is impossible?

Those dead bones we mentioned above are about to jump out of their graves, and cry out: How could we become dead bones if we could have lived forever? Let's travel back!

On what basis was science born and grown? Who can do a repeatable experiment to show how to make people go back in time?

Should mainstream science waste itself on taking folktales or fantasy as subjects for research published in prestigious journals?

How ridiculous!

Below we discuss some basic concepts about time, and then turn to Hawking's influential A *Brief History of Time* on the description and application of Space-Time light cone. Thereafter we discuss how Einstein and Hawking cooperated in creating "scientific" time travel, introducing mystery and fantasy into modern physical science in the name of science. Einstein was the initiator, and Hawking the communicator.

The Concept of "Time" that the Author Agrees With

As usual, we should first introduce various wonderful theories about time by our predecessors, but we will not waste space to introduce them here. In the information age, readers can access them by just searching. My books *Fading Modern Cosmology*, *Clarify Big Bang*, and others have detailed narratives and reference lists readers can refer to. The following presents the author's own understanding of time. Let's start with the most basic analysis.

Think about it, what is time? There is a sense of elusiveness that cannot be accurately described. Why do we feel this way? It's because the word "time" depicts different concepts. Let's subdivide them.

We believe there are **two faces of time: scale time, and the trajectory time of the existence of things**, simply referred to as **existence time**.

Our concept of time derives, first and foremost, from the natural phenomenon of our planet's rotation; in brief, standard Earth time is based on the length of time it takes the Earth to rotate once.

Application time is tied to the application of scale time. It is measured with a specified time scale. How is it measured? Of course, it is not space, which by definition is empty. What does it mean to measure space with the time scale? Is it measuring "nothing?" Therefore, it is to **measure things that exist in space with the scale time!** Is there a problem with this passage

Think carefully. If you agree with the above statement, then there is no reason to tie time and space together, because space and time have no direct relationship.

An area of subspace and time are related through things that exist in this space.

The time concept we defined consists of the following two parts:

- "Scale time (time)"
- "Existence time (time)"

To clarify "time," we first need to clearly state what "existence time" and "scale time" are. "Existence Time" is the time used in philosophy or daily work to describe the passage of process, the transformation and evolution of events, and the lifespan of everything. "Scale Time" is used for practical purposes in science, in the physical world, and so on as a tool for measuring existence time, just as a meter ruler measures length. It is very important to clearly distinguish between "Existence Time" and "Scale Time. ", If we cannot clearly distinguish between them, and do not understand their differences, we will never really know what time is.

This distinction is not only a morphological division, but also an essential division.

Scale Time

"Scale time" is a measurement tool we use to measure the "existence time" that describes the sequential process of the origin and development of an object.

In terms of application, "scale time" is strictly a tool for unit time measurement, a standard measuring meter. It is only used to measure existence time and does not belong to "existence time" itself. The "scale time" we commonly use now is just an immutable scale, as we have defined it, which measures the progress of things in strict accordance with the sequence of development, evolution, and end of things.

The traditional definition of "scale time" is that it strictly follows the provisions of the measurement unit, in accord with the development of objects, from their origin to their end; a strictly irreversible sequence. This is the definition of time that ordinary people generally accept. Indeed, almost all of humanity uses this kind of "time" to communicate with.

Our **unit of scale time**, of course, is based on the relative movements of the Earth and the Sun, which provides the year, month, day, hour, and other time units.

If we encounter an alien who has just arrived from a distant planet, he will not know our concepts about time scales, such as year, month, day, and so on. Because it is the Earthling who defines them according to the relative movement of the Earth and the Sun, they are native products. A complete cycle of the Earth around the Sun is a year, the Earth self-rotating one round is a day, and then the year is divided into months, and so on. Aliens are not familiar with the Earth. At the beginning of their arrival, they don't know these relationships. Say "one year" to an alien, or "a light-year," and he will not know what you are saying.

If we encounter an alien who has just arrived from a distant planet, will this being know our time scales, such as the terrestrial year, month, day, and so on. Because it is we on Earth when they arrive, they don't know these relationships. Say "one year" to an alien, or "a light-year," and the alien will not know what you are talking about.

But we do when speaking with each other. When talking about the measurement unit of "scale time," we Earthlings understand what the other side is saying.

Question: What time is it now?
Answer: Twelve o'clock.

This concise answer contains a part of the definition of the "time" concept, such as a day is divided into 24 units called hours, but it also implies the content of the day, week, month and year. If the answer is "Sunday, February 28, 2021," it is even more wonderful, as this brief answer also implicitly contains the long historical background of mankind.

The meaning of the definition of the concept of "time" of the Earth is closely related to "scale time." Terms such as years, months, days, etc. not only represent the concept of time, but also act as a recognized scale to measure and use. The more mankind evolves, the more precisely the time scale is defined. On the contrary, if thousands of years ago, somebody said "0.001 seconds," no one would know what he was saying. At that time, there was no demand for using such a short time.

To review: we people on Earth have a common understanding of the definition of "time", and this definition has its own strict and irreversible connotations. From an origin point, time evenly, sequentially measures the order of an object's progress — the most fundamental basis for our ability to communicate with each other.

If someone wants to change the definition of time and turn it into a more mysterious, dazzling object, then he can do so, but before this definition is universally accepted and used for communication, he can only keep playing it with himself.

For instance, if we want to define a concept of time that includes 100 dimensions, no problem! We are free to do so. We are also free to write an article on this fascinating concept of time, and have it published. But publication on its own, whether in a scientific journal or popular magazine, does not mean that this concept of time is reasonable. If by chance, the public adopts it and uses it to communicate, and the world operates in accordance with it, then my definition gains the reason it needs. The opposite, of course, is that very few readers pay attention to it and even fewer make use of it.

So, if someone wants to scale time with a variety of arrows or make it higher-dimensional, print it in a book, and publish it, he can only say that he is giving his own time definition. It does not mean that his definition will be accepted by the public, It does not mean that the public will accept his definition, adopting this concept of time, and therefore, he has no way to prove that he can put the existence of objects into numerous historical events so that they really exist there, for example, using a time machine to pop in and out of historical events.

One only needs the definition of time to see that time travel is a farce of thought. We have already refuted such nonscientific time travel in the chapter "Relativistic Literature" in the book <谁有权谈论宇宙 Who Should Talk about the Cosmos>, published as early as 2005.

When scale time is used to measure objects, several important concepts preside: "moment" and "time interval." "Moment" refers to a certain point in time, such as now, noon, and so on. A time interval can be determined by comparing two different moments, the length of which is the value determined by these two moments, such as 8:00 to 12:00.

Existence Time

"**Existence time**" is "time" itself in our usual sense, and refers to the time of a thing and its historical trajectory," which is abbreviated as "existence time".

"Existence time" can only be described sequentially by "scale time"

But why would we put it this way?

According to the understanding of space we discussed in the previous chapter, ultimate space clearly is too large for us to understand. We cannot find the origin time and final end time of ultimate space. Since all subspaces are part of ultimate space, we cannot use them to determine the origin and end of ultimate space.

Subspace is a part of ultimate space. In reality, it is not separated from ultimate space, although we distinguish it as such. Therefore, a subspace, like ultimate space, cannot be measured by time.

Why, then, do we divide subspaces? Because we want to describe something in them. We isolate a subspace we're interested in to trace back an occurrence, change or end of the subspace division: what happened and how it happened. Time in subspace is inseparable from the existence of events or things in subspace. So, we naturally know:

Existence time is only connected with things in space and has nothing to do with space. Everything in the universe, even thinking about the universe, has different intervals, from its occurrence and development to its end. If we want to describe something, we need to know the specific time trajectory of these different time intervals relative to the "now" moment.

Recall, though, that **space has no time attribute,** nor is it directly related to existence time. **Time acts on things** in subspace, **not on the subspace** itself.

From a philosophical point of view, existence time is an "absolute background" of the existence of objects and is an indispensable element in describing objects, from their origin to their end. Everything in the universe is performing in this context. It is in all objects that exist in the universe, but it never deliberately shows itself. It is unique, omnipresent. When you need it, it appears; when you do not need it, it is hidden. It can be said that it is such a special background material that it is ubiquitous and everywhere.

Existence time must relate to things in the subspace. **Scale time** is a measure of the sequential progress of the existence time of the things contained in the subspace.

Here we need to pay attention. Existence time involves all objects in space, so local time variation cannot become a universal law.

After Einstein's notion, this pseudo-god concept of his, are we any closer to understanding time and space, particularly in regard to time and space conversion and time dilation.

We believe that Einstein's "time dilation" binds time and space together rigidly, and directly spreads confusion. It also does not express the kind of data that can confirm it. In the end, it cannot accurately measure the time of any object.

Because nothing in the universe is absolutely static, for every object and event, there are numerous far or near, visible or invisible objects in relative movement with it. Each pair of objects or systems are in relative movement to one another, and the systems they define are in relative movement to one another, including subsystems and the system within the subsystem and the system layers.

But according to Einstein's theory of relativity, each relative pairing will affect the time of the object or event. Why? Because the Earth is relative to the Sun, the Moon, the Milky Way, the collision of the cosmic edge of a nebula... In the universe, everything is moving relative to other objects. Whether we are willing to admit it or not, the Earth is moving, so everything is relative to the Earth, or relative to the Earth in terms of movement or static (it's doubtful that Einstein's so-called static system exists in the universe). So, according to Einstein's theory of relativistic time, these work together in a variety of different ways to finally produce what kind of result? Up to now, none. It can only be said that what Einstein gave us was intriguing enough to fool us into believing it.

Einstein's mistake in the theory of relativity relates to what can be called "time dilation" and his assertion that it could be achieved. But in fact, it cannot. First, if time changes occur in all local sites, measuring the time of the universe is impossible. And if the speed of an inertial system slows down relative to another system, this also includes atomic electrons slowing down. Where does this energy to change things come from? And light, gravity, air, and bacteria ... Does their aging slow down?

Even if we can identify a "time dilation" phenomenon, the result is measured out and compared to the standard "scale time" of an object within a system. It can be more scientifically explained that the objects within the system are affected by the relative movement: the time speed of some passages has changed, the aging rate for some people or animals becomes slow, and so on. It needs to be studied from different relevant aspects, such as material science, medicine, etc.

The above remarks on Einstein's theory seem quite superfluous now. Although they are intriguing enough and can invite further consideration, they do not accord with what happens in reality; that is, the relativity between independent systems can only be carried out through thoughts, describing the movement of the system no longer needs to consider the influence of the relativity of the system, because for a system with 2 **independent** relative reference bodies that can only be ideologically relative to each other, it is impossible for there to be any substantial physical or chemical interactions between them.

Criticism of Time Travel

Modern Astrophysics is mixed up with superstitions and confusing things related to time travel. And this kind of thing, fueled by celebrated scientists, such as Einstein and Hawking, is driven today by modern physicists who prefer consensus to independent thinking, has occupied the minds of modern people in the name of "science."

Here is a true story that can clarify what's going on. It concerns a scientific project and the reception of a letter from a dead Las Vegas casino boss.

Some years back, Professor Zhiyong Wang, who was the head of the delegation of the first Chunhui program of the Ministry of Education, as well as an associate professor in the United States at the University of Nevada, told us this

A famous casino owner died, and though his heir suspected that the dead boss's mistress killed him, no evidence was found.

This heir, who absolutely believed in science, remembered that Einstein promoted the idea of time travel. If Einstein knew that time travel could happen, there would be no problem in sending and receiving letters from those who died a short time ago.

He approached professors at the University of Nevada and offered them $3 million to study how to communicate with the dead. The purpose was to send a letter to the dead casino owner and receive a response from him proving that he was murdered by his lover.

Several professors at the University of Nevada were so enthusiastic that they wrote a research project plan at rapid speed, in which there were meant 1, 2, 3, 4..., research steps A, B, C, D...etc. It seemed the solution to the case was just around the corner. Their results would make killers all over the world much easier to catch.

The main theoretical basis that these professors used was the past light cone described by Hawking, according to Einstein's space-time theory. They thought it would enable them to transmit information – the aforesaid letter — into the past.

Although when Professor Wang told this tale to me, he did it as a joke – and it does seem quite funny – the research went ahead without results. Money and time were spent on a useless research project.

The question posed by the research was clear enough: how can one send information from the present to the past for a particular person who died recently: the casino owner?

Those doing the research, which failed so miserably, did so as modern cosmologists; facilitated by the light cone theory of past events.

Could this project be successfully completed with the help of past events?

What exactly is the light cone of past events, and can it spread information to the past? Let's study this a bit more carefully, even though it reiterates discussion in the previous chapter.

Einstein directly applied Minkowski's space-time mathematical model without any proof or explanation, combined with a simplified version of the three-dimensional space-time light cone obtained from this model. The combination enabled him to theoretically consider how current events could propagate into past space-time. This kind of behavior, which has no experiential or factual basis, and only exists according to mathematical models, is a typical example of the kind of research pursued today, and as discussed in the book, *Lost in Mathematics*.

If this is the basis of a new theory with such far-reaching consequences – altering the course of human history, destiny, and cognition — then must be thoroughly and scientifically proved. It must be theoretically true and practically verified.

However, Einstein and his followers, represented by Mr. Hawking, used a careless attitude, and "similarly" accidentally spread the doubtful "time travel theory."

As can be seen from Mr. Hawking's writings, he later found himself doing the absurdity of propagating time travel through math in <*A Brief History of Time*>, so 13 years later he published <*The Universe in a Nutshell*> to quietly correct this mistake. However, the bad influence had already spread over a large area, and this quiet and hidden correction attracted few people's attention. To this day, **time travel** defined by the past light cone has become an orthodox theory of relativity.

For example, we mentioned in Phys. Rev. D 97, 124038 published on June 15, 2018, the article entitled "Time travel using degenerate metrics", which discusses how to travel through time and space in a well-known mainstream academic journal.

People promoted Einstein's joke and Hawking's negligence and quietly corrected the "science" theory, which was officially presented in the elegant hall of mainstream magazines.

Despite being ridiculous, it remains!

Isn't this ridiculous "science" that needs a thorough clarification?

From the perspective of practical applications in the real world, the past and future space-time light cones are wrong

1. Introduction of Space-Time Light Cone

The original work on light cones is the mathematical paper "Space and Time" by Minkowski, collected in <*The Principle of Relativity*>, while its popular description of the space-time light cone comes from Mr. Hawking's <*A Brief History of Time* >and <*The Universe in a Nutshell*>.

Minkowski's mathematics can, of course, easily, and perfectly combine space and time, making people feel the beauty and power of mathematics. But when Hawking told us how mathematicians applied such pure, beautiful mathematics into the real world, we finally came to understand that those same mathematicians had ignored the real world, so fascinated were they by their mathematical abstractions. And while this made the space-time cone appear beautiful, it also masked the errors involved both in its application and as a lesson for further mathematical thought about space-time, which we have discussed.

We won't talk about abstract mathematics because it is always right. In the world of abstract mathematics, there is no right or wrong, and the mysteries do not need us to understand.

What we care most about here is this: when mathematics becomes a model applied to the real world and which application affects our lives, we must investigate whether or not the math, however beautiful it seems, is properly applied, or just made up a new myth with it?

First, we determine the rationality and applicability of the mathematical model and then discuss its application. Whether or not a mathematical model can be applied to reality, and whether or not it helps us to understand, explain, or solve real-life problems or causes confusion.

The space-time light cone is a typical example of using beautiful mathematics and ending up with chaos, errors, and ugliness. Simply put, it is as if a person born in 1950 became one year old in 1949, five years old in 1945, and attained the ripe old age of 70 in 1880, and was living in the Qing Dynasty.

Let's take another look, then, at Hawking's narrative of the space-time light cone in *A Brief History of Time*, published in 1988, and see how he popularized the ridiculous idea of time travel for readers who have faith in modern science.

Minkowski's four-dimensional space-time mathematical model is tailored for relativistic effect, and the light cone is a three-dimensional view after replacing the one-dimensional space with the time dimension. Mr. Hawking explained a lot of space-time light cones in popular language. Because our book is also trying to use popular language to explain the universe, Mr. Hawking's narrative is heavily borrowed in this book.

In his book, Hawking explained the event light cone as follows:

"It follows from that if a pulse of light is emitted at a particular time at a particular point in space, then as time goes on it will spread out as a sphere of light whose size and position are independent of the speed of the source. After one-millionth of a second the light will have spread out to form a sphere with a radius of 300 meters; after two-millionths of a second, the radius will be 600 meters; and so on. It will be like the ripples that spread out on the surface of a pond when a stone is thrown in. The ripples spread out as a circle that gets bigger as time goes on. If one thinks of a three-dimensional model consisting of the two-dimensional surface of the pond and the one dimension of time, the expanding circle of ripples will mark out a cone whose tip is at

the place and time at which the stone hit the water. "

The above paragraph told us. starting from "now," how water waves propagating over time will form a cone at some point in the future. The future light cone is formed in a similar manner but in a two-dimensional space with the time interval, as it increases, represented as well. (See the left section of Figure 9.1 below).

In the next paragraph Hawking told us how the light cone, in fact, imitates the water wave to form at some point in the future; that is, the future light cone. Then, along a mathematically symmetrical trajectory, how did Hawking "similarly" let the light wave propagate in the past, referenced by the past light cone in the lower right part of Figure 9.1. Here we can see a typical example of a master **gone astray, led by the beauty of mathematical symmetry.**

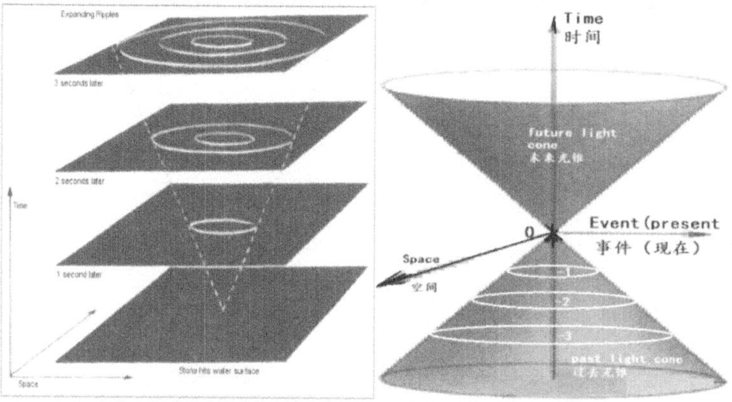

Figure 9.1 Left: Standing now and looking at the future. A stone hits the water, from <A Brief History of Time Figure 2.3>. Right: Event light cone used to explain relativity, also from the same book Figure 2.4>

Hawking said: ***"Similarly**, the light spreading out from an event forms a (three-dimensional) cone in (the four-dimensional) space-time. This cone is called the future light cone of the event. **In the same way,** we can **draw** another cone, called the **past light cone, which is the set of events from which a pulse of light is able to reach the given event.*** " (Bold added by the author.)

According to the above description, Hawking drew the complete space-time light cone in the right frame of Figure 9.1. It consists of a future light cone in the upper part, a current event point in the middle, and a past light cone in the lower part.

However, in the sense of application, the above picture is almost entirely wrong. Let's take stock of these errors one by one.

Hawking's last sentence says, "In the same way, we can draw another cone, called the past light cone, which is the **set of events from which a pulse of light is able to reach the given event.**"

Boldface marks the part of the sentence that is difficult to understand. Now let's try to understand the correct meaning of this sentence: the past light cone is a collection of past events. From this past light cone, what is a "light pulse"? How can it "reach the given event"? What is the given event? We guess that it should be the current event in Figure 9.1.

In this way, we have two questions left:

1. What does "the past light cone is a collection of past events" mean? How can we draw the past light cone based on this? What past events do we refer to?

2. What is "a light pulse"? Who or what emits light pulses? How does it move to "be able to reach a given event?"

Finally, if the "light pulse" originates in a laser or is similar to such light, it is impossible to spread out in the form of a photosphere.

Because readers couldn't understand the last sentence, and because Hawking said, "*Similarly, **in the same way**, we can draw another cone called the past light cone*," readers of the Chinese translation, including me, and readers from all over the world who don't understand the original English text, including Hollywood screenwriters, automatically filtered out the second half of Hawking's description – it is incomprehensible — and understood the past light cone in a way similar to Hawking. This is why readers thought that Hawking painted the light cone into the past.

We will only comment on Hawking's past light cone "in the same way."

2. Hawking's Light Cone of the Past – A Time Travel through the Scientific Source of Fantasy

Hawking apparently did not think carefully about this issue when he wrote *A Brief History of Time*. Like most theoretical physicists, their interest in mathematical models far outweighs the reasonable applications they can be used for. Hawking summed up the whole process of drawing the past light cone with a simple sentence: "***In the same way,*** *we can* ***draw*** *another cone, called the* ***past light cone***."

The "in the same way" drawing method is shown in Figure 9.1, starting from the time of "now (that is, 0 on the time axis)" and propagating through three oval sections. In the first oval, the radius of light propagation is 1 light-second; that is, one light second in the past (with the coordinate -1). This continues sequentially until three light seconds in the past (with the coordinate

-3). Continue in this fashion to draw an oval for the nth light second in the past, and we obtain the past light cone in the lower half of the space-time light cone on the right side of Figure 9.1.

The **craziest** thing here is: it is only when the light of the event spreads **from the present to the past** that this past light cone can be drawn! No other "same method" can draw this past light cone! The reader can pick up a pen and try it.

In this way, the past light cone in the space-time light cone in the right of Figure 9.1, in the name of science, theoretically initiated a "scientific" space-time journey through the past.

In other words, Hawking "similarly" invented time travel to the past.

Please recall the date when time travel became popular?

This is the "scientific" origin of time travel.

And to top it off all of this is completely mathematically correct. The beauty of mathematical symmetry is revealed once again.

But when applied to reality, it is completely wrong.

For example, we have started our car: now. In terms of time, we can only drive to the next minute after now, to an hour from now, to tomorrow, and to the future, all of which is covered by Hawking's future light cone. But this car cannot drive into the past, whether a minute or a day in the past.

By the way, can the car head to both the future and the past at the same time?

Mathematically, Hawking "similarly" drew the past light cone. How simple and crisp!

But in real life, the past light cone cannot be drawn "similarly" to the future light cone!

Events that happen "now" cannot spread to the past! This is the consensus of mankind. It is a truth that has not changed for ages. It is an important basic concept for us to distinguish the past, the present, and the future. It is a natural law that human beings must abide by.

If a drastic change is to be made to this natural law, doesn't it require complete verification?

But contemporary cosmic physicists, led (perhaps hypnotized) by the symmetrical beauty of mathematics, have casually and arbitrarily changed the truth that humans share in regard to time. Hawking himself casually used the method of depicting the light cone of the future to "similarly" draw the light cone of the past, completely ignoring the rules of nature and the basic principles that human beings abide by.

In Hawking's description of the past light cone, he also uses vague terms, such as "similar" and "same," in order to define the theoretical basis of time travel to the past. It is a model with two contrary values: correct mathematics but incorrect application. Due to the widespread dissemination of the book, the possibility of time travel seems to have been proved scientifically. But after thinking about it for some13 years, Hawking repudiated this possibility. In his book, *The Universe in a Nutshell*, he described the past light cone from six new angles. But each of them is still problematic.

Hossenfelder, the German theoretical physicist we referred to earlier, said in her book *Lost in Math*:

Theorists are convinced that their elegant equations and elegant formulas contain higher truths about nature. And these "too good to be untrue" theories have plunged the field into a dead end.

Doesn't the light cone here fully prove her words?

Writing here, we can't help but be sad for Giordano Bruno, the sixteenth century philosopher and scientist, who was burned at the stake by the Church for his "heretical views." To adhere to the truth of science when pitted against pseudo-theology can entail risks, as we know. Has that kind of situation returned? Perhaps it has and we haven't yet recognized it, especially in regard to the Big Bang theory (its own form of "scientific" pseudo-theology.)

Time travel, which caters to the psychology of people hunting for novelty, has also spurred an expansive set of particularly interesting themes for blockbuster Hollywood films. It has spread throughout the world at the speed of light. And the great success of Hawking's *A Brief History of Time* has facilitated and spread the ridiculous and unthought-out "time travel" to the extreme! The book was published in 1988. Since then, the story of the time traversal began to spread magnificently, and there is a tendency to gradually ascend to the halls of magnificence.

Probably, the "scientific" description in *<A Brief History of Time>* about the light spreading towards the past is too absurd, the originator, Hawking himself should have received many different opinions and criticisms from real scientists. After spending a lot of time thinking, in the book *<The Universe in A Nutshell>*, published 13 years later, in 2001, he **re-explained the past light cone** using a lot of pages. This is a disguised approach to correcting his own mistakes, and it is exactly the same as Einstein's method of correcting his confusion between "simultaneous" and "synchronous. "

Nonetheless, even with his correction, Hawking was not able to avoid problems. Indeed, the past light cone he re-drew and explained in *The Universe in a Nutshell,* although denied backward time travel, is again erroneous and difficult to understand, cannot help one to correctly understand the relationship between time and space and correct wrong ideas. As a result, time travel sustains its position in orthodox cosmological theory.

Let 's take a look at all this now in detail. After thinking about it for more than ten years, Mr. Hawking re-described the past light cone from six new, if still problematic, angles in "The Universe of Nutshell", but each has its problems.

Through careful comparison of his description of the light cone in the two books published more than ten years apart, we can see that his thinking completely changed. He gave up a favorite idea: time travel. Unfortunately, theoretical physicist that he was, he did not, at the same time, clarify the relationship between the mathematical model of space-time and the light-space cone. Not only is it in conflict with the narrative in *A Brief History of Time*, but it also quite distinctly differs from basic scientific principles that he should have followed from the get go.

Please go back and review Hawking's past light cone in Figure 9.1 (from *A Brief History of Time)*. Then let's look at the new past light cone that Hawking redrew in *The Universe in a Nutshell*.

3. 13 Years Later, Hawking's Modified Past Light Cone, but Still Has Many Mistakes

We are not saying that Einstein's view of time and space was wrong, but that Hawking's view of time and space was wrong. The main reason is that Hawking has expressed his thoughts in two popular books, <*A Brief History of Time*> and <*The Universe in a Nutshell*>. *Although* his expressions were quite different from beginning to end, both of them were wrong about the past light cone!

From a mathematical point of view, this is such a simple thing that Hawking is confused about. Perhaps he confused himself.

In addition, Hawking also introduced concepts, such as black holes. It's important to discuss the issues here clearly. If something so mathematically simply goes wrong for Hawking, it is doubtful that he can get more complicated things right!

Of course, first, we must thank Hawking for revealing, in his several popular science books, his real thoughts and intentions in such a way that readers of all kinds could discuss the issues at hand from a common starting point, without having to refer to complex mathematical equations.

Earlier, we introduced in detail the past light cone in *A Brief History of Time,* and the errors involved in its creation. Although the cone represents a visual image of the theoretical basis of the story that Hawking presented, It exemplifies an example of correct mathematics and wrong application; and it is also a typical example of wrong mathematical application due to wrong physical model.

The description of the past light cone in <*A Brief History of Time*> was introduced in detail earlier. That section defines the theoretical basis of the story. It is a typical example of mathematical correctness and wrong application.

Hawking then gave us a completely different past light cone, as represented in Figure 9.2 below. He did so after he gave us a completely different past light cone in (Figure 9.2).

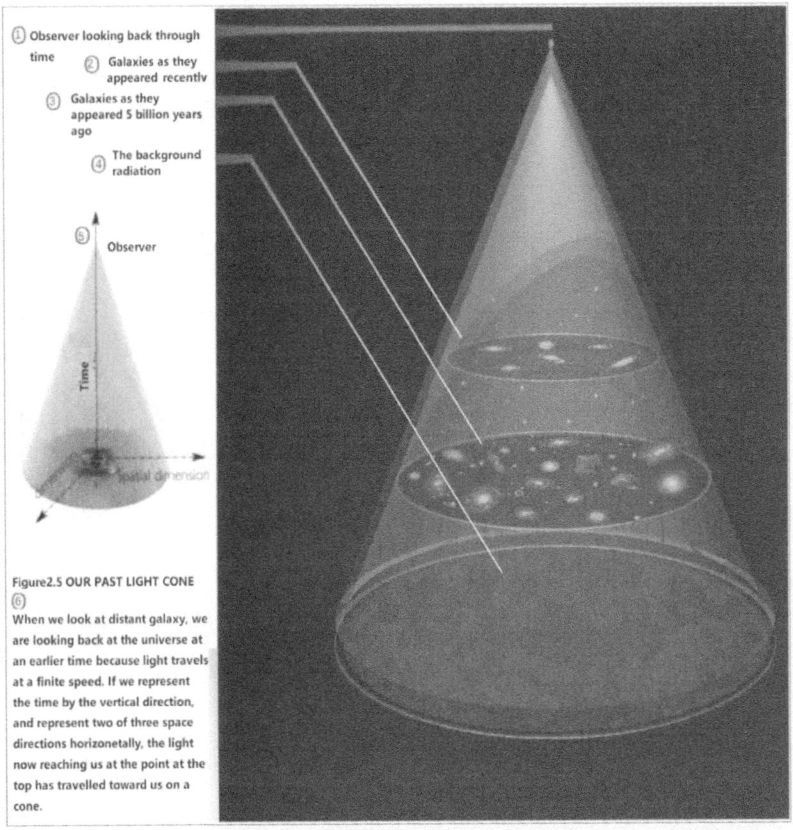

Figure 9.2, Hawking's past light cone and its interpretation. (The original Figure 2.5 in <The Universe in A Nutshell>.)

Thinking about it for more than a decade. The differences between the new and the old show that he quietly abandoned the theory of time travel.

Hawking specifically introduced this past light cone in <*The Universe in A Nutshell*> after thinking about it for more than 10 years. The differences between the new and the old showed that he quietly abandoned the theory of time travel. (Figure 9.2)

Figure 9.2 shows Hawking's past light cone and its new interpretation. The text label on the left side of the figure is a description of the figure on the right, and the right part of the figure is a past light cone with some galaxies drawn. The left part of the figure is the 6 annotations for the 6 small portions of the picture in the right figure. The circled numbers in the figure ①, ②, ③. ④, ⑤, ⑥ are added by us to make it easier to explain clearly.

Figure 9.2 is full of divergences from the past light cone in *A Brief History of Time* yet includes various common-sense errors. In his new version, Hawking used six descriptive labels, each of which, in our estimation, contains errors. For convenience's sake, we refer to the numbers in the circles on the labels in the figure. picture to mark the six labels in order.

① Observer looking back through time

This is the vertex of the light cone where the observer is located. This label shows that Hawking has completely abandoned his time travel, the absurd idea of the light or matter propagating into the past.

This is the vertex of the light cone where the observer is located. Note that we added those circled numbers ①, ②, ..., ⑥ in front of the labels to make it easier to explain.

In *A Brief History of Time*, the vertex of the light cone is "a four-dimensional space of light scattered by an event — a cone of vertices formed in time."

In *The Universe in A Nutshell*, the vertex of the light cone becomes the point where the observer looks back through time. The events no longer spread from here to the future or, "similarly," to the past. This apex no longer has anything to do with the occurrence and spread of the present event in *The Universe in A Nutshell*, but only the "now" moment, which is the time when the observer receives light from the past.

The above demonstrates that Hawking has completely corrected the fuzziness of his theory of time travel, as he initially conceived it. It's a pity that, despite the correction, his colleagues did not take to it as they should have. Almost two decades later, the possibility of time travel still has currency as a viable topic in theoretical physics.

The above fully demonstrates that Mr. Hawking has completely corrected the fuzziness of his theory of time travel. It's a pity that he quietly made corrections like this and did not purify the people who had been "similarly" drawn through the cone of brainwashing! Almost two decades later, time travel is still sitting in the hall of science! And there is a tendency to time travel more and more cheerfully.

And his vertex is also wrong. Because any galaxy spreads its light sphere bigger and bigger from the past to the present and reaches the "now" moment, it cannot be a "point", but a light sphere. This is a common-sense error.

② Galaxies as they appeared recently

③ Galaxies as they appeared 5 billion years ago

We will discuss ② and ③ together. Because it is necessary to compare their respective expressions to find problems easier.

The label is precise. But it also hides an error. This is another classic example of error without going through the brain, which tells us that sometimes even big mathematicians make extremely naive mistakes in applying their mathematics! The beauty of mathematics was rudely destroyed in this application. This reminds me, in its way, of a story. In the period of the Cultural Revolution in China – approximately 1966 to 1976 – middle school students noticed and criticized that their electro engineering professors didn't know how to install an electric motor, and criticized him for it. But in fact, this is entirely possible, because such a professor is engaged in the theory of electricity and is not an electrician. It seems funny to us and perhaps seemed funny to the students, but there is nothing to make a joke about. The same holds true for mathematicians and their application models. Application is key, and a key factor in determining whether or not a beautiful mathematical application model is correct.

In Figure 9.2, this all becomes obvious after comparing label ② pointed "Galaxies that have recently appeared" with label ③ "Galaxies as they appeared 5 billion years ago" pointed. Apparently, Hawking has concluded that, in the past light cone, galaxies closer to the observer were younger.

Did we misrepresent Mr. Hawking's intentions? Let's carefully compare the labels ② and ③ with the corresponding images in Figure 9.2 and combine them with the distances from the observer in the figure, to identify the truth carefully.

What do you think? Think about it before you try to answer the question.

Why is it wrong? Take, for example, Abell754. Experts tell us that it is 800 million light-years away from the Earth and its age is 300 million years. The distance between the Sun and the Earth is 8 light-minutes, and the age of the Sun is 4.6 billion years. Why is the Sun so close to us but its age is so much older than Abell754, whose distance from us is billions of light-years?

Such mistakes should not can be made in a popular science book. We really don't know how to comment.

Will the most recent appeared galaxies be younger galaxies? Of course not. A galaxy's age is 10 billion and one years old, away from the observer 10 billion light-years, then, for the observer, these tens billions of years old galaxy has just appeared for one year.

Do recently-appearing galaxies get closer to the observer than later-appearing galaxies? The galaxy in the example above, which has just appeared

1 year, is ten billions of light-years away from the observer.

And the age of the Earth and its distance from the observer can be discussed from another angle.

In ③, Galaxies appeared 5 billion years ago. Mr. Hawking used his new past light cone to represent "galaxies as they appeared 5 billion years ago," with many galaxies in a cross section at 5 billion light-years, which is again totally wrong.

In the light cone drawn according to the mathematical model, the cross section at any moment indicates the maximum distance that the event light can reach at that moment. For a galaxy located at X years, the time required for its light to travel to the observer is X years (without considering other factors such as the expansion of the universe).

Therefore, the galaxy located on the "cross-section at 5 billion light years" may be 1 year old, 100 years old, 100 million years old, 5 billion years old, 10 billion years old ...; and the galaxies at "now" moment may be observed in this cross section must be 5 billion years old or older.

On the other hand, it has been said before, the "galaxies that appeared 5 billion years ago" may be located at any distance from the observer, not necessarily as shown in Mr. Hawking's picture, only on the cross section of the light cone 5 billion years away from the observer. It's like the Sun is 4.6 billion years old, but it's as near as 8 light-minutes from the Earth. According to a 2013 study by Bond et al, HD140283, about 13 billion years old, is 190 light-years away from the Sun.

④. The background radiation

The background radiation at the bottom of the light cone on the right of Figure 9.2 indicates that the cosmic background radiation is located at the edge of the Big Bang universe. And if the background radiation is only at the edge of the universe (the bottom of the light cone), it is impossible to receive such a weak microwave near the Earth. It would be even more ridiculous if Mr. Hawking believed that the microwave background radiation "detected" by the space probe came from the edge of the universe! This has been analyzed before.

Moreover, according to NASA's orthodox theory, the microwave background is distributed almost uniformly throughout the universe. If Mr. Hawking believes that the bottom section of the light cone represents the entire universe, then he is very wrong.

⑤. Observer

In general, unless otherwise specified, according to the most basic convention, the intersection of the three coordinate axes is the origin of the coordinate system. In Figure 9.2, the current time (i.e., time = 0) in the left picture is located at the bottom of the light cone, which represents the beginning of the Big Bang universe. The observer's time coordinate should be 13.7 billion years after the Big Bang. However, the vertex time of the light cone in the large picture on the right in Figure 9.2 should obviously be set at the vertex, which is the "now" of 0 time. It is not a serious scientific attitude that a scientist should have to obfuscate the origin of the timeline in the two sub-pictures on the same page with the same figure without explanation.

⑥. Our past light cone

The last sentence of the description at the bottom left of the picture states: "When we look at distant galaxies…our rays reaching far at the apex come from a cone." The Chinese translation of this word is "conical surface," but the original English version is "cone." The difference is significant. We take the comments in the original English text.

Here Mr. Hawking spoke without going through his brain again…

Regardless of whether it refers to the light of a single galaxy, multiple galaxies, or even all galaxies, "now" will not be just a point.

The light of a galaxy travels in a sphere, not a light cone. A light cone is an image in which the light sphere in three-dimensional space is simplified and compressed into two dimensions, and then the largest possible propagation range is sequentially drawn in chronological order. But the spread of starlight is not in a cone of light, but in a light sphere.

To sum up the above 6 points:

1. The original meaning of the space-time light cone is to express the four-dimensional space-time relationship of light traveling with time in three-dimensional space. But because humans can only draw three-dimensional views, they compress a space dimension, and it becomes a three-dimensional image of two-dimensional space plus one-dimensional time. When light travels in a sphere, there is no way to draw light to the past. Readers, please try it out. Therefore, in the past, light cones could not exist in true four-dimensional space. Only after changing to a three-dimensional plane-time image can we draw an image of light spreading with time.

2. Einstein, Hawking, and other astronomers of the physical universe for hundreds of years, on a three-dimensional light cone diagram with two-dimensional planes and one-dimensional time without further thought, have drawn a past light cone with mathematical symmetry. It is ridiculous for Hawking to make himself the originator and defender of non-scientific time travel. After discovering his mistake, he did not admit the error generously but modified the past light cone from a new perspective for sophistry.

3. However, since light cones did not exist in the past, it is impossible to

defend past light cone seamlessly. This leaky new past light cone negates the time travel to the past, but it does not and can't explain why the light transmitted by the celestial body from a distant place in the past to the observer at present will shrink to a point. Of course, if we can barely explain the new past light cone from the perspective of information decaying with distance, etc., then is it still the light information describing the propagation of the event in space-time?

In Hawking's new past light cone, one thing is certain: he once again denied his wrong description of the past light cone in <*A Brief History of Time*>, which also completely deviated from the space-time mathematical model of Einstein and Minkowski.

4. What Should the Correct Light Cone Look Like?

Among the existing three-dimensional light cones of time and space, only the "Now" point is correct. Because light cannot travel from the present to the past, a past light cone is a fabrication. In addition, when the light from a celestial body reaches an observer, it cannot be an unchanged point. As far as the light cone of the future is concerned, it is only an expectation that hasn't yet happened.

So, the only symmetry here, from Minkowski, Einstein and Hawking, is their use of a problematic light cone of space and time, and its adoption by numerous other scientists, with only the "now" being correct, and everything else wrong.

What then should a more reasonable space-time light cone look like? See Figure 9.3 below.

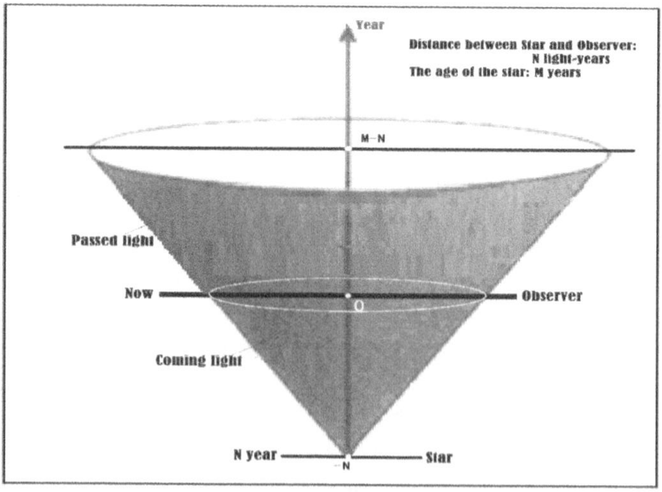

Figure 9.3 The correct new space-time light cone,

Please compare this light cone with Hawking's light cone in Figure 9.1. In the new light cone, the past light cone no longer exists. When the light of the star reaches the observer's position, it is a spherical surface, not a point. The only place is the point when the stars have just been born and have just begun to emit light. Note that from the observer’s point of view, his location is "now" at time 0; but for a star, the light emitted from its location is also emitted at the "now" time, and it has a radius of N when the light reaches the observer. The spatial axis is not drawn in the figure.

After the light is emitted from the star, it propagates to space in a continuously increasing spherical shape with the progress of time. In Figure 9.3, assume that the distance between the star and the observer is N light years, and the age of the star is M years.

Take the observer's location as the "now" moment, that is, the origin of the time axis at time 0. Any time unit can be used as its unit value on the time axis. In this figure, it is assumed that the unit of the time axis is "year". Three points are marked on the time axis: the -N point where the observed star is N light-years away from the observer, the "now" moment where the observer is at the zero origin, and the entire time from the birth of the star to the present luminescence; and (M - N) point, where M is the age of the star and N is the distance between the star and the observer.

Again, compare with the old light cone on the right in Figure 9.1. In the new light cone, the past light cone no longer exists, and when the light of the star reaches the observer's position, it is a sphere and cannot be a point. When the star has just been born and has just begun to emit light, it's just a point. Note that from the observer's point of view, the observer's location is "now" on the time axis: time 0; but for a star, the light emitted from the star's location is also emitted at "now" time. The spatial axis is not drawn in the figure.

In the new light cone, the cross-section perpendicular to any point X on the time axis indicates that the propagation range of light at that point is a circle with X as the radius. It can be seen that at the position of the star -N years, the stellar light has just begun to emit, and its propagation range is zero. When the stellar light reaches the observer after N years of propagation, the propagation range of the stellar light is a circle with a radius of N light-years.

The origin 0 where the observer is located is the watershed of the stellar light propagation. Before this point, the starlight has not reached the observer yet, they are propagating to the observer, so they are "Coming lights" for the observer; the star lights passed the origin, have passed the observer, and have left. The observer has seen them, but will never see them again, so they are "passed light".

Since the age of the star is M years, the entire propagation range of stellar light is a photosphere with a radius of M light-years, which is the coordinate of (M-N) years on the time axis.

In fact, Figure 9.3 depicts only the situation where M is greater than N. Think about it, if M is less than N, then what should Figure 9.3 look like?

There is also a problem left to be considered: in terms of star luminescence, the location of the star should be the origin of the new light cone, because the spread range of the star here is 0, and the light emitted here is emitted at the "now" moment. But in such a case, only one star can be drawn in one light cone diagram. Therefore, the observer's location is selected as the origin of the time axis. In this way, all the stars that need to be expressed can be drawn into one new light cone.

In order to understand the above description more intuitively, we use the data of the Sun as a substitute for the star in Figure 9.3, as shown in Figure 9.4 (of course it is a schematic, the size is completely out of proportion):

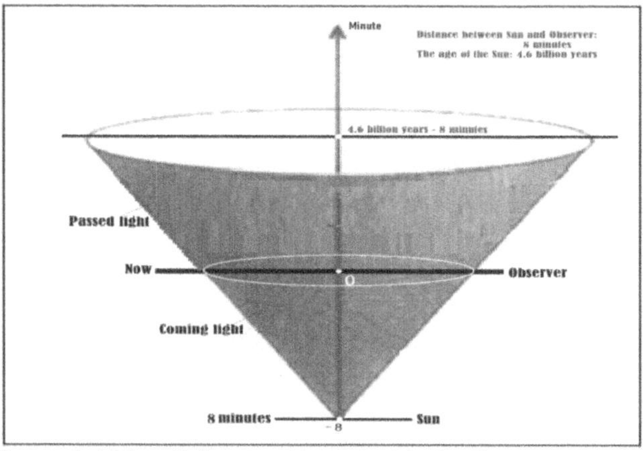

Figure 9.4 New light cone drawn with the Sun as an example

When the age M of the star and the distance between the star and the observer is N light-years, when the relative values of the two data M and N are various, their new light cones vary. They are depicted by the three small images in Figure 10.4 below.

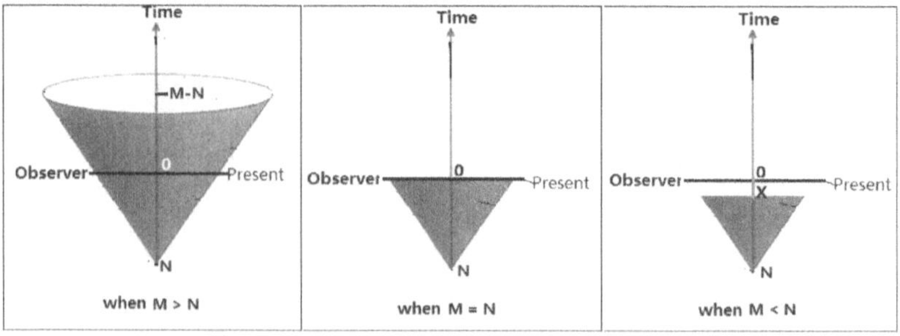

Figure 9.5 The three shapes of new light cones with the distances are N light years between the observer and the star, and the lifetime of the star is M years

In Figure 9.5, there are three new light cones, each distant N light years from the observer. M indicates the lifetime of the star as M years.

When M > N, like the Sun, the starlight already passed the "now" and the picture frame describe most of the star situation,

When M = N, the star light just arrived "now" or observer.

In the third case, when M < N, the light emitted by the star has not reached the observer, so the observer does not see the star, it is only the result of normal scientific reasoning.

Chapter Summary

We have discussed why Hawking's use of light cones in his two books, while different, with one correcting its predecessor, remain incorrect. It's shocking to consider how application of such a simple and ready-made mathematical model can have so many problems, and how so many scientists still use it blindly.

I'm relieved that although Mr. Hawking accidentally "similarly" became the originator of ridiculous time travel in his carelessness, after thinking for more than ten years, he completely denied the "scientific" basis of time travel. Although this change was quietly obscure, it was the right choice after all. And although this new explanation is full of fallacies, it is not so absurd when compared with the theory; after all, it is normal to apply beautiful mathematical models to the more normal mistakes made in the universe, although these mistakes are relatively low-level.

After comprehensively studying the application of the space-time light cone, other than by its corrected example shown in Figure 9.3 and 9.4, we have found that it is almost useless, except for the identification of the "now" point. To be more specific, the future light cone is something happening after "now." It belongs to the category of prediction. But there is nothing about prediction in this space-time light cone. In the past light cones, because light traveled backwards in time in this model, it did not exist in the real world. It just presented a "scientific" endorsement to time travel.

Compared with our four-dimensional space-time curve introduced earlier, it is clear at a glance which is right and wrong.

Tracing back to its origin, the space-time light-cone is derived from Minkowski's "Space and Time" thesis. After Einstein's usage of it (however baseless it was), and Hawking's reluctant application, it has been widely circulated in the field of astronomy and cosmology.

Minkowski's paper is a beautiful mathematics paper, the description of the symmetrical beauty of the space-time cone is very charming. If you do not carefully explore the rationality of the application in the real world, no problems will arise.

This kind of mathematical model, which obviously violates the rules of the real world, has dominated our scientific thinking now for over a century. People cannot help but ask why: Why does this happen?

The further discussion of this issue will be presented later as a summary of the book.

Part 4

Calling for Creation

How do human beings view the universe and understand the universe?

What kind of outlook should we have on life, the world, and the universe?

Looking back at the past hundred years, although we have been trying to understand the true face of the universe scientifically, as it turns out we have in large part failed, the truth ever receding from our grasp.

In order to counteract that regression as we can, we have also written this book for the layman. In it we have identified issues that require a new perspective. This requirement encompasses the pseudo-theological nature of the Big Bang theory, various other problematic theories in contemporary astronomy and cosmology that support it, and the strange phenomenon of treating such mathematics as reality while overlooking the errors involved when applying mathematics to reality.

In brief, and based on the following two reasons, we feel it important to carefully reconsider consensus scientific beliefs about the universe, based mainly on the following two reasons:

1. The acceptance of the Big Bang pseudo-theological theory has largely marginalized, if not prevented, the kind of independent critical thinking necessary to consider other opinions.

2. The apparent lack of innovative awareness and pioneering spirit in regard to the scientific understanding of the universe points to one conclusion: something has gone seriously wrong – including with a large number of NASA astronomical observations where easily identifiable mistakes have been made and which we have discussed previously.

A large number of astronomical observations of NASA are pointed out to be simple mistakes that everyone can easily judge; Is there something wrong with the contemporary "scientific" worldview led by NASA?

Therefore, the question naturally arises: Why has this occurred? Is there a problem with the contemporary "scientific" worldview headed by NASA?

Take the author's personal experience of nearly 20 years of research, thinking, and writing as an example.

With the publication of several books and more than ten related research papers published in the journal *Matter Regularity*, one would think that other mainstream scientific journals would be interested in the perspectives expressed. But all of them were rejected by all kinds of journals. For example, when I submitted one of the papers to the *International Journal of Theoretical Physics*, it was rejected with this scant explanation: "*we have decided not to accept your offer. The paper did not undergo technical review and is not being declined for any technical error.*" Is this a normal scientific argument?

It is the common attitude of mainstream scientific journals to treat papers similar to ours, which criticize the Big Bang Theory, as unpublishable. In effect if not purpose, consensus option must be preserved, whether or not its truth value holds up under criticism. Interested readers can go to the various journals that publish papers on astronomy and the universe, and they will be impressed.

Of course, scholars must have their own views like those on the simple issues pointed out in this book, but those will be equally unpublished. And researchers need to publish papers to maintain their academic status and jobs. If the innovation paper could not be published, they can only write papers that follow the trend.

But even if you don't agree with us about the creation of the universe, why don't you dare to listen to us and discuss the differences? Hopefully, this book will prompt debate, and in the process rejuvenate our capacities to apply mathematics correctly.

We abandon the scientific god who imprisons ideological progress!

We call on the God of Creation who drives history forward!

We need a truly scientific worldview cosmology!

Know this as well: the main points of this book are derived from several other titles, including "*谁有权谈论宇宙 Who Should Talk about Cosmos*" published in 2005, *Debate of Light and Dark,* published in 2017, *Fading Modern Cosmology,* published in 2020, and *10 Questions to NASA*, published in 2021. About20 years deep thinking, reasoning and summarizing has led to the crystallization of perspectives expressed in this book.

Chapter 10

New Cosmic Model Framework

If Tao can be discussed, it is not a common Tao. If a name can be named, it is not a common name. Nameless is the beginning of heaven and Earth; named is the mother of all things.

— "Laozi · Dao De Jing"

The night sky is fascinating, psychedelic, deep, colorful, hazy and full of thousands of years of dreams!

During the day, we seem to have little to do with the universe. The intense light from the Sun only allows us to see the surrounding landscape clearly.

At night, faint starlight can lead us to a place as far away as billions of light years.

Daytime is like a kind gentleman; He is easy and deep.

Night is a mysterious story, ancient and long.

Daytime is like intranet, which we can use at home or at work, connecting both; nighttime is like the internet, letting us sightsee over the entire world.

The night sky has also been the subject of recording. To this day, countless space telescopes transmit countless observation data that we store in massive databases. But how best to use this data, so that ordinary folks can understand it?

We have discussed the credibility of the data collected by space telescopes in previous chapters. From very distant celestial objects, be they stars or galaxies or clusters of galaxies, we have no way to obtain true reliable data. Data with low credibility? Yes, there is much of that. Simply, there are too many definite and unmistakable errors involved in the interpretation of such data by astronomers and cosmologists.

Let's take for an example the processing errors made on research data received from distant celestial bodies, specifically on the generation of new stars after nebulae collide. We refer particularly to the errors made about Abell 754, the galaxy cluster in the constellation Hydra, which we have discussed previously. Interpretations of Abel 754 are shot through with obviously ridiculous errors, and full of fairy tales.

Scientific fantasy and reality: just by endlessly discussing dark matter, dark energy, black holes, the Big Bang, wormholes, etc., as much in scientific as in popular literature, has made too many of us believe in their existence despite the fact that they are theoretical constructs. We are catering to an era – our own – that requires the colors of fantasy.

Someone is keen on parallel worlds. But according to science fiction movies and novels, it should not be called parallel world but eternal world. You enter that world at any time, everything in that has lived continues to live, which means that this world abides in a static eternity without the passage of seconds, minutes or hours. Excuse me: You were a minute ago and you are now, at the same time. How amazing! As you walk you also follow yourself; as you recognize where you are you are also there in the future, no matter how far in the future that is. What kind of story is this?

Many similar reports we have read on the collision of nebulae that generate new stars hundreds of millions of light-years away, based on different research observations make us question the entire event. Why did these mammoth nebula collisions only happen hundreds of millions of light years away?

Let's go back to the Big Bang web page of NASA's Web site at the beginning of this book (Figure 0.1) to see how it is said there.

In Figure 0.1 NASA established two major pillars to support the Big Bang theory.

One is Einstein's curved time and space from general relativity: But space bending is only a local disturbance; how can it be extended to the entirety of space created by the Big Bang? A stellar light wave is infinitely huge, an object blocking it from our point of view is relatively very small. How can such a small object interfere with such a huge light wave? Where does the energy to do that come from? How can it form curved space-time, supporting the Big Bang theory?

The other pillar is the cosmological principle, which states that the universe appears isotropic on large scales and that matter is uniformly distributed, as evidenced by the cosmic microwave background. The truth, however, is something else: According to the needs of a myth that the explosion of an atom created the universe, experts envisioned its result: a microwave background evenly distributed throughout the universe. Thus, in the vicinity of the Earth (humans are only be able to active near the Earth - within 1 light-year), scientists detected microwaves then capitalized on their discovery by telling us that it encompasses the entire universe and is evenly distributed. The size of the universe is approximately 13.7 billion light-years. From less than 1 light-minute to 13.7 billion light-years is quite a leap. So much so that only magical logic could make it seem real.

Wasn't it just a lack of imagination and no ability to find the source of the received microwaves? And then expanded the local finding to the whole universe? Then made it as a whole sky map?

Near the Sun and other stars in space the microwaves are certainly not evenly distributed. How far away then from luminous celestial bodies can the uniform distribution of microwaves begin? Has anyone studied it? Have you ever thought about this problem? If there is no basic model available to us here, which would help us solve this problem, what models are there to help us solve other associated problems, such as Olbers paradox or how to quantitatively describe "dark" or what the scope of the impact of light emitting objects is?

In the universe, there are so many large hidden celestial objects we cannot see. How is it then that we can chart across so distant an area of space, and with such uniformity, microwave "background" radiation?

In my book "*誰有權談論宇宙 Who Should Talk about Cosmos* published in 2005, I have simply and literally depicted several cosmic models, including an extremely huge model, where the Earth is as dust on a giant's hair. And an extremely small model, where planets are only grains of sands in the desert of an immense celestial body, from which we can depict, however much we observe it, just a few lines of sight over a sand dune. The ancients, too, had their cosmic models. They told us the tale of Huang Liang's dream, where people live, and whose whole lives begin and end inside the space of the time spent eating a meal. If you note in this dream that the flow of time is too fast, you can also note the same when living in it.

Einstein's theory of relativity is only a product of pure thought, without any power to support the grand and heavy universe the Big Bang pseudo-theology created. Of the Big Bang theory itself, several unexplained questions remain: why was there a Big Bang? What exploded? What is there outside of the space created by the Big Bang?

It's obvious, isn't it: According to the number of questions that cannot be answered, the Big Bang model is far less reasonable than that of Newton's universe.

The valid data domain of the astronomical observation data

This valid data domain is used to calculate and explain the information received by telescopes from far distant celestial objects.

It includes an **effective time domain of celestial object observation data:** For a celestial object with a distance of D light years, all the information that a telescope receives from the celestial object **occurred before D years.**

If we know that the age of the celestial object is Y years, and the celestial object's maximum coverage area is S light years, then **the ages** of all the events we observed from the celestial object are between D years and (D + Y + S) years.

Such a simple definition will help us to avoid so many embarrassing wrong results.

The essence of the definition does not focus on the wrong ages in many research reports, that is only a superficial phenomenon. It's essence reflects that the astronomical data processing methods already made common mistakes.

It tells us that to roughly determine the time of occurrence of events on a distant celestial object, we must first know the age of the celestial object and the distance of the celestial object from Earth. Then and only then can we calculate the minimum age of the information that is received from the celestial object. If the size of the celestial object is known, we can calculate the maximum age of the information received from the celestial object. All events related to this celestial object can only occur between the minimum and maximum age.

We have gone through a limited number of astronomical observation reports and research papers. In almost all of them, the time-related data about the celestial object or events, when checked by the Valid Domain of Observing Data of a Celestial Object, is generally incorrect.

This highlights the urgency of re-testing and discovering new calculation methods for handling astronomical observation data.

How to correctly describe the relative time and space between two systems

A typical example of a misrepresentation is Mr. Einstein's special theory of relativity. Take a light ray and a moving rigid rod. They have no contact and no way to contact at all. Mentally put them relative to each other, take a look, think about it, and calculate, then the light ray can make the size of the rigid rod shorter, and the time becomes slower. This is not science, but pseudo-god magic. For another example, if the atomic clock on the spacecraft is imagined to be relative to the ground clock, the clock on the spacecraft will become slower. So, if they are not relative, how do the atomic clocks go? Why are these two clocks relative to each other in your imagination (can you think of any other way to make them relative to each other not through your imagination?) Would the speed of the clocks change? What kind of thinking is this? It's a pseudo-god thinking.

This kind of metaphysics masquerading as science has been dominating our thinking for more than a century!

So, how should we correctly describe the time and space between two relatively moving systems?

First, we must make it clear： two independent systems that do not have any connection or even communications, they do not interact with each other.

How can we proceed correctly if we want to measure or determine the time and space within another system then?

In fact, humans have responded to this question with exactitude. For example, the division of time zones on Earth is the right way to correctly process relative observations between different systems.

The time zone system meets the various needs of mankind, without disrupting time and space.

One of the applications of this is that humans use basically the same 'digital time representation' to meet their need to get up at dawn and go to bed after dark. Here the phrase “nine to five” has the same meaning for working people all over the world. It means that when the Sun rises people get up and go to work and when the Sun is about to set people return from work and rest. This is the application change relative to the Sun at different times in different time zones, but in fact there is only one Greenwich Mean Time.

People also apply this method — not changing the time itself — to the observation of cosmic objects. However, this method has not yet seen completely specific provisions and descriptions. Let us discuss it in more detail to avoid the simple but universal error that Hawking made, for instance, with his "The past-future light cone" in *A Brief History of Time.*

We define this cosmic timeline as shown in Figure 10.1.

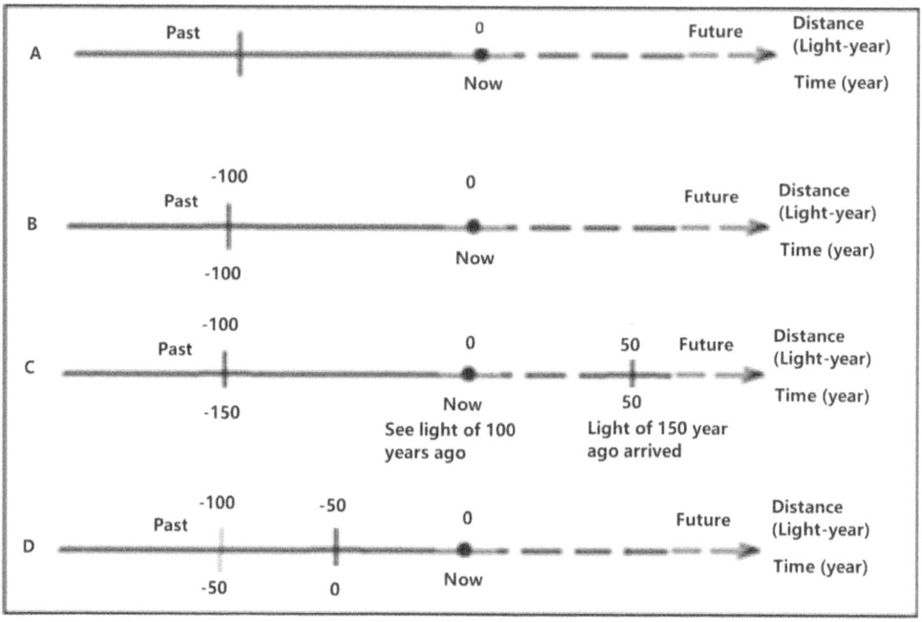

Figure 10.1 Universe space-time axis.

Taking frame A in the figure as an example, the solid line represents 'past', the dotted line represents the 'future', the solid point (current time value = 0) represents the 'now' and the arrow represents the direction of time. The upper part of the axis represents the distance, the unit must be the light-year, and the lower part of the axis represents the year at which the light emitted by a celestial object reaches the corresponding distance.

Figure 10.1 is a simplified diagram of the distance and time used to calculate the same from a distant celestial: How distant that object is and how long it takes light to reach an observer. If the scope of the event is simply extended to three dimensions as the event light cone described by Hawking, then the correct event light cone description is different from Hawking's. Readers can try it. We leave the specific description to the chapter about the depiction of the four-dimensional space-time world line.

Let's consider the following situations in the figure:

Frame A: Definition of the number axis in the figure. The solid line represents 'past', the dotted line represents 'future', the solid point (current time value = 0) represents 'now' and the arrow represents the direction of time. The upper part of the axis represents the distance, the unit must be the light-year, and the lower part of the axis represents the year at which the light emitted by a celestial object reaches the corresponding distance

Frame B: The celestial object at 100 light years from the observer. The light emitted by the celestial object 100 years ago arrived at the observer "now" at 0 year.

Frame C: The celestial object at 100 light years from the observer. Study the light emitted by the celestial object 150 years ago, what the observer now sees is the light emitted by the celestial object 100 years ago, and the light emitted by the celestial object 150 years ago has already passed by the observer 50 light years forward. This is the past-future event , the light emitted by the celestial body in the past 150 years ago has already crossed the observer and reached the future space of the observer for 50 light years. If there is a spaceship 150 light-years away from the luminous celestial body, then the observers on the earth and the observers on the spaceship can see the luminous celestial body at current time. However, what the earth observer sees is the light emitted by the celestial body 100 years ago, and the spacecraft sees the light emitted by the celestial body 150 years ago.

Frame D: The celestial object is about 100 light years from the observer. The light emitted by the celestial object 50 years ago cannot be seen by the observers.

The above expression, in fact, regards the entire universe as a space that has a cosmic time scale. The assumption of the "now" time is the time where the observer stands. The time of a celestial object is the time of where the event happened, and then calculate the relative time difference according to the distance between the celestial object and the observer, and finally, the relative time that the celestial light emits event information to the observer is calculated according to the distance between the celestial body and the observer. Please compare the relative time from the same celestial object in Figure 10.1 with the relative time that the event information is reaching the "now" of the observer.

Celestial Age Indeterminable Law

We define the law of Indeterminable Celestial Age as follows:

The age of a celestial body is undetectable; only its lower limit can be determined.

In Figure 10.1 of the previous section, starting with an event that says "the star 100 light years away has been emitting starlight for 150 years," the law of astronomical age indeterminacy can be easily introduced.

From Figure 10.1, we know that when the age value of a star exceeds the distance between the observer and the star, the information of the star observed by the observer is far beyond the value that the observer can possibly identify and imagine. The minimum age of the star can be determined by the distance value calculated from the observer and it's basically the same as the age value with unit "year" in light years. However, the true age value and maximum age value of the star cannot be known by the observer, and it is generally much greater than the minimum age value.

For example, in the figure above, a star originates from a luminous event 100 light years away from an observer. The star that began emitting light 150 years ago, is located 100 light years away from an observer, so the minimum age of the star is 100 years old. It started to emit light 150 years ago, but the light received by the observer at "now" moment is not emitted by the star 150 years ago, but 100 years ago. According to the light information received, it is only known that it must have been emitted 100 years ago, and the light emitted before it, like 150 years ago, or 200 years ago, has already passed the observer by, entering the future zone, and is no longer visible.

In general, as shown in Figure 11.1, if we replace the numbers in frame C with general letters, then Frame C reads in this manner: The time of the observer is at "now," which is located at the origin 0. The star is D light-years from the observer. The star began to emit starlight X years ago (X > D). What the observer is seeing now is the light emitted by the star D years ago, and the light emitted by the star X years ago has passed the observer by (X-D) years and is now at (X-D) light-years away. In other words, the light emitted by the star before D year has passed the observer by and has entered the "future," as noted in the figure. The situation where the past light of the star propagates in the observer's future space can be named "the Future Starlight that Has Passed".

Building a Macroscopic Existing Framework

The so-called scientific physics Big Bang universe model, neither has sufficient scientific basis, but also lack of grand imagination, nor can reasonably justify. Therefore, there is a need for a more reasonable general view of how to look at the @@ where we are existing in.

However, the above sentence — "a more reasonable general view of how to look at just what kind of universe (@@) we exist in" – coupled with another need – "to explain how to look at our @@" – is not complete. It lacks an accurate object "@@." As contemporary cosmology was flooded with a variety of strange concepts, no suitable noun could be found to serve as this object. "@@ = Big Bang" of course cannot be used, "@@ = the universe" has been polluted as dark as the Big Bang"@@ = space" was stirred up zig zag by Einstein. And the nouns themselves need to be explained.

If Tao can be discussed, it is not a common Tao. If a name can be named, it is not a common name. Nameless is the beginning of heaven and Earth; named is the mother of all things. So, we had to first add a noun to the object @@. And then build a seemingly more scientific and reasonable description of what we are in @@. (But again, encountering no words is available dilemma)

In this framework, even the Big Bang model itself, if there is, can be more perfectly, more scientifically and more reasonably described.

Build a framework to describe the surroundings'@@ scientifically and rationally

We define this object as "**macroscopic existence**," i.e., '@@=macro exists'. So, the above sentence is "to build a framework to scientifically and rationally describe the macroscopic existence around us."

Macroscopic existence is the meaning of everything; "everything" includes space, time, the universe, all of existence: material, conceptual, or any other form. And then, based on the various models and concepts (such as space and time concept, the law of existence of things, etc.) established in previous chapters, all the possible things as the basic composition, we have the materials that are required to build this macro-existence framework.

The words' Macroscopic existence' is appropriate or not can be discussed later, its meaning is 'everything' as above said.

Let's use it to describe the model in our thinking. It's clear and easy.

Our large structure of the framework of macroscopic existence should be like this:

First, there are the two basic elements involved: space and things.

Space is the ultimate space that we defined before. How big it is, we never know.

We are used to talking about the universe in a problematic way: we talk about it, but not where it exists. Where, for example, does the space created by the Big Bang come from, where does it go, what is there outside it? We have no way to respond to these questions with any credibility. So, we must first determine the container that accommodates our macro-existence. It is the boundless ultimate space, referred to as space.

Nor do we completely know what ultimate space contains. If we appeal to past and present science to help us here., we don't get very far. For example, according to Copernicus's large structural principle, this space should contain all celestial bodies, roughly evenly distributed, however far beyond our understanding they are, and however boundlessly ignorant we are. Perhaps, from another perspective, there is the so-called chaos in space, Pangu from which opened up the universe. But what "chaos" is, we don't know. There should be something else we know or do not know in space. How a celestial body exists as such, independent of other celestial bodies; or how it shares a commonality: light waves and gravitational fields.

Events or Things: events or things that exist in space are in accordance with the Existence Theorem of Objects.

There are two elements that define macroscopic existence: space, and events or things in the universe that obey the laws of the existence of things.

Let us first check if using these two elements to describe can include our known macro-existence or not? Space is the place to accommodate events or things, if there is no space, then events or things do not exist. Events or things are in motion, all events or things only exist at the top terminal point of the world line, that is, "now" moment.

This is the objective basic element of macroscopic existence. It is indispensable. We do not need more. We have no ability to know more.

Now, in this objective macro-existence, we add in our subjectivity; that is, the concept of human understanding of macroscopic existence.

The **universe** is the subspace in the ultimate space, and the edge of this subspace is the maximum range that humans can observe.

Everything that man can perceive is the universe's space inside the ultimate space.

Everything in the space of the universe is in accordance with the law of the existence of objects. Recall the law of the existence of objects: an object is born, evolves, and arrives at the present. Despite an expectation of future movement, it exists only in the "present"; this moment. "The past" is composed of the memory of the objects that have existed, like the traces of a boat across the water; the "future" is the trend of object development, like the route that the boat is going to sail; "now" is the place where the boat is, forever at its "current moment."

From the macro point of view, are these two elements included in the macro environment we are living in? Is there anything missing?

When we discuss the **basic elements of the macro's existence**, we must be careful not to miss anything, making the macro in harmony. For example, if there were no space nothing would not exist; if there were no objects, why would we talk about the existence of objects?

Maybe is something missing: how about the time?

Someone will ask: where is the time? Is time not one of the basic elements of macroscopic existence? Past, present, future are all time elements, right?

Time is not really one of the elements of the formation of the macroscopical.

The past, present, and future is used to describe the theorem of existence of objects. If we do not need to describe things, we do not need to use those past, present, and future elements of time. Without describing things, space itself and the space to accommodate things does not change. So, time is a concept, a measure that people make for communication.

According to the previous discussion, things only exist in the "now" moment. So, "time of existence" becomes simply "now". Other times of things, like the past and future of things, become the scales for measuring and describing the occurrence, development and change of things. Its main function is to describe and measure the existence process of things, rather than the existence of things themselves, so it is not regarded as the basic element of macroscopic existence.

Time is a constantly changing measure that dynamically describes the development of things.

With the above macro existence framework, we can build a clear and scientific concept of the universe.

Big Bang: If there was a Big Bang, some kind of material explosion, then it happened in the ultimate space. We identify the place where the smoke from the Big Bang explosion reaches as the Big Bang subspace. As the distance from the explosion center increases, the mechanical effect of the blast wave weakens. The volume of the Big Bang space is constantly changing over time. If we think of it as a nuclear bomb exploded over the desert, then we know that the volume of the Big Bang first expands at high speed, and then with increasing distance from the center of the explosion the rate of expansion reduces, gradually cooling, and finally at a certain distance the expansion stops.

If the Big Bang was similar to the example just presented, then we can identify where it came from, where it went, and how it progressed. What's there inside the space, what's there outside the periphery of the Big Bang, what specific thing is there in the end, what exploded, and how it happened – was it an automatic explosion or a kind of Creator let it explode — we do not know.

The universe is the subspace that humans can observe, perceive, or reason out, with everything accommodated inside it. This definition is one of the many different definitions of the universe.

Unfortunately, in contemporary cosmology and popular argument, the universe and the Big Bang space are so mixed up and confused with each other, that many things remain unclear. If in fact there was a Big Bang, and the Earth was born from the aftereffects of the Big Bang, then, when we define the universe, naturally we must consider the relationship with the space created by the Big Bang. Let's do that now.

In the early days following the Big Bang, its subspace was still relatively small, and subspace of the universe, of course, was larger than it. But today, the expansion of the Big Bang subspace caused by the Big Bang, has made the Big Bang subspace larger than the cosmic space we can observe, and the cosmic space was smaller than that of the Big Bang. Figure 10.2 is a macro-space diagram that assumes that the Big Bang has occurred

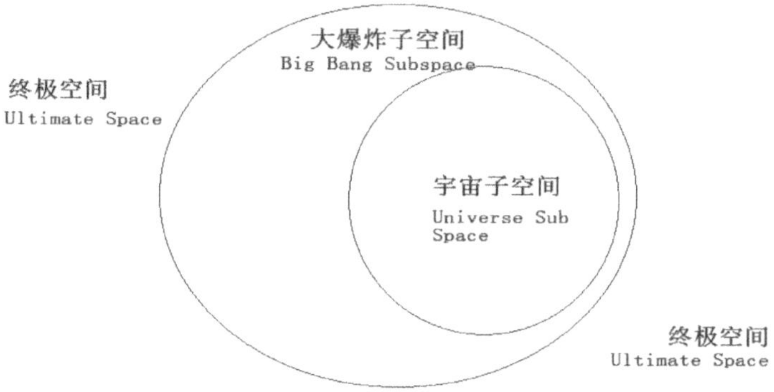

Figure 10.2 the macro-space structure assume the Big Bang has occurred

NASA released observations of cosmic data, the whole sky map, etc., that put the Earth center of the universe. But according to NASA data, the Earth's age is more than four billion years, and the age of the universe is 13.7 billion years. Even the Big Bang theory cannot put the Earth at the center of the universe.

Figure 10.3 Macroscopic spatial structure assuming no Big Bang occurred

Figure 10.3 removed universe subspace from Figure 10.2. This is the figure show the big cosmic structure without Big Bang.

If there is no Big Bang, then cosmic space, as shown in Figure 10.3, is the largest subspace humans may observe in the ultimate space (Figure 10.4). The radius of this subspace is determined by the absolute observable radius.

From another perspective, from the finite observable radius theory of celestial objects, the Earth is at the center of the observable cosmos, with its finite dimensions. This again provides proof of the correctness of referring to the hidden systems of celestial bodies when seeking to determine how celestial phenomena occur if from another aspect.

It is also possible to infer the size of the universe from known observations of astronomical data. Using the celestial object data provided by NASA and combining the hidden celestial object theory, we construct a new sketch map of the distribution of celestial objects in Figure11.5.

Figure 10.4 uses the astrometric data provided by NASA. The supergiant galaxy, IC1101, is by far the largest object found, and we can calculate that its light's limited propagation radius is 477 billion light-years, which means that the light of IC1101 is likely to spread to 477 billion light-years. We use it as the theoretical maximum distance man can observe at the present stage, and from here draw Figure 10.4.

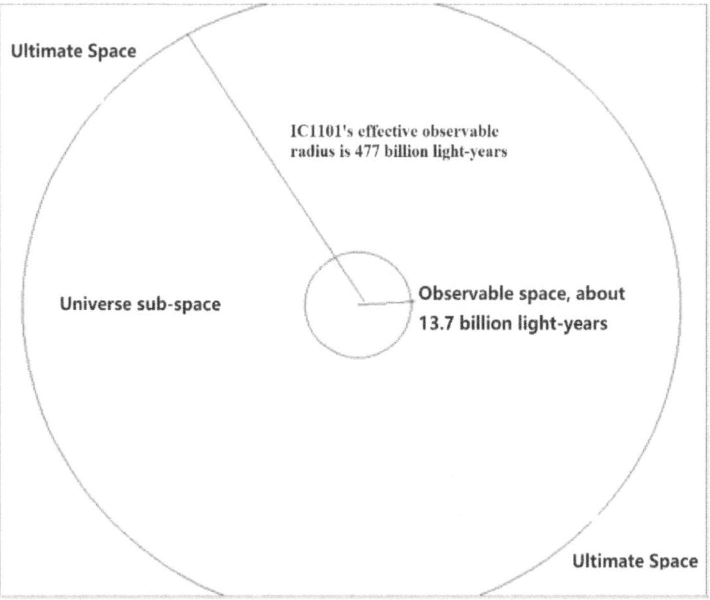

Figure 10.4 According to the IC1101 light propagation radius of reasoning macroscopical universe imagination picture

Comparing IC1101's effective observable radius to its life-time comparison, where is its starlight arriving? This is a topic that can be further explored.

The IC1101 also gives us a hint: can the light of the huge celestial objects that lie far beyond the observable universe be similar to, or the same as the light from IC1101; thus becoming an invisible possible source of microwave background radiation – if we cannot find a source near Earth?

There is no research either that can yet explain how the Big Bang, if it acts as a barrier, can block what happens to IC1101's light when it crosses it. Do you know what the current astronomy and cosmology related thinking or explanation is?

Similarly, according to the cosmological large-scale theory, there should also be many galaxies on the fringe of the Big Bang, including celestial objects like the Sun. What will happen when the light from all these celestial objects that we haven't known shines on the edge of the Big Bang?

In contrast, the size of the universe of 13.7 billion light years given by cosmologists in accordance with the Big Bang theory is too insignificant to reach the 0.03 of the farthest distance that the light of IC1101 can reach.

It is now possible to outline a general picture of the entire macrocosm in my head:

The ultimate space is enormous and boundless, and we never know where its borders are. Any attempt to completely understand ultimate space is futile. Maybe only God can know!

The universe is the largest subspace within the ultimate space we can observe. The universe we can understand is currently limited to a relatively small range of observable radii, within 13.7 billion light-years, as certified by NASA. And while even this measurement is inexact, it basically reaches the limit of what we know, and can know, about the universe.

Human beings are banned in less than a light year area, completely without the possibility of interstellar adventure; human sight cannot see through the thick veil of distance.

The universe we can reasonably maximize is determined by the relative observable radius of the largest luminous object. What we know now is defined by the data for IC1101 in Figure 10.5. This can be changed, but up to now, there may not be too many breakthroughs. As far as IC1101's data is concerned, the ratio of the size of the universe that humans claim to have known to the limit range of IC1101's light propagation is also very small. We are confined to an insignificant region relative to the universe. The proportion of the universe known to us is a fraction of less than 0.000001 percent. And those invisible celestial objects are hidden celestial objects (we cannot see them), so there is no room for a large proportion of "dark matter"!

In addition, objects in the universe may be distributed in different ways, perhaps according to different distances between them, in patterns we can recognize but are not sure about. We do not know other things too: whether or not the universe will transform the form and appearance of its existence beyond what we can perceive

Cosmic material - celestial distribution scheme map

We will use the above macro-space structure, coupled with the material - that is, the distribution of celestial bodies, to make a cosmic astronomical distribution diagram Figure 10.5, in which the most important thing is adding the hidden celestial objects distributed inside and outside the universe to it.

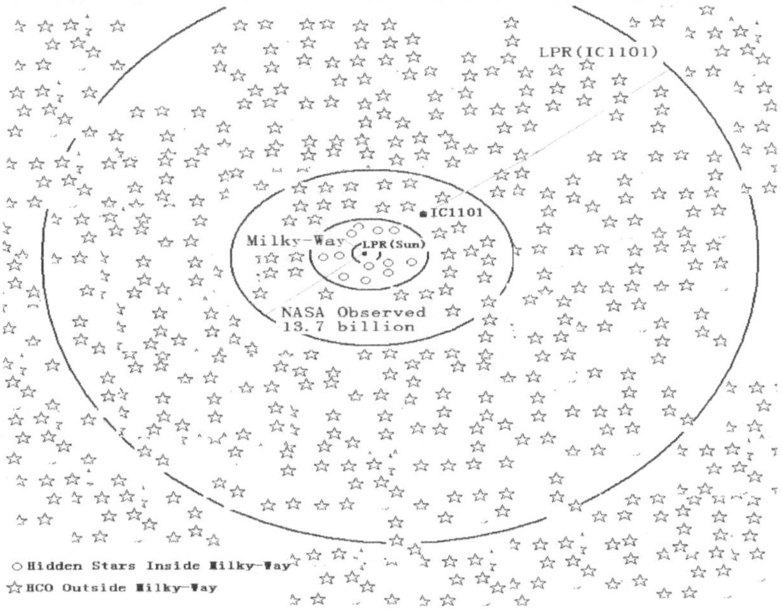

Figure 10.5 The distribution of cosmic hidden celestial objects, where the large circle is the visible radius of the IC1101, the middle circle is NASA's identification of a Big Bang of 13.7 billion light years, the small circle is the radius of the Milky Way, and the smallest is the solar system radius.

Of course, these are not proportional with exactitude. Small circles represent observable celestial bodies, while small stars represent hidden objects. From the perspective of macroscopic existence, or from the imagined universe based on the visible radius of IC1101, the number of celestial bodies that humans can observe is extremely small.

As can be seen from Figure 10.5, hidden celestial objects abound in the immense space between at least 13.7 billion light-years from the Earth to 477 billion light-years from the Earth.

If mainstream science finally realizes that there are possible models other than the Big Bang and begins to accept our point of view, then it will also accept the concept of macro-space. At the same time, it will admit that the universe has materials unknown to us, and that macro space is filled with all kinds of celestial bodies, chaos and other materials, we will find that the hidden celestial object is the main component of macro space.

Because of the immensity of the universe, the maximum range we can observe relative to the space we cannot perceive is so small a subspace, that what exists outside it, what must exist outside it, are hidden celestial objects. Even if the Big Bang happened and generated the subspace we can observe, the quantity of dark matter – if it exists – in relation to ultimate space, is equally a minute quantity.

Chapter Summary

We have established a simple macro-existent model, and now we can talk more clearly about the macroscopic world.

When we know what space is and how it has evolved, equally so time, we can avoid the ridiculous space-time conversion we are plagued with. When we look at the more magnificent picture, we can easily answer those problems we have yet failed to answer, including the Big Bang, space, the universe, and the relationship between them. For example, if there was a Big Bang, we can now respond to its associated questions, such as: what was there before the Big Bang, what is there after the Big Bang, how the Big Bang will be in future, and so on

Chapter 11

Why Don't Aliens Communicate with Humans?

Why don't aliens interact with humans?

In Chapter 3 we discussed hidden celestial objects. We also mentioned that we can try to discover alien civilizations. With more and more incidents related to aliens appearing in different media, people naturally ask: why don't aliens come into contact with humans?

We really do not want to answer this question. Because we are also a member of humankind, we hesitate to repeat an answer, which is both humiliating and frustrating. On the other hand, if we did answer the question, perhaps it just might prompt us and our leaders to wake up and begin our self-salvation?

To put all of this into its proper context, consider a few other questions first:

Why don't we humans communicate with ants?

Why don't we humans communicate with rats?

Why don't we humans communicate with snakes?

The answer is that we humans consider ourselves more powerful than ants, nobler than rats, and more glorious than poisonous snakes!

Is that true?

Ants are weak, and we can arbitrarily kill a few without paying it any mind.

The rat is dirty, spreads disease, and hides in dark corners.

Poisonous snakes are sinister and ferocious, but if they bite us, we can die, so we must always be weary of them.

We are noble, so why should we communicate with these lowly animals?

But are humans really noble? Is this why aliens disdain us and are not willing to contact us?

In the eyes of aliens, what exactly are we?

Do you know?

Let us tell you, in the eyes of aliens, human beings are

The weakest, filthiest, most ferocious animal!

Weakness is not only physical, but also spiritual; the soul is dirtied by ruthless, evil means, including how we treat other humans and animals.

We invented ferocious weapons for mass murder. We inflicted on other humans the cruelest punishments we could imagine. How many millions of poor people have suffered as a minority of us feasted? How much blood was buried under the ruins as arms dealers counted their banknotes

For such human beings, for those who live richly off others who live poorly, selfishness and profit lust are nearly instinctual. The result: society can only be maintained by law and power. Why would any aliens, more evolved and civilized than we are, no wish to make contact with us? We who have shed so much blood to keep what we desire and steal what we don't possess. How many of those highly evolved and highly civilized aliens are willing to approach us who have such a dirty body with terrible thinking?

We humans have also aimed our aggressions at nature, exploiting her whenever and however we can. Human beings are destroying nature unscrupulously, destroying nature for the small interests in front of them. This increasingly decadent society is bound to be retributed by nature. All kinds of raging diseases are just the tip of the iceberg. (This article is copied from my book <Clarify Big Bang> in 2018. Today, looking back after the COVID-9 ravages for more than a year, how does it feel?)

In the immediate Russo-Ukrainian war that is killing humanity on a massive scale, we see the dirty fingers of an ugly man who has the ability and may press a button to fire a nuclear war and destroy the world. Why do humans give such power to such ugly people? Aliens will not associate with such terrible people!

Yes, we have also had saints, masters, nobles, and pure souls. But today we are driven more by money than by our saints. We no longer need masters, and we no longer generate them. A smile is a hypocritical cover-up, and purity is ridiculously naïve! The evolution theory we advocate is cruel, a scientific excuse of "the weak should die, and the powerful should survive."

The French existentialist, Sartre, said that hell is other people.

A Chinese idiom puts it another way: "If one is not selfish, heaven will ruin him, and Earth will destroy him."

Despicable human beings, according to their own ugly acts, would impose the law of the jungle on aliens. Yet in most literary and cinematic works that deal with alien contact the tables are turned and it is the aliens who are evil. In the famous science fiction novel *三體 The Three-Body Problem*, in order to survive the Earthman, caught up as he is in the law of the jungle – kill or be killed — must fight against aliens.

Actually, it's ridiculous. Comparing the potential ability of aliens to the current ability of human beings, and aliens crushing a few people is like people killing a few ants; and enslaving humans is like humans enslaving cattle and sheep. We are more likely being observed like white mice rather than the aliens' enemy.

If we want to live justly on our planet, in the solar system and beyond, we must celebrate what makes us noble, and win back the honor, integrity, and loyalty that characterizes us at our best. And we must strive to join the universal family.

I feel God's forgiveness and vision of love.

I feel the Buddha's grand, compassionate expectations.

I feel Allah's encouraging, loving voice.

To overcome evil thinking and behavior, be sincere and kind, worthy of heaven and Earth!

Then, maybe one day, aliens will be willing to shake hands with us.

Will aliens massively invade the galaxy or even Earth?

This kind of sad science fiction is pure grandstanding without value other than by its implicit expression of fear. Does this expression also contain within it concern about our destiny?

Look, in the Milky Way distances are calculated in thousands of light years, right?

How long does it take for electronic waves, light, or any information to travel over 1,000 light-years?

How long would it take a human spacecraft to cross such distances?

Are we even capable of it?

Considering the eating, drinking, and communication on the road, we have several issues that need to be clarified:

What is the farthest distance we humans can now reach?
What is the farthest distance humans already have reached?
What level of technology is required for humans to reach this distance?
How big is the gap between the levels of technology required to fly across the 1,000 light-years, and the technology currently mastered by humans?

When these questions are clarified, we will find how difficult our star path to the heavens is. And how big distant even the outer solar system is for us, despite our desires to explore it. That is a place that only God can reach.

After clarifying these questions, we can think in reverse: what kind of technological level must the aliens who can reach the Earth have mastered enabling them to cross such great distances?

With their technological level, can human beings fight against them?

Some people would shout excitedly: then I would rather die than give up and fight.

Unfortunately, there is no struggle, no need for any struggle.

Put yourself in the picture and imagine that people in primitive society may compete with ants for their property resources, while people who have entered a civilized society today will hardly have the idea of snatching when they see ant nests in the wild. Unless someone's wickedness is at fault and sabotage.

For aliens with highly developed technology, imagine what humans can offer them? What do they need from humans?

Nothing!

With the technological level of aliens, it is so easy to turn a stone into gold or a drop of water into gold.

Do they need to go to war with the earthlings?

They would buy the Earth with gold mountains or silver mountains or anything human wants. How about this?

Aliens don't communicate with humans because they look at humans as we look at ants. Just like humans like tame puppies and kittens, aliens don't like dark, cunning, and vicious people!

Wake up, poor people!

Are you happy to fight to the death and fight for power and profit?

God have pity!

God have mercy!

Dare to ask: Where Is the Star Road?

The star road is long, but the road in our hearts is also far, far away.

The star road is fuzzy and confusing, and it seems impossible to traverse.
The heart twists and turns, but it can freely get to the horizon of space.

In recent years, human science and technology have triumphantly, rapidly developed, but cosmology is deep in the pit of an atom's Pseudo-God Big Bang Theory and cannot extricate itself.

Looking back at Figure 1.1, NASA's official webpage on the Big Bang announced that the Big Bang theory has two main pillars: one is Einstein's general theory of relativity, and the other is the principles of the universe.

But can these two seemingly scientific pillars really be able to support the pseudo-god universe of the Big Bang? We have meticulously shown that those two pillars are too fragile and cannot support the weight of the normal universe. The former is only an explanation of some of the disturbances that occur in the small corners of the universe and cannot be extended to larger spaces; the latter is the blind impression of blind men touching an elephant, imagining that they can see microwaves in every corner of the universe, when, in fact, they only can see local microwaves around the Earth.

A ray and a rod are fully independent. Then, with Einstein, they composed a relative system. With various dazzling steps and abstruse mathematics, Einstein convinced himself and others that the movement of the ray could change the length of the rod, and change the time on the rod. No mathematical magic of this stature has we humans seen before. This cannot but say it is the greatest success of mathematical magic of mankind. The celestial body radiates into the space around the light source in the shape of a larger and larger light sphere and is regarded as a light ray that can be distorted by a relatively very small planet. It also walks along a so-called time-space geodesic line, with various myths related to it.

The scary thing is that more than a century has passed, and most of us, including mainstream astronomers and cosmologists, still regard such fantasy as scientific bible. Why has no one ever asked, "How can light affect the length of the rod and the time on the rod?" "How can a sphere of light be taken as a ray of light?" Those scientists who initially opposed the theory of relativity were only concerned with Einstein's mathematical derivation. No one starts from the fundamentals of a ray and a rod to query the fantastical nature of the theory of relativity.

The author asked that "*Is Mathematics Old*" in <*誰有權談論宇宙 Who Should Talk about Cosmos*> published in 2005, gave answer "Math is old" in <Debate of Light and Dark> published in 2016. Through more than ten years of thinking, the author has been vaguely aware of this problem, but still needed take a few more years before the author suddenly realized clearly about this fundamental question of the theory of relativity, that is, the theory of relativity is based on a mathematical model of fantasy, because the movement of a ray cannot have any effect on a rod, and there is even no way to make "relative" between the ray and the rod. He realized that the most important thing in applied scientific research is not the problem of mathematical deduction, but the scientific and rationality of the research model. In a field that is difficult to empirically study the universe, because it is difficult to verify, it is easier to go astray due to model errors.

We can't consider the fact that a ray and a rod can't influence each other in a completely independent relative model. Instead, we always go through the mathematical formula to make a fuss, which fully demonstrates that human scientific and logical thinking has stiffly involved in the production line of industrialization and mechanization, it is necessary to be wary of "math is old"!

We also built a very rough new cosmic model that conforms to basic scientific doctrine, and it can be used to describe the new universe we see and imagine, according to our superficial understanding and the fruits of scientific progress. If the Big Bang had ever occurred, or if it could appear in the future, the model could also be used to explain where it comes from, what happened after, what was outside it, and so on. But these ideas are based on the knowledge of mankind. We cannot, and we have no way to prove it.

And so the Big Bang model follows suit. And even if, for argument's sake, it did occur or will occur, the model used to identify it should also explain where it came or comes from, what happened or will happen afterward, what was or is outside it, and so on.

Humans will never see the true face of the universe!

Humans can never see the true face of the Creator!

Compared to the grandness of ultimate space, the universe in which we exist has become insignificant, and our vision as we look into the far distance is full of confusion.

This is the road that we humans must travel when exploring the universe and all its wonders. We are not gods, we are not omniscient, but humble explorers with limited resources for research and for planning the future of mankind and seeking long-term existence.

Judging from the all-wrong research about distant celestial objects discussed before, contemporary cosmic science is fundamentally unable to deal with distant astronomical information received by telescopes. This requires us to do a lot of new research.

If the mainstream cosmological sciences and astronomy sciences do not turn back in time in the smoke and confusion of the Big Bang, it will only lead mankind further and further on the road of wrong understanding the universe.

The scientific community must face up to its limitations, from its capacity to understand the true face of the universe, what can be done, what cannot be done, and what cannot be known. Only in this way can we progress, and hopefully ensure our survival.

Only then can there be true reverence for the Creator!

First, we must find the correct method of dealing with distant astronomical observation data, reinterpreting a large number of observations so that we can attain the kind of clarity we currently lack.

And then we must come to a reasonable assessment of how best to continue exploring the distant universe.

It can be said that for contemporary cosmology, reasonable data analysis, observation, results, and theoretical summaries have just started!

Chapter Summary

From theory to practice, modern astronomy and cosmology have spent a lot of money to make observation instruments beyond imagination.

But experts have yet to consider whether or not their research facilitates our understanding of the universe or confuses us.

Experts and scholars don't seem to think seriously: Does such effort and profligacy promote humanity's understanding of the universe or confuse us about the universe? Does it give us a clearer picture of the universe, or does it make things more confused to get caught up in illusory telescope images?

We need to have enough reverence for the universe and the Creator!

As long as they are trapped in the mud pit of the Big Bang theory, humans will never be able to approach the true face of the universe.

We need new models of the universe to broaden our thinking and correct our understanding of the universe.

We need to re-understand ourselves so that humans can survive in the extended family of the universe.

Where is the correct human star road?
Right in the way of our hearts.

Chapter 12

Is It Possible for Humans to Migrate to Planets Outside the Solar System on a large Scale?

If the Earth is damaged by humans so that it cannot be inhabited, is it possible for humans to abandon the Earth and immigrate to other planets outside the solar system on a large scale?

It's impossible!

Because the speed we can master is too slow and our lifetime is too brief, we are unable to overcome the immense distance between the stars, and thus cannot emigrate to other planets outside the solar system.

Please take good care of the Earth because it is the only place humans can live safely.

Meaningless Reports From NASA

NASA, year after year, publishes news that new "habitable" planets have been discovered. Such a description can easily be misleading if it encourages us to wonder if it is indeed ink "possible to migrate to" such planets. Did NASA append to these discoveries any suggestion that they were planets we could reach? No.

Figure 12.1 shows one of the numerous similar reports of the search results obtained in Baidu online with the key words "外星移民 alien immigrants." We can see in the figure that the report embellished and misinterpreted "habitable planet" to mean that "migration to another planet is not a dream" – that is, the dream of human migration to planets beyond the solar system. Here, no longer is it just a fanciful dream. Rather, migration to the planets can be really achieved.

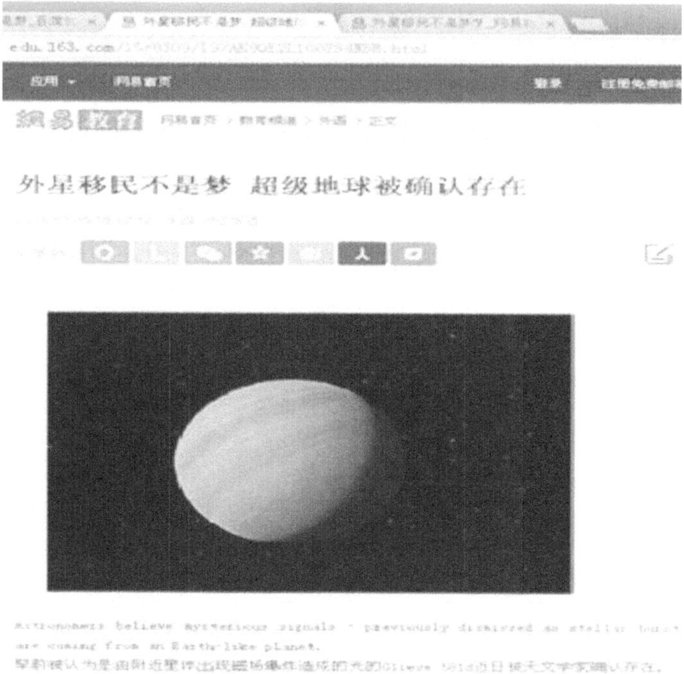

Figure 12.1 Chinese news reports "migration to other planet is not a dream"

These kinds of reports, which exaggerate the possibility of human space travel beyond the solar system, are typical responses to new planetary's discoveries in Chinese media. If the planets are there, why can't we travel to them? Won't new technologies enable such a trip at some point in the future? Although we cannot completely blame the news media, it's all really caused by NASA repeatedly announcing the discovery of habitable planets — and by the difficulties in translating one language into another on this topic.

So, all the ordinary people who are not scientists truly believe that cosmologists are able to lead us in our time of need to successfully flee from the Earth.

We too, believed that immigration to a "habitable planet beyond the solar system was possible.

It seems that migration from the Earth is not a dream, ah!

But, in fact, it really was, still is, and will ever be, a pure dream!

And in the future, even if we count future years in ten-thousand-year units, it would still be a pure dream – if we continue in the current way of thinking and insist that the current cosmological theory about the universe is correct.

Brutal fact:

So far, outside of the solar system, humans have not found a single planet that we are truly able to migrate to!

According to existing cosmological knowledge and theory, there are also no prospects for us to get there.

So, please be patient and follow our discussion. There yet is hope.

Minimum conditions of a planet that human beings can migrant to

If humans want to migrate to a particular planet from the Earth, the planet must meet at least two conditions:

1). It must be habitable.
2). It must be reachable.

To immigrate to a planet, it first needs to have a habitable condition. If not, of course, we will not immigrate there. If we immigrate to a star, like the Sun, is it not a rush to commit suicide? And, indeed, the number of discoveries of habitable planets beyond our solar system is climbing.

In accordance with the standards of human needs on Earth, cosmologists have found many planets that we can live on. The closest are only about five light-years from Earth while the ones far away are more than two thousand light-years distant. There are countless habitable planets farther away, but cosmologists no longer overemphasize them because of their distance.

The real problem is that the planet must meet the second immigration requirement: **humans can arrive there**. We humans must be able to travel there.

Cosmologists usually talk about billions of light-years, painting maps of the sky that span billions of light-years, and discussing what happens and why at that immense, unimaginable distance. Their lofty vision ignores things just a few light-years away.

Unfortunately, I'm just a little person who is only concerned about the immediate reality that occurs within a small distance from me.

We just wanted to figure out: how far is one light-year?

Cosmologists have told us that asteroids might hit the Earth, or that the Earth will be destroyed in the future. But do not worry, cosmologists have given us a good way out. April 18, 2013, NASA again announced its discovery of three "new Earths." In addition to the Earth, in the vast universe, there are also Earth-like habitable planets. International media was excited. As NASA put it, the Kepler Space Telescope observed two extrasolar planets that are "the most like Earth" and "may be the most suitable for human habitation." The planets are "moist, warm like Hawaii, cold like Alaska," and 1200 light-years away from the Earth.

Being a coward, I'd have to prepare myself well in advance of traveling there. So, I'd first calculate how much food I'd have to bring.

Such a calculation, Wow! My eyes were sent straight!

Guess how long it takes to fly there, OK?

Currently, on NASA's fastest spacecraft, it would take an immeasurably long time. If, on that spacecraft, **one light-year requires 37,500 years to cross**, how many years would we need when multiplying that by the distance to the desired planet?

Well, to get to the nearest habitable planet, just five light-years away, it would take 187,500 years!

To get to the newly discovered "moist and warm like Hawaii" habitable planet, it will take 45,000,000 years!

Ten thousand years as one unit, yeah!

If someone said to me, "I'm sure you'll be the richest man in the world 1000 years later," I wouldn't be very thankful and, perhaps, would like to punch him in the mouth.

Does it make sense to repeatedly talk about what might happen tens of thousands of years later to a person with a lifespan of one hundred years?

Anyway, we must remember the following brutal fact:

So far, and in the foreseeable future, we will not possess the technology, even to travel one light year's distance. There will be no ability to move people out on a large scale.

Mankind has dreamed of migrating to a habitable planet outside the solar system. Might this dream come true?

To migrate to a habitable planet outside the solar system on a large scale is a dream that we have had for some time. But it is only a dream and there is no hope of making it true.

1. Relocating to an extrasolar planet has been the dream of mankind.

The famous theoretical physicist, Stephen Hawking, has pointed out that mankind must migrate out of the solar system to escape the fate of extinction. An asteroid impact on the Earth, or the threat of nuclear war, sooner or later can destroy humanity as will the Sun when it dies. The solar system is too close to the Earth, so humans must find habitable planets outside the solar system.

NASA Aeronautics and Space Administrator Michael Griffin has said that, in theory, the planet's species are unlikely to survive forever. He has told us: We have strong evidence that, on average, every 30 million years, the Earth's species will face a large-scale destructive event. One day, we must immigrate to an alien world, but we do not know when that day will be.

2. What are the barriers to be overcome when migrating from Earth?

The first hurdle is to establish livable standards for humans and to find those planets that meet those conditions.

The second hurdle is to reach the livable planet. In brief, to be able to migrate to an alien world outside the solar system, we must create vehicles that can travel as fast as the speed of light.

The third hurdle involves living on an alien plane. Currently, on the International Space Station, in weightless conditions, the astronauts have cultivated more than 100 kinds of crops, fruit flies, spiders, fish, and other animals. If this technology can be applied to the livable planet, our problems of survival can be easily solved. Yet our ability to survive on a lengthy space trip presents its own issues. A French scientist discovered that, in weightlessness, important structures of living cells cannot properly form, which means that humans cannot live and thrive long in weightless or near weightless conditions.

After reading the above description; what impression does it give you? Do you not feel that "immigration to an alien planet is not a dream," without a question mark?

However, the fact is, even after thousands of years, that this is very likely an unattainable dream!

Of course, technological advances will increase the speed of our spacecraft. But it still will not be enough to overcome the distances and time involved. We can't send a large number of humans into space and fly many light years.

Based on our current situation, our ability to immigrate to an alien world seems hopeless. Yet, if conditions on Earth change so that our survival hangs in the balance, another possibility arises. Dr. O'Neill at Princeton University believes that the best approach here is to build a city in space, so that some number of humans can relocate there, at least until conditions on Earth are more favorable for us – and we can return to our home planet.

But, again, please note the significance of this sentence: "Based on our current situation, humanity's ability to immigrate to an alien world seems hopeless."

This is the cruel truth we are facing:

Humanity has an urgent need to immigrate to extrasolar planets, but human beings have yet to find a safe planet we can truly immigrate to.

Which planets are humans able to migrate to?

Astronomers and cosmologists should not be giving us boundless hope that we can migrate to planets outside the solar system when we need to survive, then fall back to its improbability, when mapping the immense distances that we would need to cross.

However, Heaven must be a way.

If we change our way of thinking, broaden our focus, and update our theory of the universe, we very well may be able to find a planet, when the Earth would be damaged too hard to survive, we are able to immigrate to there.

This place is hidden in the celestial bodies that we are going to propose in later sections.

If aliens exist, where are their bases?

In fact, this problem is basically the same problem of finding a planet that can migrate to.

We know that while some people do not believe in aliens, others do. Some have said they met the aliens, and some have said they had dealings with aliens.

In California's capital city, Sacramento, a group of people occasionally meet who claim to have had dealings with aliens.

My close friend saw a very strange UFO but did not see aliens. I'll relay his story later.

Let us tentatively assume that aliens exist. Then there are interesting questions that naturally follow:

Where are the alien bases or homes?

Alien aircraft take off from where?

Alien spacecraft from Earth fly back to where? They should have somewhere to go.

In fact, there are many issues involved, including whether or not there are aliens at all. If no one can give a reasonable explanation of where the alien bases or homes are, then it indirectly denies the existence of aliens.

However, if we can give a reasonable explanation to the question of where the alien bases are, it will greatly enhance the possibility of the existence of aliens.

Often, people will say: our telescopes, including space telescopes, which can observe stars that are hundreds of millions of light-years away, have not found any alien bases, and therefore, alien bases do not exist; or aliens should only be present in very distant places, such as a hundred, thousand, million light-years away, so we cannot see them through our telescopes. The following discussion on the Hidden Celestial Objects will tell us that this is a completely wrong idea.

In ancient Egyptian pyramid texts, we often see many astronomical stories that tell about aliens from distant places of the universe. For example, the Titans came to Earth after "millions of years" of travel in the dark without air to breathe; the god of wisdom, Naxos, who counts the number of stars in space and measures the land on the ground, has a magic power that can make an already dead Pharaoh live for several hundred more years; the eternal god Osiris traveled for millions of years to get here. In addition, there are many puzzles in Scripture, like "millions of years" and "one hundred million of millions of years," and so on. These are telling people that aliens in ancient Egypt came from distant planets.

However, from another point of view, if the aliens who frequently appear on the Earth come from faraway bases, then what kind of speed should the alien spacecraft travel at so that it can freely haunt the Earth?

We already know that, with the fastest spacecraft we currently possess, travel over one light-year takes more than 30,000 years to complete. If the alien base is hundreds of light-years away, even if the alien aircraft fly to Earth at the speed of light, one Earth visit would take hundreds of years.

Some people might assume aliens have extremely long lives, with one of our centuries equating to one alien year. This may be possible, but our reasoning does not base itself on such a myth. We make our assumptions with mankind as a reference. Suppose the basic physiological conditions on the alien planet are not that different than for humans on the Earth. So, a century is a relatively long time.

Some people would say that the aliens come to the Earth through a black hole, or a wormhole, etc.

We hereby solemnly declare: this book does not use any of these seemingly "scientific" ideas, for which there is no scientific basis and which have never been proven with a scientific experiment. They exist only in the theory of contemporary cosmologists, Hollywood star war blockbusters, or the fantasy fiction of so-called "theory." Such theories are distinctly different with the assumption of the possible existence of aliens.

Specific to this chapter, time travel through wormholes or black holes exists only in the mythic story of theoretical cosmologists. There is no convincing scientific experiment or observational facts to support these ideas. Even the proposed existence of wormholes or black holes itself is still not supported by solid scientific experiments or any observations but is a result of bold, indirect speculation based on a number of observed phenomena, which we discussed in detail when talking about Hidden Celestial Objects.

However, the numerous eyewitness testimonies since ancient times of alien spacecraft may be said to have a proven track record over the years. It can be said that the research on aliens exists and there is a solid basis for designing a scientific observation project to capitalize on this research.

Everything this book describes or will describe is based on empirical scientific experiment or observation of the facts, not unfounded speculation and imagination. It is not a purely theoretical speculation, but a combination of theory and facts.

In fact, according to existing theories in physics, reaching the speed of light is an impossibility. Why? Because Einstein said, at the speed of light the spacecraft will be converted into energy as will its astronauts. In our discussion on the theory of relativity to come, we note that "near light speed" is an ugly concept. The orthodox astronomical community does not agree that the speed of light can be changed or made slower. But when in need, they use the ugly concept to quibble over numerous topics.

Therefore, if the alien bases are really in distant space, it is almost equivalent to the fact that aliens basically do not exist. Because it would take them too long to make a trip to Earth. There should only be occasional visits from alien spaceships from distant bases on Earth, and they may stay for a long time after each visit.

But the history and current recorded facts are: alien spacecraft frequently haunt the Earth, and there are many different types of space aircraft. Some of these vehicles are common, such as a disc-shaped UFO; some are rare, like a square or oval shuttle.

It is concluded that, not only are there occasional guests to Earth from very distant depths of space; more aliens are coming from our near neighbors.

That is, according to an analysis on UFOs in human history, the conclusion is: if aliens exist, then some alien's bases are very far away from the Earth, but some are not far away and perhaps, just in our neighborhood. Some may even be in the solar system, or near the solar system.

This is an argument that overthrows public and expert knowledge.

It is generally believed that, with telescopes, we basically have clearly seen the area of space very close to the Earth (such as within a few light-years or tens of light-years); only very distant celestial space cannot possibly be seen by powerful instruments like the Hubble Space Telescope.

Unfortunately, this perception is totally wrong. Indeed, it is entirely possible that the most powerful human telescope cannot see celestial objects that are very close to us.

Fortunately, if we promptly recognize this, then it is possible to solve the problem of finding alien bases. More importantly, in this new way of thinking, searching for a suitable planet for humans to inhabit is not just a dream but a real possibility.

This chapter will clarify this point of view in a popular way. For specific mathematical proofs, please refer to the author's related books.

Through the process of searching for possible extraterrestrial bases, and the associated process of looking for a planet for humans to immigrate to, it can be seen that, with regard to the problems of the universe, even cosmology experts are also prone to cognitive bias. Correcting these deviations can bring great benefits to humans.

The estimated range in which alien bases possibly exist

Let's start with a reasonable range estimate on where alien bases exist to further explore some of the problems about finding alien bases.

First, we can compare the historical record to the Earth's existing space technology, the pace of development of science and technology, and so on, to make a rough estimate of the level of technology aliens must possess to reach us, for example, about how advanced the alien spaceship is and what its speed is.

From the comparison of sightings of alien spacecraft in history and currently, and from technological progress on the Earth, to speculate, the pace of the speed of the alien spacecraft is now probably a hundred times that of the fastest human spacecraft speeds, a thousand times or 10 thousand times; such a large range estimate, should be somewhat reasonable.

From the historical record, since a few thousand years ago, there are reports of alien flying saucers landing on Earth. But during that time, their technology was not so advanced, so their early aircraft would cause a huge ring, a loud sound, and radiant light.

More contemporary reports on UFOs present a different point of view, indicating that when UFOs appear, the sound and light they generate has diminished. Equally so their size, which is now smaller. The UFO my friend witnessed, for example, was rather small and almost silent in flight.

There are many historical records of UFO sightings. In China, from the Qin Dynasty to the present, various yet uninterrupted records tell of UFOs that came to the Earth. We have excerpted the following:

In the Jin dynasty mythical masterpiece, "Notes of Pickup Oblivion Stories" Volume of Tang Yao, there is the following account: "Notes of Pickup

Oblivion" Volume of Tang Yao, there is the following account: After Yao became Emperor for thirty years, a huge ship appeared in the West. There was a light at night on board, and the people at seaside called it "貫月槎 Ship Go through Moon." On board were flying fairies wearing white feathers. This story happened in 2327 BC, dating back 4,000 years ago.

In the chronicle book "資治通鑒 Governor Mirror of History" written by Sima Guang in the Song dynasty, there are 17 suspected alien astronomical records and related activities. For example: in the Jian-yuan second year of the Western Han Dynasty, in the summer in April, a star like the Sun came out at night. This is a record of the object as bright as the Sun that appeared in the sky at night.

Song scientist Shen Kuo's book "夢溪筆談 Dream Creek Essays" also recorded UFO stories. In the Yangzhou region's lake, there was a strange big bead with a clam shell shape, a typical UFO shape. It emitted intense light, stayed there for more than 10 years, and switched its residence between three lakes, as residents witnessed. This may be an account of visitors from distant space, where it is not easy to come from, so they stayed longer.

Figure 12.2 "*赤焰騰空* Red Fire Soaring into the Sky" is the work of Qing Dynasty painter Youru Wu in his old age, in 1892. Figure 12.2 is considered to be a detailed and vivid UFO sighting report.

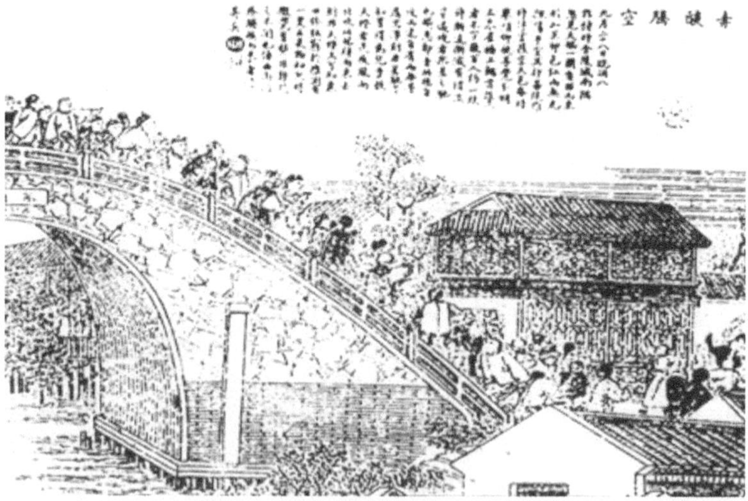

Figure 12.2 Red-fire vacated FIG.

The screen diagram of "Red Fire Soaring into the Sky" depicts the scene when, on the Nanjing ZhuQue Bridge, pedestrians crowded like clouds to all look at the overhead sky, eager to watch a group of gleaming flames.

In the inscription at the top of the screen, the painter wrote:

"September 28, at eight o'clock in the evening, Jing Ling (now Nanjing) south, a group of fire blankets the sky from west to east, shaped like a giant egg, red light without shining, floating through the air with very slow movement. At that time, the floating cloud hides the sky, the heaven is darkened. For heads looking up, the sight is very clear. More than hundreds of people stand on tiptoe on ZhuQue Bridge, looking for time upon the sight. With increasing distance, the brightness decreases. Somebody said there was a meteor in transit, but that should be fast and appear instantly. But this ball from near to distant, gradually disappeared, like stagnating water, so we knew not a meteor. Some guessed that there were children putting lanterns into the sky, but at night, the storm was a north wind, and the ball turned east, so we know it was not sky lanterns. Testimonials guessed a lot, but have no way to speculate an accepted result. An old man said that, at the beginning, he could hear a weak sound that one had to be quiet to hear. It flew from outside of the south gate, giggled, and then carried on in a different direction! Hey, strange!"

In this detailed and vivid illustration, the time it took the fireball to pass, its location, speed, size, color and intensity, and the number of people who witnessed it, were all clearly described. It is a qualified academic observational report.

In recent years, newspapers, magazines, television shows, and media networks have been filled with a steady stream of sightings and reports on alien encounters and alien spacecraft.

My friend told us that he saw a UFO like this:

It was on the evening of September 9, 1997 7:30 pm, on the way back from taking his son to a piano lesson. A few hundred meters from his house before a traffic light, suddenly he saw in front of him about four or five hundred meters away a page from the newspaper blown strongly across the rooves of several houses. He felt very strange: why was the wind so strong tonight? Reaching his home, he saw in front of the house an object like a piece of newspaper about eight meters above the power line pole. Its body was round with a slight yellow light, making the material look more like a yellow flag swinging in the wind. He realized that he was encountering a UFO and was shocked and hurried back to the house to get his camera. Meanwhile, the UFO had changed into a flag-like shape. It quietly rose into the sky and flew off to the west. While its light was not bright enough to be captured on film, after a long time in the air, it could still be seen in the clear sky.

The UFO my friend saw was both small and silent. If you assume that an existing spaceship on Earth has a similar shape and speed as an alien UFO from thousands of years ago, the idea that now, after thousands of years of evolution, the aircraft is a hundred thousand times faster is a logical inference.

Using this rough estimate to measure vehicle speed of those UFOs that frequently appear on the Earth, the alien base should be near the Earth, from less than a light-year to just tens of light-years away. According to our estimation of possible alien spacecraft speed, the aliens would be able, with no difficulty, to depart from the base and frequently visit the Earth.

Their base might also be not very near the Earth. Otherwise, aliens would take Earth as their roots and often come in droves.

The Dogon people of Mali, Africa, are still a tribal society. But the tribe has the legend that they are from the third star of Sirius, a planet of Sirius. The planet is covered with water and is eight to nine light-years from Earth. This is our estimate of the alien base location that is relatively close. So, there may be a base or a hometown for aliens.

If this alien spacecraft is from a place that is not too far away, why hasn't the powerful Hubble Space Telescope found the planet where the alien base is? Why hasn't it found the ancestral home of the Dogon people near Sirius?

With respect to the broader point of view, might there exist places near the Earth that even the Hubble Space Telescope cannot see? Yes.

There is also a considerable number of such planets that even the most powerful telescopes cannot see.

If this assertion contradicts our given concepts, which we have inherited, then so be it. It is no doubt a fact, and we will prove it below.

People do not see these planets mainly because the vast majority of them cannot be seen. There are others that we should be able to see but, for various reasons, have not yet discovered.

Still, a few should be identified but haven't been so far. One possible cause is that the alien base is masked by aliens using stealth technology to shield it from our detection.

Another possibility is that, in the years after the invention of the telescope, people have slowly focused more and more attention on distant space, and the most advanced telescopes do not focus on near space, so some objects in the vicinity that should have been found years ago still have been overlooked.

More likely is that those hidden objects, which we discussed in chapter 3, although indeed in our neighborhood, cannot be seen.

So, when we focus the most powerful telescopes on near-Earth space, it is possible that we might find some previously unseen planets close by. In this regard, we have a very optimistic attitude because many of these hidden celestial objects may not be alien to us and may not be blocked from being seen from us. But due to our own reasons, we lose sight of our near neighbors.

So, inevitably, there are many celestial bodies that exist far and near but that have drifted away from our current vision. Our research needs further approaches to find them.

Those alien-infested bases near the Earth should be from some Hidden Celestial Objects relativity near the Earth.

Hidden Celestial Objects are a Possible Source of the Cosmic Microwave Background

It is possible that hidden celestial bodies are a real source of the cosmic microwave background. Maybe just a few objects, maybe several groups of objects, but there is this possibility.

Because microwaves of unknown origin were detected near the Earth, and because the source of this microwave could not be found, it was attributed to the remnants of the Big Bang. Because satellites have been used to detect such microwaves near the Earth for several years, it is said that almost the same microwaves are everywhere in a universe that is approximately 13.7billion light-years old.

This topic is left to be described in detail in the discussion of the whole sky map of the cosmic microwave background.

I remember that in one of NASA's early papers there was the suggestion that microwaves near the Earth could have been emitted from two invisible celestial bodies. The article is accompanied by a diagram describing this situation. Later, I searched for a while, and then put it down without finding it. Interested readers can continue to look for it.

Love the Earth and look for hidden planets that can migrate on a large scale

Mankind's large-scale migration to extrasolar planets seems hopeless. If migration to planets in the solar system becomes possible, how can it benefit those of us on Earth?

Human beings wantonly destroy the Earth's environment and squander the Earth's resources. Now that we understand that human beings have nowhere to go and nowhere to escape to, we should wake up to protect the Earth and the planet where we live.

Please cherish our Earth.

Chapter Summary

The topic of looking for alien bases is only introduced to make the writing interesting. Its real meaning is this: if the people on Earth do not take good care of the Earth but think that they can leave the Earth and run away, it is a big mistake.

With our current technology, we cannot even travel one light-year distance. How can it be possible then to migrate to the outer solar system on a large scale as Hawking called for? Hawking's appeal made people think that human beings can escape the Earth when needed.

In fact, large scale migration out of the solar system is completely impossible!

Don't listen to the bigwigs fooling around.

Human capacity is too limited. And they don't know it, they don't know that they are really very small creatures in the universe.

The Creator created the universe. When it is unwilling to show its world to people, no amount of human effort can change the fact that humans cannot see the true face of the universe.

Let's take good care of the Earth we live on.

Let us change our behavior and thinking, and strive to be a qualified member of the cosmic family.

Chapter 13

Math Is Old

The title of the first chapter in my first book, *谁有权谈论宇宙 Who Should Talk* About *Cosmos* published in 2005, was a question: "Is Math Old?" At that time, we intuitively felt that mathematics seems to be aging, and the creative vitality of mathematics gradually declining.

Mathematics is old.

It is my understanding now, after 18 years of thinking. We still think it's telling the truth today in 2022. In this final chapter of this book, my response to the question noted above. It comes, with some minor changes, from the section, "Mathematic is Old" of my book *Debate of Light and Dark*.

Today, after more discussion with friends, we and several mathematicians exchanged views. We consider the statement, "mathematics is old," as a basic fact.

A fact that is at stake.

Mathematics is the foundation of natural science, a symbol of natural science.

If mathematics is old, is the construction built on it, natural science, old? Are we no longer able to create with them the revolutionary breakthroughs we expect? Or are we at that end point, as science journalist, Hogan, has put it in his book *The End of Science* "Today, science has revealed the basic facts of man and the universe, from the nature of scientific work has been completed on the terms in almost all 'pure science', the major discoveries have been found, and no major incidents of amazing."?

There is a series of ideas worth exploring.
First, is mathematics really old?
If mathematics is really old, why is it old?
If mathematics is really old, is science old?
Do we have a way to rejuvenate mathematics?

Climbing trees and planting trees—Is Mathematics old?

We often do not know what the real reason is that makes people old.

As a person ages, his mental life is gradually reduced, his pace and external performance increasingly falters, and his actions become more and more sluggish.

In fact, why aging is a difficult problem is tough to say clearly.

The same molecules make up our cells; why are these "old" while those are still young? On the molecular level, what kind of mechanism is this?

The idea that "math is old" is also inferred from its external performance.

The main manifestations of mathematical aging are as follows:

The development of mathematics is stagnant, and the fundamental revolutionary development of mathematics has not progressed for nearly a century.

After simple mathematical problems are gradually resolved, mathematics is powerless when dealing with a complex world. The delicacy of mathematics leaves it powerless in front of the changing, rough, real world. Tough to use sophisticated mathematical tools to describe the wild and varied system, only a fine play on the practical system! Only the free universe left a space to play with exquisite mathematical formulas, the rest are bluntly replaced by **computer numerical analysis** when dealing with real-life issues.

In this light, the application of mathematics is increasingly going astray. There are many examples of rigorous yet unrealistic applications. The mathematical myth played by a rod and a ray has ruled human scientific thought for a hundred years. Mathematically proven myths in mainstream scientific journals have thrown the human mind into chaos. People can't tell the difference between myth and reality, science and story. The scientific truth-seeking spirit, its practical use of experiment and verification are ever more obscured by the noble mask of mathematics. Can time go back? Are there countless parallel worlds? Where is the reasonability here? At least myths and tales are true to their source? It's not as credible as paradise, it's just a few "scientific" labels rudely sticked on...

Let's take a look at a short story that shows how mathematics is getting old.

The protagonists are of two sorts: coaches and students.

The story takes place in a forest.

The trees of this forest were planted some long ago, some hundreds of years ago, and the youngest is one hundred years old. Some trees were planted by the elderly, but some were also planted by people less than twenty years old at that time.

The coaches begin to teach the students at a very young age.

This is a modern, extremely formal education.

When the students are young, the coaches tell them how great, how excellent and talented those are who planted a tree in this forest. Slowly, the coaches teach the students to look at these trees in awe.

As they grew older, the coaches started teaching them tree-climbing skills.

These trees have a rich fruit of inexhaustible wisdom.

Some children are not good at climbing trees; they do not have tree-climbing abilities. And while they are still encouraged, after multiple failures, they are sent for other training. These students spend the rest of their lives overlooking the forest from afar, with respectful or disgusted moods.

Other children are very clever, with very good tree-climbing abilities, and they are left to slowly climb this or that tree.

There are exotic flowers and fruit trees, with long branches, and vines entangled with them. The children grow up in the world of this forest, climbing higher in the trees, as high as they can. They build houses in their trees, and even connect the trees with vines. It is tireless, hard work. And they do it until they have unconsciously grown old!

Year after year, the size of the forest is the same, and for the last hundred years, no more trees have been added.

Generation after generation, no one thinks to climb down from the trees, perhaps to plant a new tree.

The forest is mathematics; the trees of the forest are the branches of mathematics founded by the predecessors: this one is thousands of years of geometry, that there is a long history of algebra. Here is a little, younger tree for group theory, and there is the small, young Boolean algebra tree.

Almost every tree was planted at least a century ago by math masters.

In the past century, we humans were busy climbing trees!

Humanity over the past century has not planted a new tree!

Indeed, over the past century we have not created a new mathematical discipline!

Figure 13.1 is the cover of "A History of Mathematics."

In this book, after 1900, the recorded history of mathematics and major groundbreaking mathematical achievement is blank!

Figure 13.1 Book cover of "A History of Mathematics". In this more than 700-page book, published in 1968, and reprinted in 1989 and 1991. It reveals after 1900, the recorded history of mathematics and mathematical achievement is totally blank.

Are we growing more and more stupid?

Have we failed to produce talented mathematicians over the last hundred years?

Of course not!

Why is mathematics old? Is it due to natural or human factors?

If mathematics is old, is science also old?

Do we have a way to rejuvenate mathematics?

Can we let science never age?

We can study several areas, including educational, economic, moral, cultural, and even science itself, from the status quo, from history, from various development angles, etc., to seek answers.

No doubt, the aging of mathematics is due to the aging of our creativity; and the aging of human creativity and human teaching methods are inextricably linked.

During the Cultural Revolution in China, I wrote a little piece of comedy, "The Third Shoot," for a propaganda team in the countryside. Because of its humor, it achieved great success. But when, for a new comedy, I sought more humorous material, no matter where I searched, I didn’t find anything to laugh about. In retrospect, I realized that, during those years, people did not smile in their thoughts in those years, so it is hard to break established thought to write interesting pieces.

When people who grow up in an environment where unified thinking is the norm and independent thinking is restricted, are they usually incapable of recognizing just how imprisoned their thinking is? When one's mind is tied to the industrial transmission belt and cannot break through, he can only lament that "The invention of mathematics has been completed, the science is over."

Forest Guardians

In the mathematics forest, the tree of "algebra" is already a millennium old, and that young tree "Boolean algebra" just around a century old. In this forest, most are unaware that there are guardians. They are familiar with every tree, and every tree they love. They enthusiastically introduce people to each tree, and are keen to train children to climb the trees.

When did those guardians actually appear?

It is not possible to pinpoint a specific time, but the approximate time is when the Industrial Revolution was basically completed. They brought the mechanization of the Industrial Revolution into this forest, and began to manage it with mechanized efficiency, making full use of the forest for human wealth increase. Thus, Countless applications are generated from the forest, and dazzling technological achievements emanate from here.

However, the guardians only concentrate on caring for each tree, seeking only to guide each child to climb a tree. They only allow children to climb trees near the forest; they only encourage children to climb the tree to pick fruit. At the same time, they unintentionally promote industrialized standards to educate their children by, and the children train in a near uniform manner. The result: not a single child entertains the idea of planting a tree. And if by chance a child does come up with this idea, the guardians, who set the rules, filter it out

For the last one hundred years, no new tree has appeared in the forest.

How can humans face such an issue?
How can we, once again, plant new trees?

A few suggestions follow. Let's remember as well.

Here, we just threw out some rubble, made from our ability to see, our ability to think of problems and phenomena, and to try to give limited answers. We only hope that this piece of debris can arouse a few jades of shock, and have some resonance in the thoughts of interested people so that they can fuse together to seek possible answers and solutions.

This is disturbing the status quo, but also an exciting challenge. The rise of science and technology is our opportunity: perhaps new saplings of mathematics and scientific theory will grow into forests cultivated by the younger generation!

This is possible as long as we can grasp the opportunities available to us by infusing the spirit of education to make creative inventions.

"End of Science" and the rising opportunities of science and technology

In 1997, *Scientific American* senior writer John Horgan published a book entitled *The End of Science*, after visiting scientific leaders in more than ten fields. The book presents the following idea: today, science has revealed to humankind an understanding of the basic facts of the universe, and scientific work discovering the big concepts has ended. In almost all the "pure sciences," the major discoveries have been made, and there will be no revelations of equal stature.

In 2006, on Mid-Autumn Festival Day, I queued in the payment line of the Berkeley Bowl grocery store, and picked up a *Discover* magazine. In it I was surprised to find the same John Horgan, after one decade, picking up the old topic in an article entitled "The Final Frontier." (Figure 13.2)

In the article, Horgan said: "Over the past decade, scientists have announced countless discoveries that seem to undercut my thesis: cloned mammals (starting with Dolly the sheep), a detailed map of the human genome, a computer that can beat the world champion in chess, brain chips that let paralyzed people control computers purely by thought, glimpses of planets around other stars, and detailed measurements of the afterglow of the Big Bang. Yet within these successes there are nagging hints that most of what lies ahead involves filling in the blanks of today's big scientific concepts, not uncovering totally new ones."

Figure 13.2 Horgan published in *Discover* magazine the article "The Final Frontier"

Yes, we were surprised that, a decade on, there was not a single scientist who could convince Horgan that his assertion was wrong. For Horgan, there was overwhelming proof that the end of science was at hand.

His main argument is simple: in all areas of pure scientific theory—mathematics, physics, chemistry, etc.— innovation has not come. In other words, in all scientific disciplines, in the scientific forests, he did not see any people today planting a tree!

Like Horgan, we also believe that the progress of scientific theories is at a standstill, despite discoveries, and the dizzying variety of new applications born from them. In fact, they are just filling in and completing the existing scientific theoretical gaps, but there is no fundamental basic scientific theory that has made a breakthrough.

Mathematics has not produced any new branch, although the number of mathematics researchers has increased exponentially. Equally, there have also been no new revolutionary scientific theories offered to us, although the number of scientific researchers has increased exponentially as well.

Recent scientific achievements have been focused on getting fruit from the trees planted in previous efforts, and not on planting new trees.

However, we do not believe that this pause in the creation of theory is because we have reached the end of science. As long as we dare to face the fact that we humans have lost the ability to plant trees, we might realize that there are yet places in the world for the cultivation of a new forest!

Solemn classical science and frivolous modern science

We do not know when it began, but science is becoming increasingly non-serious, floating noise.

There are many man-made factors involved, such as telling right and wrong not by truth but by academic faction, and judging good or bad as the authorities deem. And this ugly phenomenon is derived from the complexity of modern science itself, and the difficulty of using black and white to determine yes or no without ambiguity. Our arrogance doesn't help either. We do not possess the kind of insight we believe we possess. There is even the arrogance of human beings who think they have insight into everything, and they don't admit that we, as a species, have come to a turning point. Sometimes the manpower is indeed exhausted with our effort.

Where are we headed?

The first is to treat the attitude of the speed of light.

Einstein asserted that the speed of light is constant. Michelson–Morley and many similar experiments "proved" this assertion. However, if the light is so fine that their quality is only one millionth of elementary particles or ten millionth, by what means can we humans distinguish between details of the light? If the light is transmitted in units of distance, say, of ten thousand light-years of distance and then starts to weaken as it approaches us, do we get "tired", can humans have any means to simulate the same with the same expanse of time and distance? Light travel in a unit of ten thousand light-years? In the Michelson–Morley experiment and other similar experiments, the experimental time duration of the light traveled was maybe even less than a year; that is, less than one light-year distance, right? It is not possible to simulate a hundred light-years or more in a lab.

Previously, we discussed in detail earlier, human-observed starlight is the statistical result of multiple photons, which must be "tired" and observable near the relative observable radius.

Another example is "dark energy" theory. What, in fact, is it? We just don't know. Does it even exist other than in theory? We bet it doesn't. Is this thing in itself a very strange thing with a very unnatural theory? I have a bet with NASA because I basically do not think that dark energy exists. We'll see what kind of magical means cosmologists use to find or create a little dark energy. If it has to account for 96% of the matter in the universe, as they surmise, it can never be dark energy. The reason for this, we have fully discussed earlier in the book.

In truth, advances in modern technology have well exceeded advances in theoretical innovation, and the power of capital driven talents tends to accumulate money. So, willingness to engage in some research fields with little research funds are essentially rare. In this case, an amateur's illusory research of the universe is simply surprising!

A friend of mine did an experiment. He sends the first draft of my book, *Debate of Light and Dark*, to an editor. His comment was: if this man can write such boring stuff for a decade, he is definitely an eccentric

I listened to the remark and could only respond with a wry smile. I feel very strange to myself, ah! I spent a decade of my spare time on this and did not earn a penny—sick, ah? What kind of worldview is that? What kind of universal view? Can you be more stuck in reality?

My glorious image is ruined by this book, ah!

In Horgan's book, *The End of Science*, the first example he used was cosmological science! Why was cosmological science the primary basis of his argument? Should we not consider this carefully? Who impeded its progress?

We believe we have told you in this book in your hands.

We think that discerning what is right and what is wrong when considering the universe — what we understand about it and how we understand it — is more meaningful to human progress. If our work can give the field of rigid untouchability little impact, people will realize that since a hundred years the human predicament has been that no new trees were cultivated. So, by discussing the end of science topics that can impact authority, people can realize the plight, to have excited desires that want to jump down from the old trees to plant a new tree, and even build a new forest. It will be a wonderful thing!

Then, too, there is this hope, which we have stressed, even if implicitly throughout this book: our hopes in the younger generation, in the next generation of scientists and mathematicians to come. May they accustom themselves not to standard rules and thinking but to independent, critical thinking. May we guide them to create new theories that are practical and verifiable, and which then become the great standard-bearers.

We do not need a science and technology God to imprison scientific progress!

We call for God who does creation that advances science!

To Astronomical and Cosmic Scientists

We know you are the best in the world and are working hard.

However, we would like to ask: What is the ultimate purpose of your work? Isn't it to allow humans to understand the universe clearly? Have you achieved it?

When we saw those much-publicized incorrect reports like that you discovered "an infancy 30-year-old black hole at 50 million light-years away," our first reaction was that experts must have gone crazy to make such a simple mistake! You have been describing a completely wrong universe to us!

What we don't understand is: How could such simple and intuitive errors persist for decades? Shouldn't you reflect? Shouldn't you listen to the opinions of us ordinary people? The various points of view we point out in the book are trying to answer this question for you.

Please don't think of yourself as the god of science, or the messenger of god. The myth that the universe was created by the Big Bang of an atom that you have supported with scientific methods for many years not only insults science, but also tarnishes God.

The contemporary astronomy and cosmology circles should really reflect on: whether to continue to adhere to the theory of the pseudo god of the Big Bang; or to truly respect science and admit that you yourselves are not the god of science, and recognize the road to the end of the universe is still far far away. Start from scratch, make good use of taxpayers' money, and truly scientifically analyze how much humans can know about the universe and how deep they can understand it, and what are humans impossible to know. Only in this way can humans clarify the correct research direction in the future, and will not lie lazily and uncreatively under the shadow of the senior masters, tinkering and making the papers!

As taxpayers, we want to ask if our money is being used correctly? Please either refute or apologize to the questions raised in the book.

Chapter Summary

Aging is a slow, subtle process, particularly lightly noticed during adolescence. Only time can unconsciously decay the cells of the body and erode the vitality of the human body. It is only when time has painted our temples with frost, that we can realize how short life is.

Compared to humans, any one particle of the universe belongs to eternity!

The mechanisms of modern society have used up the land available for planting.

Thanks to the rise of the internet, so that human knowledge can be shared beyond time and space, in this era, there is likely to grow master genius so that the century of mathematical decline may be stopped now.

At the same time, the internet has become a major social tool, and only a handful of strong-willed young people exist who are not captivated by social media or digital gaming networks. But these people are not necessarily really geniuses.

So, the mediocre becomes the mainstream of the society, and for universal social communication, pleasure becomes fashion. In humans, this pleasure of mediocrity leads us towards a more mediocre existence.

Ugh!

Mathematics is old!

Postscript

Where does the star trail meander toward?

Ups and downs to the deep universe.

The path of science is extremely uneven, and the mysteries of the universe are harder to pry into.

For more than 18 years, we have collected a lot of questions about the universe and the world. It seems that they are scattered, but when we gather them together, we see a faint and coherent vein. This leads to an interesting question:

Why do experts, professors, and ordinary people ignore the simple, low-level, and large number of questions such as the propagation time and distance conflicts raised in this book? Why let those obvious mistakes exist in cosmic scientific research for a period of tens or a hundred of years?

There should be many reasons.

One question we hear most often is: Are people all over the world wrong? Are all academics, experts, and professors wrong? Just you correct alone?

Isn't such a question that scares people away and down?

However, it is obvious starting from Chapter 1 of this book, isn’t the whole world wrong?

In fact, I have been subconsciously seeking this answer since 2005 in Chapter1"Is Math Old?" of my book *Who Should Talk About Cosmos*, and "Math is Old" in *Debate of Light and Dark* published in 2006. I initially realized that the reason is like this:

It was the unified education model in the era of industrialization, training people's thoughts into the same mode.

Of course, there are many aspects of specific issues, which can be extensively discussed from various aspects such as science, philosophy, psychology, and society. Interested readers can read my books <Fading Modern Cosmology>、<Debate of Light and Dark>.

But one thing needs to be mentioned in particular: the incompetent thought of worshiping authority in the scientific world. Now these academic circles engaged in astronomy and the universe must obey academic authority and have been trained to worship great men like Einstein since childhood, so they dare not step beyond the thunder pool.

We once saw an American physics textbook published in the 1970s, which stated that the special theory of relativity can basically be used to determine its correctness, while the general theory of relativity has many questions. Now, these promising next-generation scholars use general relativity as a life-saving weapon, and use it wherever they are needed. Throughout the decades, no new ideas have emerged in the entire academic circle of astronomy and physics. They just use the brilliance of advancement in space technology to waste taxpayers' money to make newer and more advanced telescopes, although these instruments are of little use.

Why must it be subordinate to academic authority? Because whoever doesn't follow the authoritative way of thinking, whose paper cannot be published, whoever will eventually have to leave behind. For example, NASA has published that a 30-year-old black hole has been discovered. If anyone writes an article against it, the paper will definitely not be able to be published. But if one follows the trend to write supporting articles such as argumentation and data analysis. It is easy and sure to publish! Over time, what new things can be done in the university? Don't you see, many Nobel Prizes in favor of the Big Bang have been given out! Has any paper against the Big Bang won any awards? So, why go against it? How good to be a stable scholar.

Why do I dare to write this, because I am retired, I am in the United States, and I have the Internet!

The prospect of astronomical academia is confused and unclear.

Winding star road, fuzzy and nothingness.

My Ambitious Desire

This book has a deeper attempt: we hope this book can give alert to people to start to straighten out the distorted relationship between philosophy and technical science. This is what I have been working on for more than a decade.

Hawking said in his book <A Brief History of Time>: *"Up to now, most scientists have been too occupied with the development of new theories that describe what the universe is to ask the question why. On the other hand, the people whose business it is to ask why, the philosophers, have not been able to keep up with the advance of scientific theories. In the eighteenth century, philosophers considered the whole of human knowledge, including science, to be their field and discussed questions such as: Did the universe have a beginning? However, in the nineteenth and twentieth centuries, science became too technical and mathematical for the philosophers, or anyone else except a few specialists. Philosophers reduced the scope of their inquiries so much that Wittgenstein, the most famous philosopher of this century, said, 'The sole remaining task for philosophy is the analysis of language.' What a comedown from the great tradition of philosophy from Aristotle to Kant!*"

Below is what I pointed out in "谁有权谈论宇宙 *Who Should Talk about Cosmos*" in 2005:

Here, we see that, thanks to the brilliant achievements of science and technology over the centuries, and the wealth that science and technology has brought to human society, the scientific and technological method has become an instinct of people's minds: where any technical science works, the philosophical thought of the humanities will wither or even die out, and it will automatically and willingly cut itself off from this field that is involved in science and technology.

The above passage has been written for more than ten years since then, but the situation is getting worse. Philosophy really retreated from fundamental questions about the universe. However, this is completely wrong. For more than a decade, we have constantly been pointing out that the current science and technology of the universe cannot help humans correctly understand the universe, but contemporary scientists have not recognized or are unwilling to acknowledge or even dare to discuss and criticize.

To put it simply, at the speed of the current manned spacecraft, the improvement of human beings is not only subject to various limitations in technology and humanities, but also subject to Einstein's relative theory of "moving ruler shorter". Any distant celestial body humans observed with telescopes cannot be determined whether it is a real celestial body or just the remnant light left by the celestial body that has been destroyed. Humans can neither travel to explore the universe outside the solar system with the help of science and technology, nor can they see the real universe with telescopes. So, why does the philosophy of finding the truth of everything succumb to the physicists who are only followers of Hawking and of Einstein's mathematical formulas??

Because Mr. Einstein's principle of relativity cannot be applied between systems that maintain absolute simultaneity and systems that maintain relative simultaneity, physical theories cannot defeat philosophical models. Relativity and absoluteness are the two sides of everything, even in relativity or physics.

In today's artificial intelligence information age, what new and fundamental breakthroughs have been made in mathematics, philosophy, and basic theory? We have seen a series of great achievements in applied science, but we have hardly seen any great progress in the fundamental nature of thought, philosophy, mathematics, and especially, in the field of astronomical sciences of the universe.

In the post-industrial era, we must break through the century-old traditional thinking that keeps us in thrall, and reconstruct the current academic model of mathematical supremacy. Only then will we be able to properly explore and understand the true face of the universe.

We hope that some latecomers will open up a road to the sky on this difficult wild trail!

Why do I keep writing this book?

I searched and chased my dreams in the universe, sometimes happy, sometimes anxious. Thanks to the gravity-free space on the US side, I am just rolling and bewildered instead of bleeding and bruising and even dying.

I wanted to clear the fog that covered my eyes and ears, and seize the touch of the golden light penetrating the dark; but I was surprised to find that I was shaking the whole world and offending the gods of the universe sciences of the past and present.

Though Living in such a prosperous and lively information age, I found that my voice could not arouse even an echo.

But what's the big deal?

I firmly believe that time's flow will wash away all hypocrisy. Truth is still shining even after a century.

It's a special surprise, too, that you were able to stick with the reading up to here.

Thank you, for your appreciation, or your criticism.

Recalling the book's writing, it seems like I'm still in a dream. I really did not believe I could stick to it until today. I had hard times trying to hold that ray of golden sunshine but always realized that it was a dream of going beyond distance. Life never gave me easy kindness and writing never went smoothly. However, giving does not necessarily require gain.

However, giving does not necessarily require responding.

I tried my best to obtain recognition from mainstream science circles that study the universe, but I never got any results. A critical article that I wrote for *Nature* was promptly refused. I was amused: What kind of ideas can these childish editors have?

For me, being able to read and write freely makes me happy:

I'm thinking, I'm writing, and I'm delighting.
I tapped on the keyboard with "pa, pa" noise,
that is saying to the world loudly:
I am still unrestrained
cheerful.

Yes, I can read and write freely, what can make me unhappy?

Whenever I think of those early days when I couldn't find a book, I felt suffocated and mad. Taking my ID with me, I went to the mountain village to work as a farmer. Burning books and torturing its scholars, my ancient civilization, my motherland, almost destroyed its heritage, sweeping my smile, my sense of happiness, away.

There was an attempt, however, by workers, peasants and soldiers to criticize the theory of relativity theory, which later turned into an attitude towards science, however asphyxiating it was. As a result, scientific and academic issues became political issues, and criticizing relativity became a terrible topic to engage in.

If I had known this would happen, I would not have delved into this vortex in China. But I didn't. So I left the software college I founded for Beijing Normal University at Zhuhai in 2003 and returned to the United States in 2007. Since then, my thoughts have had the freedom to expand into the wider universe. The boundless stars attract my thinking, and the magnificent universe moves my mind across the river of time, from ancient to modern.

When my heart was bewildered by the beauty of artificial intelligence and attracted by the endless starry sky, I chose the latter, exploring the star road to release my desire to view the beautiful universe.

I gave up on the artificial intelligence software, which was in its early stages then, and pondered on just how to explore this vast star road. Then the hard work began, and my thoughts about the universe began to crystalize. Hopefully this kind of thinking about the universe will inspire us, and lead our thinking into new spaces, sooner or later.

If you look at the references in the appendix, you will know how many golden years I have condensed into crystals and presented to you in this book.

It has been more than a decade since my work began. Time is like running water, and life is like flowers, but where and to whom does the water of time carry flowers of the life?

Despite all the hardships along the way, I still want to thank God, who finally gave me the answer I have been seeking.

Although I stumbled along the way, I still have to thank God, and finally, with the help of Lǚ Wu, gave me the answer I have been looking for many years.

Enduring hardships, sticking to the end, and enjoying a spirit of pioneering – these portray my life.

Thank you for joining me to complete this twisty cosmic journey.

This book was completed during the rampant period of the COVID-19 pandemic. I wish people from all over the world would overcome the disease, keep physical and mental health, and be safe and good luck!

01.31, 2023 at Green Garden Home in San Francisco Bay Area

A thought after completing the book

While writing “Who Should Talk about Cosmos” in 2004, I began to throw my energy on the winding star road. Gradually, my rucksack got heavier, my eyes became farsighted, and my legs and waist gave me trouble. Only then did I realize that I had walked too deeply into the sky for 18 lonely years. After repeated thinking and revision, the book about the universe was written as it is today in 2022. Regardless of the result, I finally left a touch of trace in this world, and I felt worthy of living.

Looking back on my lifetime, on all its bizarre twists and turns, its bitter and sour encounters, I now know as a kind of infatuation. In a troubled world, how fragile a personal life is! I have never followed trends, insisted on my own dreams, and pursued the truth of the universe. Be able to live safely to this day, I must say thanks to my family, my friends, Buddha and God.

A farmer’s desire (excerpt from a poem that is in several remaining pages of my diary. Under big risk I kept it.)

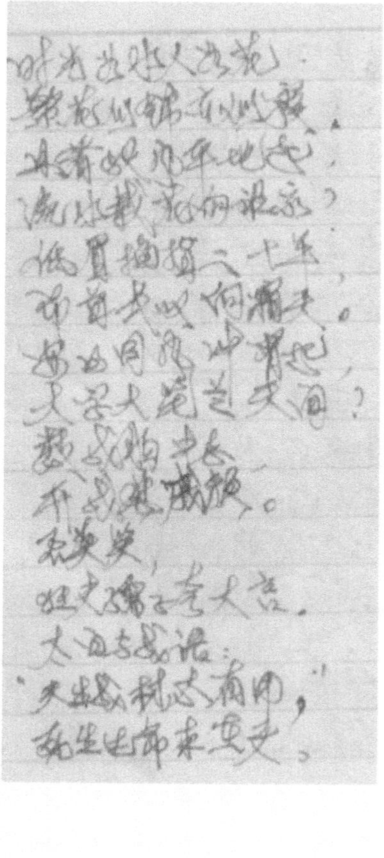

Time is like water and people are like flowers,
The flowers are like brocades and haze.
And watching the autumn wind rise on the ground,
To whom is the water flowing?
Painstakingly thinking and learning for twenty years,
Lift head high sighing to frost sky:
How can I rise with the wind,
write free words with big pen in blue sky?
Pouring dreams out from my heart,
and cheer up my sad feeling.
Don't laugh at me -
Crazy man small fellow saying exaggerated words.
Taibai told me:
"Everybody is good for something,"
Live or die, rich or poor, are in one's own hand.
January 2, 1971

Planted seedlings when I was a kid.

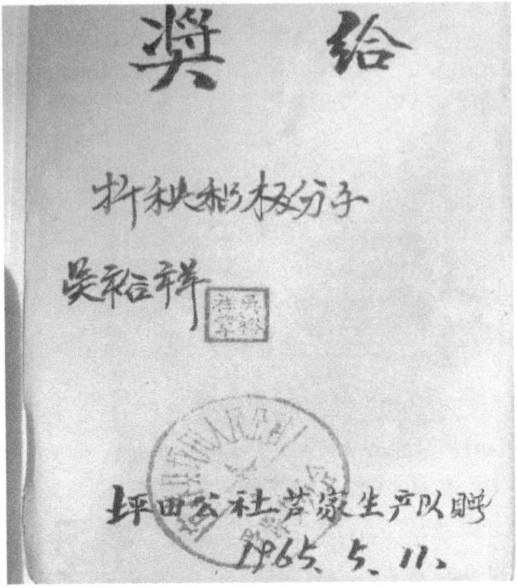

After growing up, it became a green tree.

Thanks

Look at the night sky, the brilliant Milky Way, stars flashing.
So mysterious, so fascinating.
But today, how many people really care about it ?
how many people aspire to study it?
Down to Earth is the attitude of life.
Only those who like fantasy and dreaming,
And neglect the festive fireworks and city neon lights,
care about those too far and faint stars' calling.

But I am still persistent thinking and writing,
to keep my obsession with the sky
in a certain corner of the universe as a piece of landscape
beyond the lights of the human world.

Thank you for going with us on this long journey.
Thank you for reading.

Partial References

I. Universe-related books and papers published by the author

1. Yuxiang Wu, "10 Questions To NASA", Two-W Object, 2021, ISBN: 978-1-7370977-0-9
2. Yuxiang Wu, "Fading Modern Cosmology", Two-W Object, 2020, ISBN: 9798645172633
3. Yuxiang Wu, "Clarify Big Bang", Two-W Object, 2018.2., ISBN 978 1985629974
4. 吳裕祥："宇宙大爆炸質疑 探索宇宙新模樣", Two-W Object, 2017, ISBN 978 1978495531
5. Yuxiang Wu, <Debate of Light and Dark – A 100 Year Bet with NASA>, Two-W Object, 2017.3., ISBN 978-1684197385
6. 吳裕祥 (Yuxiang Wu,), 《光暗之爭-與美國宇航局的百年期約》, 上海科學技術文獻出版社, 2016.9., ISBN 978-775439-7071-7
7. 吳裕祥 (Yuxiang Wu,), 《光暗之爭-與美國宇航局的百年賭約》, Two-W Object, 2016.9., IISBN-13: 978-1537631318
8. 吳裕祥 (Yuxiang Wu,), <誰有權談論宇宙 Who Should Talk about Cosmos >, 香港文匯出版社有限公司, 2005.2, ISBN 962-374-114-6
9. 吳裕祥, "從歷史的角度看狹義與廣義相對論的困境", 格物 USA, MATTER REGULARITY, January, 2019, Vol 19, No.1, pp1-pp38
10. Yuxiang Wu, "尋找新的宇宙觀測資料方法的建議書", 格物 USA MATTER REGULARITY, January 1, 2018, Vol 16, No.6, pp1-pp4
11. Yuxiang Wu, "The Confliction between Flying Rabbit and Dunning Rabbit Damaged the Principle of Relativity", Matter Regularity, Vol 17, Sum 84, No.6, pp14-pp17

12. Yuxiang Wu, "Can We Change Material and Time by Imagination? - On Einstein's Theory of Relativity", <Matter Regularity-> 4th, August 2017
13. 吳裕祥，"憑想像就能改變物質和時間嗎？- 評愛因斯坦先生的相對論"，《格物》2017 年 8 月第 4 期
14. 吳裕祥，"吳裕祥論壇：1. 憑想像就能改變物質和時間嗎？ - 評愛因斯坦的相對論"，《格物》2017 年第 4 期，Vol.17 sum No.82, Matter Regularity No4 2017, pp 1-14
15. 吳裕祥，"吳裕祥論壇：2. 為什麼說愛因斯坦的系統相對只能在思想上進行？"，《格物》2017 年第 4 期，Vol.17 sum No.82, Matter Regularity No4 2017, pp 15-16
16. 吳裕祥，"吳裕祥論壇：3. 空間與時間淺論"，《格物》2017 年第 4 期，Vol.17 sum No.82, Matter Regularity No4 2017, pp 16-25
17. 吳裕祥，"構造一個新科學宇宙觀的框架"，北京相對論研究快報，北相第七屆高端論壇報告會論文集，《格物》Vol.15, No.4, 2017, pp15-18
18. Yuxiang Wu, **Lü Wu**, "相對論理論蘊含的時間悖論、空間災難及三個驗證實驗設計 Relativistic Theory Contains Time Paradox, Space Disaster and Three Verification Experiment Design", Matter Regularity 2015 5, Vol. 15 sum No.71, No.5, pp 59 – 68
19. Yuxiang Wu, **Lü Wu**, '愛因斯坦先生的"動鐘變慢"悖論及"動尺變短"災難從何而來？Where did Einstein's "moving clock slow" Paradox and "moving meter shorter" Disaster Come From?' Matter Regularity 2015 6, No.6, Vol. 15 sum No.72
20. **Lü Wu**, Yuxiang Wu, "4-Dimension Spacetime Diagram", Matter Regularity 2015 5, Vol. 15 sum No.71, pp 55 – 59
21. Yuxiang Wu, **Lü Wu**, "Hidden Celestial Object (HCO)", Matter Regularity 2015 1, Vol 15 sum No.67, pp 25 - 31
22. Yuxiang Wu, **Lü Wu**, "New Solution on Olbers' Paradox: Why Sky is Dark at Night?", Matter Regularity 2015 2, Vol.15 sum No. 68, p9 - p19
23. Yuxiang Wu, **Lü Wu**, "Speed of Observed Image and the Observer-Distance Caused Star Redshift" Matter Regularity 2015 4, Vol. 15 sum No.70, pp10-15

II. Astrophysics Monograph References

【1】 SABINE HOSSENFELDER, LOST IN MATH – HOW BEAUTY LEADS PHYSICS ASTRY, BASIC BOOKS, ISBN: 978-0-465-09425-7, 2018

【2】 Barbara Ryden, Introduction to Cosmology, Cambridge university Press, 2017

【3】 Anderson R. Liddle, An introduction to modern cosmology, 3rd edition, 2015 John Willey & Sons, Ltd

【4】 Dale A. Ostilie, Bradley W. Carroll, An Introduction to Modern Stellar Astrophysics, 2nd edition, Pearson Education, Inc., 2007

【5】 D. Scott Birney, Guillermo Gonzalez, David Oesper, Observational Astronomy, 2nd edition, Cambridge University Press, 2015

【6】 王傳晉，葉彬潯，天文可見光探測器，中國科學技術出版社，2013

【7】 Steve B. Howell, Handbook of CCD Astronomy, 2nd edition, 2006

【8】 D. Emerson, Interpreting Astronomical Spectra, John Wiley and Sons, 1996

【9】 Micharl A. Convington, Pratical Amateru Astronomy Digital SLR Astrophotography, Cambridge University Press, 2007

【10】 mke de Pater, Jack J. Lissauer, Planetary Sciences, Cambridge University Press, 2015

III. Reference for Dark Energy

[1] James M Overduin, Paul S. Wesson, The Light Dark Universe, 2009, p26, p27, p32, P69, World Scientific Publishing Co, Pte. Ltd. ISBN-13 978-981283-589-5.

[2] P.S. Wesson, Sp. Sci. Rev. 44, 169 (1986)

[3] P.S. Wesson, K. Valle and R. Stabell, Astrophys. J. 317, 601 (1987)

[4] P.S. Wesson, Astrophys. J. 367, 399 (1991)

[5] P.S. Wesson and H. Liu, Int. J. Mod. Phys. D10, 905 (2001)

[6] E.R. Harrison, Nature 204, 271 (1964)

[7] E.R. Harrison, Mon. Not. R. Astron. Soc. 131, 1 (1965)

[8] E.R. Harrison, Am. J. Phys. 46, 948 (1978)

[9] E.R. Harrison, Science 226, 941 (1984)

[10] E.R. Harrison, Darkness at Night (Cambridge, MA: Harvard University Press, 1987)
[11] H. Bondi, Cosmology (Cambridge: Cambridge University Press, 1952)
[12] H. Bondi, The Advancement of Science 12, 33 (1955)
[13] M. Hoskin, in V.J. Martnez et al.(eds), Historical Development of Modern Cosmology (ASP Conference Series, V. 252, 2001) 11
[14] S.L. Jaki, The Paradox of Olbers' Paradox (New York: Herder, 1969; 2d. ed. Pinckney, MI: Real View Books, 2001)
[15] S.L. Jaki, Olbers Studies (Tucson, AZ: Pachart Publishing House, 1991)
[16] W. Thomson (Lord Kelvin), Philosophical Magazine, Ser. 6 2, 161 (1901)
[17] Bonnor, W. B., 1963, "On Olbers' paradox", Monthly Notices of the Royal Astronomical Society, Vol. 128, p.33
[18] D.D. Clayton, The Dark Night Sky (New York: Demeter Press, 1975)
[19] Boughn, W.P., & Kuhn, J.R. 1986, Ap.J., 309, 33
[20] Tyson, J.A. 1988, A.J., 96, I, p.18.
[21] F.J. Tipler, Q. J. R. Astron. Soc. 29 313 (1988)
[22] Martin C., & Bowyer, S. 1989, Ap. J., 338, 677
[23] H. Kragh, Conceptions of Cosmos (Oxford: Oxford University Press, 2007)
[24] Nikishov A. I., 1962, Soviet Phys. JETP, 14, 393
[25] Jelley J. V., 1966, Phys. Rev. Lett., 16, 479
[26] Gould R. J., Schreder G. P., 1967, Phys. Rev., 155, 1404
[27] Martin Raue, Tanja Kneiske, Daniel Mazin, 2013, First stars and the extragalactic background light: How recent γ-ray bservations constrain the early universe, arXiv:0806.2574v2 [astro-ph] 30 Jan 2009
[28] Rudy C. Gilmore,1,2, Rachel S. Somerville,3,4 Joel R. Primack2 and Alberto Dom´ınguez5,6,7,8, Semi-analytic modeling of the extragalactic background light and consequences for extragalactic gamma-ray spectra, Monthly Notices of the Royal Astronomical Society C! 2012 RAS
[29] Eli Dwek, Frank Krennrich, 10/2012, The Extragalactic Background Light and the Gamma-ray Opacity of the Universe, arXiv:1209.4661v2 [astro-ph.CO]
[30] A. Dom´ınguez1, J. D. Finke2, F. Prada3,4,5, J. R. Primack6, F. S. Kitaura7, B. Siana1, D. Paneque8,9, May 10, 2013, DETECTION OF THE COSMIC -RAY HORIZON FROM MULTIWAVELENGTH OBSERVATIONS OF BLAZARS , Draft;

arXiv:1305.2162v1 [astro-ph.CO] 9 May 2013
[31] Dube, R., Rieke, G.H., & Rieke, M.J. 1977, Ap. J. (Letters), 215, L51
[32] Dube, R., Rieke, G.H., & Rieke, M.J. 1979, Ap. J., 232, 333
[33] Cassan, A. et al. (January 11, 2012). "One or more bound planets per Milky Way star from micro lensing observations". Nature 481 (7380): 167–169. arXiv:1202.0903. Bibcode:2012Natur.481..167C. doi:10.1038/nature10684. PMID 22237108.
[34] Staff (January 2, 2013). "100 Billion Alien Planets Fill Our Milky Way Galaxy: Study". Space.com. Retrieved January 3, 2013.
[35] Deutsch, David (2011). The Fabric of Reality. Penguin Books Limited. pp. 234. ISBN 978-0-14-19696-9.
[36] H. A. Rey (1976). The Stars. Houghton Mifflin Harcourt. p. 145. ISBN 0395248302
[37] Antonaldo Diaferio, THE LARGE-SCALE DISTRIBUTION OF GALAXIES, www.mariecurie.org/annals/volume1/diaferio.pdf?
[38] Crowther, Paul A.; Schnurr, Olivier; Hirschi, Raphael; Yusof, Norhasliza; Parker, Richard J.; Goodwin, Simon P.; Kassim, Hasan Abu (2010). "The R136 star cluster hosts several stars whose individual masses greatly exceed the accepted 150 M⊙ stellar mass limit". Monthly Notices of the Royal Astronomical Society 408 (2): 731–751. arXiv:1007.3284. Bibcode:2010MNRAS.408..731C. doi:10.1111/j.1365-2966.2010.17167.x.
[39] David R. Williams, NASA Goddard Space Flight Center, Earth Fact Sheet, July 2013, http://nssdc.gsfc.nasa.gov/planetary/planetfact.html.
[40] Yuxiang Wu, 2014, 'Hidden Celestial Object (HCO)', to be published with this paper,
[41] Yuxiang Wu, 2015, 誰有權談論宇宙 Who Should Talk about Cosmos (in Chinese), Honkong Wenhui Publisher, ISBN 962-374-114-6
[42] Research Consortium on Nearby Stars, GSU (2007-09-17). "The One Hundred Nearest Star Systems"
[43] Christian, Eric; Safi-Harb, Samar. 2007, "How large is the Milky Way?". NASA: Ask an Astrophysicist
[44] Rix, Hans-Walter and Bovy, Jo (2013). "The Milky Way's Stellar Disk". The Astronomy and Astrophysics Review. in press. arXiv:1301.3168. Bibcode:2013A&ARv..21...61R. doi:10.1007/s00159-013-0061-8.
[45] Douglas C. Giancoli, Physics for Science & Engineers, 2000, p867, Prentice Hall, ISBN 0-13-021519-8

IV. Reference for Dark Matter

1. Bessel, Friedrich W. Extract from the Translation of a Letter from Professor Bessel, dated Konigsberg, 10th of August, 1844. “On the Variations of the Proper Motions of Procyon and Sirius." MNRAS 3.21 (1844): 136-41. Oxford Journals. Web. 4 Sept. 2014.
2. "Case Study: Friedrich Bessel and the Companion of Sirius." American Museum of Natural History. American Museum of Natural History, n.d. Web. 4 Sept. 2014.
3. http://home.web.cern.ch/about/physics/dark-matter. CERN. Retrieved on November 25 2014.
4. Trimble, Virginia. "Existence and Nature of Dark Matter in the Universe." Annual Review of Astronomy and Astrophysics 25 (1987): 425-72. Annual Reviews. Web. 4 Sept. 2014.
5. Rubin, Vera. "Dark Matter in the Universe." Indiana University. Indiana University, n.d. Web. 16 June 2014.
6. Kroupa, P. et el. "Local-Group tests of dark-matter concordance cosmology. Towards a new paradigm for structure formation." A&A 523 (2010): n. pag. Astronomy & Astrophysics. Web. 4 Sept. 2014.
7. Bergstrom, L. (2000). "Non-baryonic dark matter: Observational evidence and detection methods". *Reports on Progress in Physics* 63 (5): 793–841. arXiv: hepph/0002126.Bibcode: 2000RPPh...63..793B. doi:10.1088/0034-4885/63/5/2r3.
8. Dean, A. J. "Telescope Sensitivity Instrument Background & Mission Scenario." PHYS2011/2012. University of Southampton. Print. Lecture transcript.
9. Australia Telescope Outreach and Education, Resolution and Sensitivity, http://outreach.atnf.csiro.au/education/senior/astrophysics/resolution_sensitivity.html 2014
10. Skinner, Gerald K. "Sensitivity of Coded Mask Telescopes." Applied Optics 47.15 (2008): 2739-49. Optics InfoBase. Web. 4 Sept. 2014.
11. A.J.DEAN , PHYS2011/2012 - BACKGROUND, SENSITIVITY AND MISSION SCENARIOAJD 1, *School of Physics & Astronomy, University of Southampton*
12. 王傳晉，葉彬潯，天文可見光探測器，中國科學技術出版社，2013

13. Kitt Peak National Observatory, Telescope and Instrument Documentation, 2014, http://www.noao.edu/kpno/docs.html
14. NASA Sun Fact Sheet, retrieved August 4, 2014, http://nssdc.gsfc.nasa.gov/planetary/factsheet/Sun fact.html
15. P. Lancaster-Brown (1985). Halley & His Comet. Blandford Press. ISBN 0-7137-1447-6
16. "New Image of Comet Halley in the Cold". European Southern Observatory. 1 September 2003.
17. Perkins, J. S. et al, TeV Gamma-Ray Observations of the Perseus and Abell 2029 Galaxy Clusters, 06/2006, The Astrophysical Journal, Volume 644, Issue 1, pp. 148-154
18. McKee, Maggy. "Titanic Merger of Galaxy Clusters Revealed." New Scientist. Reed Business Information, 24 Sept. 2004. Web. 16 June 2014.
19. "Massive Merger of Galaxies Is the Most Powerful on Record." NASA, 23 Sept. 2004. Web. 17 July 2014.
20. QUICK-LOOK PHOTOMETRY and LUMINOSITIES for IC 1101 http://ned.ipac.caltech.edu/cgi-bin/objsearch?objname=IC+1101&extend=no&hconst=73&omegam=0.27&omegav=0.73&corr_z=1&out_csys=Equatorial&out_equinox=J2000.0&obj_sort=RA+or+Longitude&of=pre_text&zv_breaker=30000.0&list_limit=5&img_stamp=YES#AbsoluteMagnitudeLSun_0 NASA/IPAC EXTRAGALACTIC DATABASE, Retrieved 2014-11
21. "Quick-look Photometry and Luminosities for IC 1101." NASA/IPAC Extragalactic Database. N.p., n.d. Web. June 2014.
22. http://aether.lbl.gov/www/science/cmb.html, “The Cosmic Microwave Background Radiation”retrieved on November 30, 2014
23. Yuxiang Wu, “Who Should Talk About Cosmos 誰有權談論宇宙”, (in Chinese), 2015, Honkong Wenhui Publisher, ISBN 962-374-114-6

V. Reference for Cosmic Redshift

1. Yuxiang Wu, Johnny Wu, Eugene Wu, August 2014, The Limited Propagating Radius (LPR) of a Celestial Object, and Hidden Celestial Object (HCO) vs Dark Matter,
2. Jayant Vishnu Narlikar, 2002, An Introduction to Cosmology, Third Edition, Cambridge University Press, ISBN 0521 79376 9

3. Jerry Lodriguss, A Guide to Astrophotography with Digital SLR Cameras, Astrophotography Techniques, 2014
4. J. J. Condon and S. M. Ransom, ESSENTIAL RADIO ASTRONOMY, National Radio Astronomy Observatory, 2010, http://www.cv.nrao.edu/course/astr534/ERA.shtml
5. Kitt Peak National Observatory, Telescope and Instrument Documentation, 2014, http://www.noao.edu/kpno/docs.html
6. 王傳晉，葉彬潯，天文可見光探測器，中國科學技術出版社，2013
7. A. J.DEAN , PHYS2011/2012 - BACKGROUND, SENSITIVITY AND MISSION SCENARIOAJD 1, *School of Physics & Astronomy, University of Southampton*
8. Australia Telescope Outreach and Education, Resolution and Sensitivity, http://outreach.atnf.csiro.au/education/senior/astrophysics/resolution_sensitivity.html 2014
9. Gerald K. Skinner, Sensitivity of coded mask telescopes, 12 May 2008, Vol. 47, No. 15 / APPLIED OPTICS
10. A.A. Michelson and E.W. Morley, 1887, On the Relative Motion of the Earth and the Luminiferous Ether, Am. J. Sci. (3rd series) 34 333–345 (1887).
11. http://www.aip.org/history/gap/PDF/michelson.pdf
12. Higgs, L. A., 1960, The Solar Red-Shift, Monthly Notices of the Royal Astronomical Society, Vol. 121, p. 421-435, http://adsabs.harvard.edu/full/1960MNRAS.121..421H
13. John Huchra, Extragalactic Redshifts, Harvard-Smithsonian Center for Astrophysics, http://ned.ipac.caltech.edu/help/zdef.html , retrieved August 10, 2014
14. Doppler, C. J. (1842). Über das farbige Licht der Doppelsterne und einiger anderer Gestirne des Himmels (About the coloured light of the binary stars and some other stars of the heavens). Publisher: Abhandlungen der Königl. Böhm. Gesellschaft der Wissenschaften (V. Folge, Bd. 2, S. 465-482) [Proceedings of the Royal Bohemian Society of Sciences (Part V, Vol 2)]; Prague: 1842 (Reissued 1903).
15. Doppler, Christian (1846). "Beiträge zur fixsternenkunde". *Prag* (Prag, Druck von G. Haase sohne) **69**. Bibcode:1846QB815.D69......
16. https://www.cfa.harvard.edu/~dfabricant/huchra/hubble/

17. Maria Luiza Bedran, 2002, A comparison between the Doppler and cosmological redshifts, 2002 American Association of Physics Teachers
18. Geraint F. Lewis *et al.* (2008). "Cosmological Radar Ranging in an Expanding Universe". *Monthly Notices of the Royal Astronomical Society* **388** (3): 960–964. arXiv:0805.2197. Bibcode:2008MNRAS.388..960L. doi:10.1111/j.1365-2966.2008.13477.x.
19. Paul Marmet etal, 1988, A New Non-Doppler Redshift, Physics Essays, Vol. 1, No: 1, p. 24-32, http://www.newtonphysics.on.ca/hubble/index.html
20. David W. Hogg et al, Oct 2002, The K correction, Astrophysics, arXiv:astro-ph/0210394v1

VI. 4-Dimensional Image of World Line

[1] Corry, L. (1997). "Hermann Minkowski and the postulate of relativity". Arch. Hist. Exact Sci. (Springer-Verlag) 51 (4): 273–314. doi:10.1007/BF00518231. ISSN 0003-9519.
[2] Galison, P. L. (1979). R McCormach et al., eds. Minkowski's Space-Time: from visual thinking to the absolute world. Historical Studies in the Physical Sciences 10. Johns Hopkins University Press. pp. 85–121. doi:10.2307/27757388.
[3] Catoni, F. et al. (2008). Mathematics of Minkowski Space. Frontiers in Mathematics. Basel: Birkhäuser Verlag. doi:10.1007/978-3-7643-8614-6. ISBN 978-3-7643-8613-9. ISSN 1660-8046.

VII. Reference for Colour-Magnitude Relation (CMR)

1. https://hubblesite.org/contents/news-releases/2001/news-2001-22.html ,"Star Clusters Born in the Wreckage of Cosmic Collisions," Release date: Jul 19, 2001 9:00 AM (EDT) ", retrieved on May 5, 2021
2. NASA web site, "https://www.nasa.gov/centers/goddard/news/topstory/2004/0831galaxymerger.html," MASSIVE MERGER OF GALAXIES IS THE MOST POWERFUL ON RECORD", date of web publication: September 23, 2004, Retrieved on May 5, 2021
3. NASA web site,

“https://www.nasa.gov/mission_pages/hubble/multimedia/ero/ero_stephan_quintet.html, "Galactic Wreckage in Stephan's Quinte“, date of web publication: September 23, 2004, Retrieved on May 5, 2021
4. https://science.nasa.gov/science-news/science-at-nasa/2010/14nov_babyblackhole, “Youngest-Ever Nearby Black Hole Discovered”, November 4, 2010, retrieved on May 5, 2021
5. NASA web site, http://hubblesite.org/news_release/news/2019-30 , "Galaxy Blazes with New Stars Born from Close Encounter, Release date: May 16, 2019 10:00 AM (EDT) ," Retrieved on May 5, 2021
6. Robert Gryncewicz et al, " Typeset using LATEX default style in AASTeX62 Mapping Milky Way Halo Substructure using Stars in the Extended Blue Tail of the Horizontal Branch,” arXiv:2103.04438v1 [astro-ph.GA] 7 Mar 2021 Draft version March 9, 2021
7. Nobuo Arimoto-P and Tadayuki Kodamal, Origin of the Colour-Magnitude Relation of Elliptical Galaxies, Astron. Astrophys. 320, 41–53 (1997)
8. A. I. Terlevic et al, Colour±magnitude relations and spectral line strengths in the Coma cluster, Mon. Not. R. Astron. Soc. 310, 445±452 (1999)
9. Ruixiang Chang et al, Colour–magnitude relations of late-type galaxies, Monthly Notices of the Royal Astronomical Society, Volume 372, Issue 1, 11 October 2006, Pages 199–208, https://academic.oup.com/mnras/article/372/1/199/974229; https://doi.org/10.1111/j.1365-2966.2006.10826.
10. S.C. Ellis1 et al, The colour-magnitude relations of ClJ1226.9+3332, a massive cluster of galaxies at z = 0.89, Mon. Not. R. Astron. Soc. 000, 000–000 (0000) Printed 5 February 2008 (MN L ATEX style file v2.2)
11. Simona Mei et al, EVOLUTION OF THE COLOR–MAGNITUDE RELATION IN GALAXY CLUSTERS AT Z ~ 1 FROM THE ACS INTERMEDIATE REDSHIFT CLUSTER SURVEY, ApJ, in press October 10, 2008, https://arxiv.org/pdf/0810.1917.pdf
12. James Schombert and Karl Rakos, THE AGE OF ELLIPTICALS AND THE COLOR–MAGNITUDE RELATION, The Astrophysical Journal, 699:1530–1540, 2009 July 10 doi:10.1088/0004-637X/699/2/1530-1540, 2009. The American Astronomical Society.
13. Yu Wang et al, A joint Chandra and XMM–Newton view of Abell 3158: a massive off-centre cool gas clump as a robust diagnostic of a merger stage, Mon. Not. R. Astron. Soc. 403, 1909–1918 (2010)
14. Nathan Roche, Mariangela Bernardi, Joseph Hyde, Colour gradients and the colour–magnitude relation: different properties of brightest cluster galaxies and E/S0 galaxies in the Sloan Digital Sky Survey, Mon. Not. R. Astron. Soc. 407, 1231–1244 (2010)

15. R. De Propris, S. Phillipps, M. N. Bremer, Deep luminosity functions and colour–magnitude relations for cluster galaxies at $0.2 < z < 0.6$, Mon. Not. R. Astron. Soc. 10 August 2013
16. Atsushi J. Nishizawa et al, First results on the cluster galaxy population from the Subaru Hyper Suprime-Cam survey. II. Faint end color-magnitude diagrams and radial profiles of red and blue galaxies at $0.1<z<1.1$, Astron. Soc. Japan (2014) 00(0), 1–14, https://arxiv.org/pdf/1709.01136.pdf
17. Shota INOUE et el, "Search for a Non-equilibrium Plasma in the Merging Galaxy Cluster Abell 754," March 4, 2016, arXiv:1603.01396
18. Eric Gaidos et al, "Origin of Interstellar Object A/2017 U1 in a Nearby Young Stellar Association," 2017 November 8, 2017. The American Astronomical Society. Research Notes of the AAS, Volume 1, Number 1
19. Department of Physics, University of Durham, The Colour-Magnitude Relation for Elliptical Galaxies, http://community.dur.ac.uk/ian.smail/colMag/colMag_intro.html
20. Ellianna S. Abrahams et el, "Informing the Cataclysmic Variable Donor Sequence from Gaia DR2 Color-Magnitude and Inferred Variability Metrics," arXiv:2011.12253, Submitted 24 November 2020
21. Robert Gryncewicz et el, "Mapping Milky Way Halo Substructure using Stars in the Extended Blue Tail of the Horizontal Branch," arXiv:2103.04438v1 [astro-ph.GA] 7 Mar 2021 Draft version, March 9, 2021
22. Yuxiang Wu, (2021), 10 Questions To NASA, Two-W Object, ISBN 978-1-7370977-0-9
23. Yuxiang Wu, (2020), Fading Modern Cosmology, Two-W Object, ISBN 9798645172633
24. Yuxiang Wu, Chinese book <Who Should Talk about Cosmos 誰有權談論宇宙->, 香港文匯出版社有限公司, 2005, ISBN 962-374-114-6
25. Yuxiang Wu, "尋找新的宇宙觀測資料方法的建議書", 格物 USA MATTER REGULARITY, January 1, 2018, Vol 16, No.6, pp1-pp4

VIII. Reference for Relativity

[1] 愛因斯坦, 孫良英, 趙中立, 張宜三編譯, 愛因斯坦文集, 第 1 – 3 卷,商務印書館, 2018
[2] Albert Einstein, Relativity: The Special and General Theory,

Translated by Robert W. Lawson, ISBN 10:1-4200-4633-1, A Digireads.com Book, 2012 Edited by Paul AuthorS, Albert Einstein, Philosopher-Scientist, NJF Books, ISBN:1-56731-432-5,1970
[3] Einstein, H.A.LORENTZ, H.WEYL, H.MINKOWSKI, THE PRINCIPLE OF RELATIVITY, Translated by W.PERRETT and G.B.JEFFERY, DOVER PUBLICATIONS, INC.Standard book number: 486-60081-5, 1952
[4] TAYLOR/WHEELER, SPACETIME PHYSICS, W.H.Freeman and Company, New York, ISBN: 0-7167-0336-X, 1966
[5] Albert Einstein, Letters to Solovine, 1906-1955, Page 21, Phylosophycal Library, ISBN：978-1-4532-0488-7, 1934
[6] O'Connor, John J.; Robertson, Edmund F., A History of Special Relativity, http://www-groups.dcs.st-and.ac.uk/~history/HistTopics/Special_relativity.html, Retrieved 2015-08-01
[7] Brown, Harvey R., Michelson, FitzGerald and Lorentz: the Origins of Relativity Revisited,
http://philsci-archive.pitt.edu/987/, Retrived 2015-08-01
[8] Corry, L. (1997). "Hermann Minkowski and the postulate of relativity". Arch. Hist. Exact Sci. (Springer-Verlag) 51 (4): 273–314. doi:10.1007/BF00518231. ISSN 0003-9519.

IX. Paper of Hidden Celestial Object (HCO) Mathematic Classification

Hidden Celestial Object (HCO)

Johnny Wu[1], Sean Y. Wu[2]

Abstract

We named a celestial object hides from telescope observation as Hidden Celestial Object (HCO). First, for celestial light propagation, we defined the Limited Observable Radius (LOR), Relative Limited Observable Radius (RLOR), and Absolute Limited Observable Radius (ALOR). Using these definitions, HCOs were divided into two categories - Permanently Hidden Celestial Objects (PHCO) and Temporarily Hidden Celestial Objects (THCO). They contain 10 subcategories. The HCO concept is novel, easy to demonstrate, but significant. HCOs help us break the old thinking, understand the real face of the dark matter and that of the universe. We described a new Scientific Super Cosmos skeleton.

1. Introduction

The conceptualization of Hidden Celestial Objects was initially provoked by the Sirius B story. This story involves a deviation in Sirius's movement from what was expected, and the subsequent labeling of the "missing mass" that was influencing Sirius's trajectory as dark matter [1]. 18 years later, with an improved telescope, Sirius B was discovered and identified to be exactly the aforementioned missing mass that was incorrectly labeled as dark matter [2]. Following figure gives a clear concept.

This tells us that should we detect "missing mass" through gravitational effect [1-7], it is likely that said missing mass is just another celestial object that's temporary hidden from telescope's observation due to reasons such as using too weak of an observational tool, the celestial object's luminosity being too weak, and the distance being too big, and so on.

---

[1] Johnny Wu, johnnywu@berkeley.edu

[2] Dr. Sean Y. Wu, Beijing Relativity Association, yuxiangwu@outlook.com

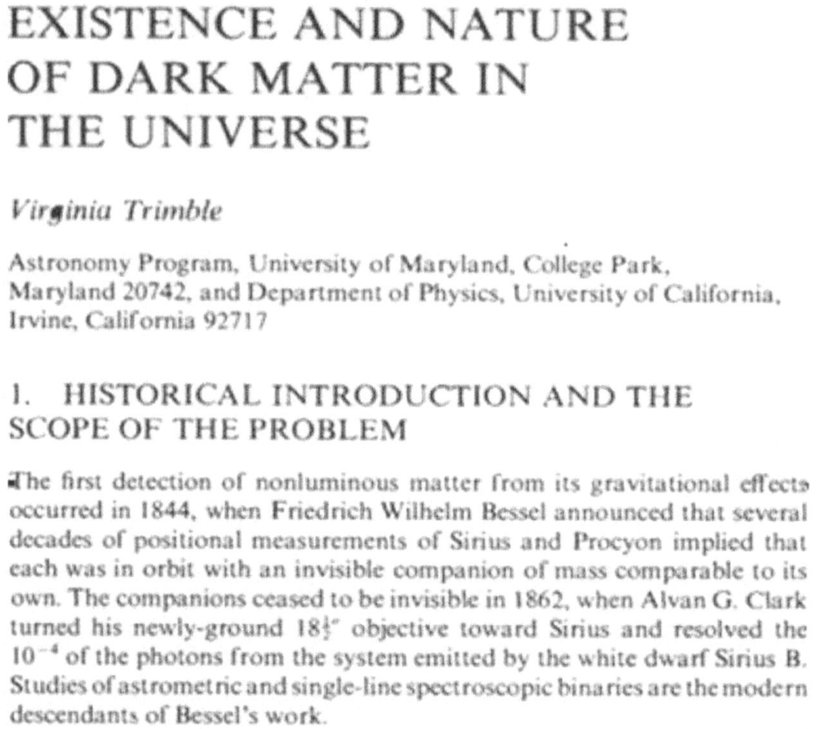
EXISTENCE AND NATURE OF DARK MATTER IN THE UNIVERSE

Virginia Trimble

Astronomy Program, University of Maryland, College Park, Maryland 20742, and Department of Physics, University of California, Irvine, California 92717

1. HISTORICAL INTRODUCTION AND THE SCOPE OF THE PROBLEM

The first detection of nonluminous matter from its gravitational effects occurred in 1844, when Friedrich Wilhelm Bessel announced that several decades of positional measurements of Sirius and Procyon implied that each was in orbit with an invisible companion of mass comparable to its own. The companions ceased to be invisible in 1862, when Alvan G. Clark turned his newly-ground 18½" objective toward Sirius and resolved the 10^{-4} of the photons from the system emitted by the white dwarf Sirius B. Studies of astrometric and single-line spectroscopic binaries are the modern descendants of Bessel's work.

Figure 1. News report about found the "Dark Matter" — Sirius B

There are many other cases that are similar to the Sirius B story as well. For example, all but 2 galaxies are hidden from view when we observe with our naked eyes. When observed through a telescope, however, many more galaxies are detectable. Comets hide and appear periodically depending on their location, and some planets, stars, and even galaxies hide from poor telescopes but are detected by better ones.

Summarizing all possible factors in cases involving observations, we arrived at the HCO concept as well as the cause of it – Limited Observable Radius (LOR), Relative LOR (RLOR) and Absolute LOR(ALOR). Below, we first derive LOR, RLOR and ALOR from basic physics principles, and then using LOR and RLOR define HCO. Application examples are then given at the end of the paper.

2. Concepts and Definitions of Limited Observable Radius (LOR), Relative LOR (RLOR) and Absolute LOR(ALOR)

Determined by factors such as instrument sensitivity, observing interfering noises, luminosity of celestial object C and distance between C and the observer, each C has its LOR(C). Beyond LOR(C), the observer obtains no valid information from C. LOR(C) possibly can be changed while any of the factors changes, in such case we call it RLOR(C). If a celestial object is located too far away that change any factor still can't be observed, it is called ALOR(C).

2.1 Instrument Sensitivity SEN_{inst}

Instrument sensitivity [8-13] is a measure of the minimum signal an observational instrument can distinguish from background noise. Let SEN_{inst} represent the sensitivity of observer the observational instrument in use where a lower value of SEN_{inst} denotes a more sensitive observer.

2.2 The Interfering NOISE on an Observer

There are different kinds of noises [8-13] that interfere with the observation result of an observer. Natural noise is denoted as $NOISE_{natural}$ that includes random (statistical) noise, dark current noise, shot noise in the source, shot noise in the sky, sky brightness, etc. Instrument and other controllable noise is denoted as $NOISE_{control}$ that includes noise sources in CCD image, read noise, systematic error, aperture efficiency filter effect, and others decrease the incoming light intensity. We take all into account, giving us the understanding that the minimum luminosity intensity that an observer requires to distinguish incoming information from noises is nowhere near zero. [8-13]

Let NOISE be the collection of all interfering noises on an observer such that

$$NOISE = NOISE_{natural} + NOISE_{control}$$

2.3 Minimum Required Luminosity Intensity f (C)

The minimum luminosity intensity f(C) that an observer requires to distinguish incoming information of celestial object C from noises $f(NOISE_{natural} + NOISE_{control}, SEN_{inst})$ is always bigger than zero. [8-13]

$$f(C) > f(NOISE_{natural} + NOISE_{control}, SEN_{inst})$$

Given that we cannot control $NOISE_{natural}$, but keep trying to lower SEN_{inst} and $NOISE_{control}$ toward 0, we place a lower bound on the minimum incoming light intensity required for C to be detected such that:

$$f(C) \geq f(NOISE_{natural} + 0, 0))$$

Let $f(NOISE) = f(NOISE_{natural}, 0)$, then the minimum threshold of intensity is

$$f(C) = f(NOISE)$$

2.4 The Distributed Luminosity Intensity I (C, D(C)) on the Spherical Surface

D(C) is the distance from C. D (C, C1) is the distance between celestial objects C and C1. L_C is the luminosity of C. **I (C, D(C))** is C's luminosity intensity over a spherical surface with radius D(C), I (C, D(C)) can be calculated as

$$I(C, D(C)) = \frac{L_C}{4\pi D^2(C)}$$

Obviously, we have

$$\lim_{D(C)\to\infty} I(C, D(C)) = \lim_{D(C)\to\infty} \frac{L_C}{4\pi D^2(C)} = 0 \quad (1)$$

For an example of the above Limit equation, take the Sun into consideration. We know the luminosity L_{Sun} of Sun is $3.846*10^{26} Js^{-1}$. [14] If we were to assign the distance D(Sun) = 1.01 million light-years, then the luminosity intensity I (Sun, D(Sun)) would be 1 photon per meter^2 per second. Even if you say that is still possible for an observer to detect, we can increase the distance such that I (Sun, D(Sun)) be 1 photon per mile^2 per second, or less… There exists no observer that could ever capture a single photon in such a wide area, let alone allow the observation of the Sun at that far.

2.5 LOR and RLOR

The LOR of celestial object C is the distance from C at which its light intensity L_C is equivalent to and indistinguishable from a function of background interference noise threshold f (NOISE). The light intensity I (C, D(C)) must be greater than or equal to this threshold in order to be detectable.

$$I\,(C, D(C)) \geq f\,(NOISE)$$

By combining the aforementioned two equations, we arrive at the equation:

$$I\,(C, D(C)) = \frac{L_C}{4\pi D^2(C)} = f\,(NOISE)$$

So, we get:

$$D(C) = \sqrt{\frac{L_C}{4\pi f(NOISE)}}$$

This distance D(C) is what we term LOR, that is LOR(C) = D(C). So, we have:

$$LPR(C) = \sqrt{\frac{L_C}{4\pi f(NOISE)}} \tag{2}$$

We define the Relative Limited Propagating Radius RLOR (C, $NOISE_{natural}$ + $NOISE_{control}$, SEN_{inst}) of a Celestial Object C on an observational instrument with sensitivity $SEN_{inst} > 0$ where $NOISE_{control}$ was not optimized to 0:

$$RLOR\,(C, NOISE_{natural} + NOISE_{control}, SEN_{inst}) = \sqrt{\frac{L_C}{4\pi f(NOISE\mathit{natural} + NOISE\mathit{contral}, \; SENinst)}} \tag{3}$$

When $NOISE_{control}$ and SEN_{inst} in RLOR are optimized to 0, RLOR becomes LOR.

$$LOR(C) = RLOR\,(C, NOISE_{natural} + 0, 0)$$

2.6 D_{now} and D_{max}

A consequence of the above is that the furthest distance we can observe is bounded under the assumptions that the luminosity of all celestial objects is below a constant L_{max}, $NOISE_{natural}$ remains constant, and $NOISE_{control}$ is optimized to 0. With this, the furthest distance D_{now} we can currently observe based on the observer sensitivity of our existing observational instruments is given by:

$$D_{now} = \sqrt{\frac{L_{max}}{4\pi f(NOISE_{natural}, \; SEN_{inst})}}$$

However, assuming our observational instruments have been optimized to the fullest so that both SEN_{inst} and $NOISE_{control}$ are zero, the furthest distance D_{max} we would be able to observe is limited to:

$$D_{max} = \sqrt{\frac{L_{max}}{4\pi f(NOISE_{natural}, 0)}}$$

In other words, D_{max} is this furthest distance this instrument will ever be able to observe; it defines the size of our observable universe. We also note that beyond a certain point, improving the sensitivity of the observational instrument to lower the noise on the instrument will not increase the maximum observable distance of the instrument. At this point, natural noises will begin to dominate the sensitivity of the instrument and further improvements may no use.

3. Hidden Celestial Objects (HCO)

We introduce Hidden Celestial Objects (HCO) as celestial objects whose luminosity intensity arrived are not strong enough to be detected by an observer. Thus, from the perspective of an observer on or near Earth, HCO cannot be detected. By defining HCO in terms of LOR, we believe it can provide a nuanced mechanism to better understand dark matter objects.

3.1 HCO Definition

A Hidden Celestial Object (HCO) is formally defined as a celestial object C whose RLOR is less than its distance to Earth such that:

$$RLOR\,(C, NOISE_{natural} + NOISE_{control}, SEN_{inst}) = \sqrt{\frac{L_C}{4\pi f(NOISE_{natural} + NOISE_{contral}, SENinst)}} < D\,(C, Earth) \qquad (4)$$

The above depends on the parameters L_C, $NOISE_{natural}$, $NOISE_{control}$, SEN_{inst}, and D (C, Earth), each of which is not static constant. Thus, the status of an HCO is not necessarily permanent - currently observable objects may transition into hidden celestial objects and previously undetectable objects can eventually become observable. By analyzing sensitivities of these parameters of RLOR, we get different HCO categories.

In following discussion, we always write RLOR (C, $NOISE_{natural}$ + $NOISE_{control}$, SEN_{inst}) as RLOR(C).

3.2 Categorization of HCO (C, Parameters, Value)

Here "Parameters" are "L_c", "NOISE", "SEN_{inst}", "D", and "Age". "Value" will be "increase" or "decrease".

With the changing Value of different Parameters in HCO (C, Parameters, Value), HCO C becomes either PHCO (Permanent HCO) or THCO (Temporary HCO). In which THCO can be further classified into 8 subcategories.

3.2.1 THCO (C, L_c, increase)

To a hidden C, if changing Parameter is luminosity L_C in (4), its Value keeps increasing over time, and eventually RLOR(C) > D(C, Earth), C transition from hidden to observable. Common examples of this class include stars that transition into a supernova state.

3.2.2 THCO (C, L_c, decrease)

To an observable C, if luminosity L_C in (4) keeps decreasing over time, and eventually RLOR(C) < D(C, Earth), C transition from observable to hidden. This commonly occurs in dying stars that exhaust their fuel.

3.2.3 THCO (C, $NOISE_{control}$, decrease)

As we lower the $NOISE_{control}$ it increases RLOR and gives us farther observation ability. For example, moving a telescope to a mountain top it will have better vision than under the mountain. We see much more stars in the sky while camping in the mountain than we see in the city is another example.

3.2.4 THCO (C, SEN_{inst}, decrease)

As we improve the sensitivity and efficiency of our observational instruments so that SEN_{inst} decreases, it increases the RLOR. Thus, celestial objects that are at the cusp of detection will become observable. Sirius B is an example of such a phenomenon, a star that was previously classified as Dark Matter, [1] and through improved telescope sensitivity, could be observed. [2]

A consequence of our RLOR definition is that the distance we can increase our observation of an object C is bounded by LOR (C, NOISEnatural, 0).

3.2.5 THCO (C, D, dec)

As celestial objects move towards Earth, we become within the objects' RLOR range. Halley's Comet [15, 16] is such an example of THCO (Halley, D, dec) while it is traveling toward the Earth. Consider RLOR (Halley, NOISE, SEN_{pupil}) where NOISE is some constant and SEN_{pupil} is the sensitivity of a human observer pupil. Although we can model Halley's movement, we can only observe it every 75 years when it comes within the human eyes' observable range. As the distance between the observer and Halley's Comet is decreasing, it is in an THCO (Halley, D, dec) state until said distance is equal to or less than RLOR (Halley).

3.2.6 THCO (C, D, inc)

As celestial objects move away from Earth, we may become outside the objects' RLOR range. When Halley's Comet is start leaving us, it turns from visible to invisible. And will become visible after another 75 years. All comets are such kind of similar THCO. This THCO possibly becomes PHCO if D keeps increasing. If the universe did expanding, there should be a large quantity of such disappeared celestial objects.

3.2.7 THCO (C, Age, young)

This is the first of the three special cases, consists of celestial objects whose age in years is below its distance in light-years from Earth. For these objects, they are hidden from our observational instruments, irrespective of their luminosity and our instrument sensitivity. We actually haven't seen it yet. It only exists in our imagination. There were many similar cases that took such THCO as observed objects that alert us to pay attention to such THCO. A related example is the so-called new-born star Abell 0754, which is reportedly 300 million years old. However, its distance from Earth is 800 million light years. Thus, its observation by NASA is a surprising contradiction to our model.[17, 18, 19] Obviously the light emitted from it needs travel 500 million light years more to hit the observer. So Abell 0754 is not 300 million light-years old. The observed information was interpreted in a wrong way.

3.2.8 THCO (C, L_C, pulse)

This THCO consists of celestial objects whose luminosity exhibits a periodicity in amplitude that is significant enough that it transitions between hidden and observable states. Pulsars, such as the Crab Pulsar, are an example of THCO (C, L_C, pulse) due to their directional electromagnetic radiation coupled with their rotation.

3.2.9 PHCO (C, L_c, 0)

If C has $L_c = 0$, then we always have RLOR(C) = 0, C is always an HCO, so C is a Permanent HCO (PHCO), All dark matter is PHCO because $L_{dark\text{-}matter} = 0$ is always true according to the definition of dark matter whose presence cannot be detected through electromagnetic means.

Another case is planets that they will not emit light, like the Earth or Moon. But they absorb as well as reflect light so $L_{planet} > 0$. This makes them also different from dark matter. They are not PHCO if they are located within some distance and can be observed.

We haven't seen much PHCO (C, L_c, 0) yet. Most of so called "dark matter" like Sirius B were found later were THCO.

3.2.10 PHCO (C, D, D_{max})

If D(C, Earth) $\geq D_{max}$, C is out of LOR(C), and will be Permanent HCO without much chance to change its visible status. Outside human limited observing range, all celestial objects are PHCO. This will be a large quantity of them.

4. HCO Application

HCO gave us many new ideas and concepts to understand the universe. Following is a few of them.

4.1 New Concept Map of the Universe

Using the concept of LOR, we can draw a conceptual picture of the Observable Universe.

First, we need to assign $NOISE_{SUM}$ a reasonable value. We can assign a rough value and use this value to aid in creating a conceptual picture of the Observable Universe. To simplify our discussion, let's just assume $NOISE_{SUM}$ is 1 photon per second per square centimeter, and also assume that 1 photon at 0.6 μm has $3.32*10^{-19}$ J, the average photon energy,

$$NOISE_{SUM\_TEMP} = 1 \text{ photon } s^{-1} m^{-2} = 3.32*10^{-19} \text{ J } s^{-1} m^{-2}$$

Using $NOISE_{SUM\_TEMP}$ we can calculate the LOR of selected objects, such as the Sun, to get a conceptual picture. Since Sun's luminosity [14] is L_{Sun}= $3.846*1026Js^{-1}$. Using $NOISE_{SUM\_TEMP}$ we have RLOR(Sun) = $1.01*10^6$

light-years. This expresses that the Sun under current condition assumption can only be observed within a distance about 1 million light years, which isn't very far.

The largest galaxy ever discovered is IC 1101[20, 21]. According to NASA [20, 21], the visual luminosity of IC 1101 is about 2.23 * 10^{11} L_{Sun}. So we have RLOR (IC1101) = 477 billion light-years.

NASA assigned their maximum observation radius to be 13.7 billion light-years. We can say that $D_{now} = D_{NASA}$= 13.7 billion light-years.

Summarizing above, we get Figure 2.

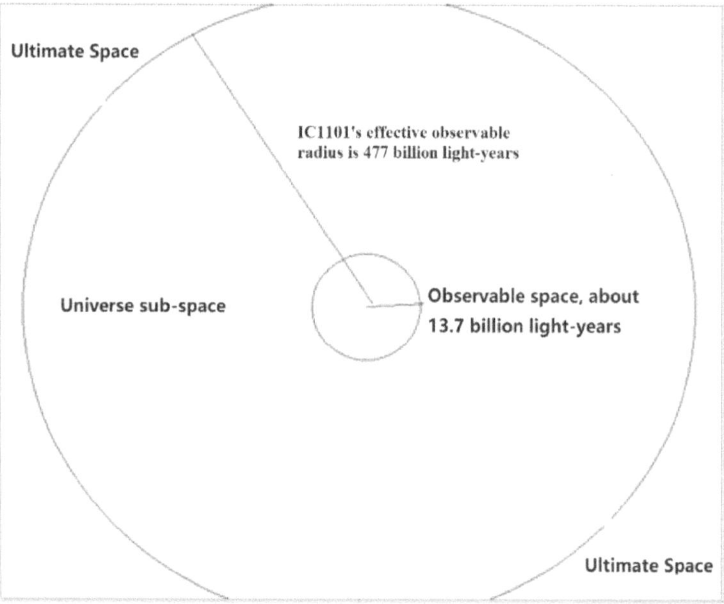

Figure 2. Conceptual Picture of the Universes According to the IC1101 light propagation radius of reasoning macroscopical universe imagination picture

From above figure, we can see that LOR(IC 1101) current is the maximum size of our known universe R_{max}, and it can be observed at each point within this R_{max}; and the observable area of NASA within D_{NASA} is much smaller than that of LOR(IC 1101). The LOR(Sun) is pretty small. Sometimes a celestial object has a large LOR but small RLOR, then the part outside RLOR but inside LOR is not part of our immediately observable universe, but part of our possible observable universe, which we call Temporary Hidden Universe.

4.2 HCO Conceptual Distribution Map

RLOR (DX Cancri (G 51-15)) is about 10 light-years. If there is a star Y like DX Cancri located 11 light-years away, we are out of RLOR(Y) and Y is a THCO.

RLOR (Wolf 359) is about 100 light-years, If there is a star X same as

Wolf 359 located 110 light-years away, we are out of RLOR(X) and X is a THCO to us.

If SunA is the same as the Sun and located 1.1 million light-years away or father, SunA is a THCO.

HCO are normal celestial objects. According to the Copernican principle, they distribute everywhere in the universe, billion light years or a few light years away.

There are observable celestial objects as well as HCO everywhere, near or far away from the Earth. In overall, HCO are much more than observable objects.

Outside D_{max} all celestial objects are HCO. Figure 2 is the conceptual map of the hidden universe, in which the normal celestial objects were not shown.

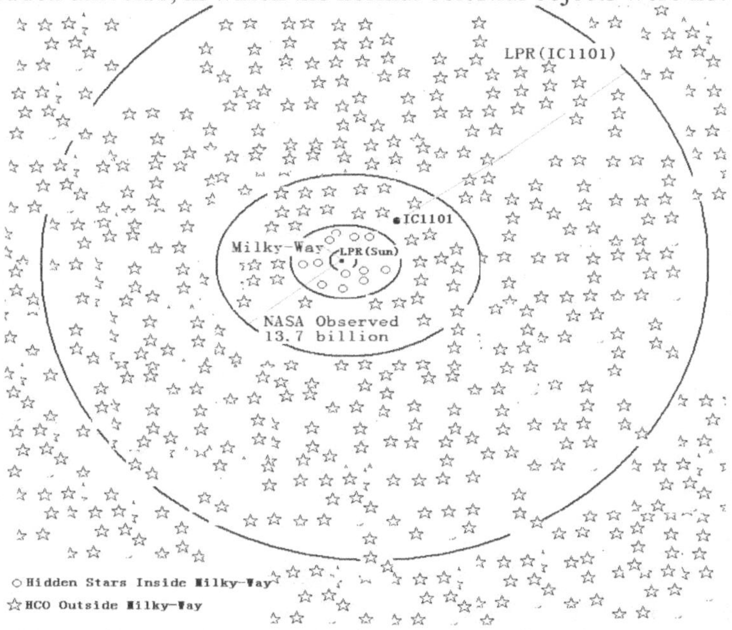

Fig.2 Conceptual Picture of known universe with Hidden Celestial Objects but without normal celestial objects. In the figure we draw LOR(Sun) the most inner circle, then Milky-Way, NASA Observed, and LOR(IC1101). Inside Milky-Way we have hidden stars and light emitting objects. Outside Milky-Way we have hidden galaxies and other HCOs. This is based on current know data of the Sun, the Milky-Way, the NASA Observed limitation, and the IC1101. The picture is not in proper proportion.

4.3 Test of Expanding Universe

The Big Bang is a dominating theory in contemporary cosmic theories, although no way to convincingly explain what is before the Big Bang and what is there outside the Big Bang Universe. The expanding Universe is major based on Doppler Redshift. But using THCO(C, D, inc) we can double check the theory: Comparing the ancient star maps to the contemporary maps with the same sensitive observer [23] (such as naked eyes, old telescopes, etc.) we can see how many celestial objects disappeared from the comparison. If there is almost none of them disappeared after hundreds or thousands of years, then we need reconsider if the universe is really expanding. If there are lots of disappeared objects, then it confirms the Big Bang theory.

5 Discussions

Our classifications have shed light on the distinction between Dark Matter and HCO. By definition, Dark Matter is defined as objects that do not absorb, emit and reflect electromagnetic waves, [1-7] and is simply a special case of HCO. Instead other HCO are just normal celestial objects like stars, galaxies, and light emitting or reflecting materials. From the case of finding Sirius B, we know that other HCO is easily to be misidentified as Dark Matter. As we seek to understand the nature of Dark Matter, we can reduce false positive identification by ensuring objects that are candidates for Dark Matter are not any other class of HCO.

"As much as 90 percent of the matter in the universe is invisible."[22] If that were true, then traveling among stars would be impossible, because no way to avoid the space ship hit on dark matter.

There are very few true dark matters because follow the same logic used by NASA on whole sky map[22], COBE never hit or detected any dark matter, and this is true to thousands of satellites running through the sky, which means there is no or very few dark matter in this sky area. The logic used by NASA on whole sky map [22] is: "fell back on the Cosmological Principle: which states that the universe, on the average, looks the same from any point."[22], we know there is little dark matter in the Universe.

HCO is derived naturally with numerous examples, like we comparing the observed results using our naked eyes from those using telescopes. It is that simple and nature. But this naturally derived HCO concept is against many contemporary astronomy thinking and research results, and will be so tough to be accepted by many people.

6 Conclusions

HCO is vital in our understanding of the Universe, in that there are many celestial objects that we simply cannot detect. HCO allows us to understand that should we be outside the LOR of some object C， even C is located very close to us, no matter how powerful and more sensitive we make our telescopes, we will never be able to detect C due to the effect of universal noises and hence C would remain an HCO to us. However, our observational instruments are nowhere near perfect, so we can still extend our current view of the Universe by improving our SEN_{inst} and $NOISE_{control}$ values to extend current RLOR, making some current

HCO become observable.

HCO give evidence of the wide existence of the invisible celestial objects that are located beyond their RLOR, and are hiding from our observation. This concept helps us further understand the dark matter. Comparing to consensus among cosmologists that "dark matter is composed primarily of a not yet characterized type of subatomic particle", [7] HCO concept is much more convincible. HCO give us brand new concept to understand the universe.

References

1. Bessel, Friedrich W. Extract from the Translation of a Letter from Professor Bessel, dated Konigsberg, 10th of August, 1844. "On the Variations of the Proper Motions of Procyon and Sirius." MNRAS 3.21 (1844): 136-41. Oxford Journals. Web. 4 Sept. 2014.
2. "Case Study: Friedrich Bessel and the Companion of Sirius." American Museum of Natural History. American Museum of Natural History, n.d. Web. 4 Sept. 2014.
3. http://home.web.cern.ch/about/physics/dark-matter. CERN. Retrieved on November 25 2014.
4. Trimble, Virginia. "Existence and Nature of Dark Matter in the Universe." Annual Review of Astronomy and Astrophysics 25 (1987): 425-72. Annual Reviews. Web. 4 Sept. 2014.
5. Rubin, Vera. "Dark Matter in the Universe." Indiana University. Indiana University, n.d. Web. 16 June 2014.
6. Kroupa, P. et el. "Local-Group tests of dark-matter concordance cosmology. Towards a new paradigm for structure formation." A&A 523 (2010): n. pag. Astronomy & Astrophysics. Web. 4 Sept. 2014.
7. Bergstrom, L. (2000). "Non-baryonic dark matter: Observational evidence and detection methods". *Reports on Progress in Physics* 63 (5): 793–841. arXiv:hep-ph/0002126.Bibcode:2000RPPh...63..793B. doi:10.1088/0034-4885/63/5/2r3.
8. Dean, A. J. "Telescope Sensitivity Instrument Background & Mission Scenario." PHYS2011/2012. University of Southampton. Print. Lecture transcript.
9. Australia Telescope Outreach and Education, Resolution and Sensitivity, http://outreach.atnf.csiro.au/education/senior/astrophysics/resolution_sensitivity.html 2014
10. Skinner, Gerald K. "Sensitivity of Coded Mask Telescopes." Applied Optics 47.15 (2008): 2739-49. Optics InfoBase. Web.

4 Sept. 2014.
11. A.J. DEAN, PHYS2011/2012 - BACKGROUND, SENSITIVITY AND MISSION SCENARIOAJD 1, *School of Physics & Astronomy, University of Southampton*
12. 王传晋，叶彬浔，天文可见光探测器，中国科学技术出版社，2013
13. Kitt Peak National Observatory, Telescope and Instrument Documentation, 2014, http://www.noao.edu/kpno/docs.html
14. NASA Sun Fact Sheet, retrieved August 4, 2014, http://nssdc.gsfc.nasa.gov/planetary/factsheet/sunfact.html
15. P. Lancaster-Brown (1985). Halley & His Comet. Blandford Press. ISBN 0-7137-1447-6
16. "New Image of Comet Halley in the Cold". European Southern Observatory. 1 September 2003.
17. Perkins, J. S. et al, TeV Gamma-Ray Observations of the Perseus and Abell 2029 Galaxy Clusters, 06/2006, The Astrophysical Journal, Volume 644, Issue 1, pp. 148-154
18. McKee, Maggy. "Titanic Merger of Galaxy Clusters Revealed." New Scientist. Reed Business Information, 24 Sept. 2004. Web. 16 June 2014.
19. "Massive Merger of Galaxies Is the Most Powerful on Record." NASA, 23 Sept. 2004. Web. 17 July 2014.
20. QUICK-LOOK PHOTOMETRY and LUMINOSITIES for IC 1101 http://ned.ipac.caltech.edu/cgi-bin/objsearch?objname=IC+1101&extend=no&hconst=73&omegam=0.27&omegav=0.73&corr_z=1&out_csys=Equatorial&out_equinox=J2000.0&obj_sort=RA+or+Longitude&of=pre_text&zv_breaker=30000.0&list_limit=5&img_stamp=YES#AbsoluteMagnitudeLsun_0
NASA/IPAC EXTRAGALACTIC DATABASE, Retrieved 2014-11
21. "Quick-look Photometry and Luminosities for IC 1101." NASA/IPAC Extragalactic Database. N.p., n.d. Web. June 2014.
22. http://aether.lbl.gov/www/science/cmb.html, “The Cosmic Microwave Background Radiation”retrieved on November 30, 2014
23. Yuxiang Wu, “Who Should Talk About Cosmos 谁有权谈论宇宙”, (in Chinese), 2015, Honkong Wenhui Publisher, ISBN 962-374-114-6
24. Yuxiang Wu, “Clarify Big Bang”, ISBN: 978-1985629974,

Two-W Object Company, 2018 USA
25. Yuxiang Wu, Lǚ Wu, “BIG BANG PSEUDO-THEOLOGY”, ISBN: 978-1-7370977-7-8, Two-W Object Company, 2018 USA

Made in United States
Cleveland, OH
05 November 2026

47669895R00217